Exercises in Environmental Physics

Valerio Faraoni

Exercises in Environmental Physics

 Springer

Valerio Faraoni
Physics Department
Bishop's University
Lennoxville, Quebec J1M 1Z7
Canada
vfaraoni@cs-linux.ubishops.ca

ISBN 978-1-4419-2222-9   e-ISBN 978-0-387-35835-2

Printed on acid-free paper.

Printed in the United States of America.   (MV)

9 8 7 6 5 4 3 2 1

springer.com

To Louine and Donovan

Contents

Preface

The study of environmental physics requires understanding topics from many different areas of physics as well as comprehension of physical aspects of the world around us. Several excellent textbooks are available covering most aspects of environmental physics and of applications of physics to the natural environment from various points of view. However, while teaching environmental physics to university students, I sorely missed a book specifically devoted to exercises for the environmental science student. Thus, the motivation for this book came about as in physics, as well as in many other disciplines, satisfactory knowledge of a subject cannot be acquired without practice. Usually students are not familiar with the various areas of physics that are required to describe both the environment and the human impact upon it. At the same time, students need to develop skills in the manipulation of the ideas and concepts learned in class. Therefore, this exercise book is addressed to all levels of university students in environmental sciences.

Because of the wide range of potential users this book contains both calculus-based and algebra-based problems ranging from very simple to advanced ones. Multiple solutions at different levels are presented for certain problems—the student who is just beginning to learn calculus will benefit from the comparison of the different methods of solution. The material is also useful for courses in atmospheric physics, environmental aspects of energy generation and transport, groundwater hydrology, soil physics, and ocean physics, and selected parts may even be used for basic undergraduate physics courses. This collection of exercises is based on courses taught at the University of Northern British Columbia and at the University of Victoria, Canada.

Each problem and its solution are self-contained so that they can be attempted or assigned independently. For students willing to deepen their knowledge of the subject, references to the literature are sometimes given in the text or the solution of the problems.

The problems are arranged by topic, although problems usually overlap two or more different categories. This should make the students aware of the fact that problems of the environment—even relatively simple ones—often involve different areas and require various techniques in an interdisciplinary approach. This is even more true for the complex problems that the environmental scientist encounters daily. To put it in John Muir's words, "When we try to pick out anything by itself, we find it hitched to everything else in the universe" [53].

This book is not comprehensive: covering the complete spectrum of topics in environmental physics would require a monumental work and most readers would have little appreciation for the more specialized topics. Many books or review papers on specific topics exist and they sometimes include exercises, but they are often too detailed for the purpose of a general course in environmental physics. The selection of topics contained in this book is to a certain extent arbitrary, as is the choice of subjects presented in most courses in environmental physics currently taught in university. However, I do believe that the essential topics common to any general environmental physics course are covered here. Rather than presenting exercises on the plethora of empirical formulas appearing in the literature on the various areas of environmental physics, the focus is on the unifying physical principles that can be applied to many different subjects.

How to Use This Book

The International System of units (*SI* system) is used in this book. Exercises are labeled with the letters **A**, **B**, or **C**. **A** denotes lower mathematical level (algebra-based) problems that can be solved without knowledge of calculus; whereas **B** indicates higher mathematical level problems usually requiring calculus for their solution. The letter **C** denotes conceptual questions that do not require calculations—these are inserted at the beginning of each chapter in lieu of lengthy review sections. Problems labeled **A** or **C** are not necessarily the easiest just because no calculus is required: they test the student's understanding and knowledge of the physical concepts and normally require more than just common sense for their solution.

The student should not browse through the solution before a problem has been attempted and a honest effort has been made to solve it. If a problem cannot be solved in spite of serious and repeated effort, the student should not be frustrated but should read and understand the solution and then review and correct his or her knowledge of the subject.

This is what exercises are for, after all, and the student will certainly learn from this process. Many exercises in the book require a sound mathematical background, and Chapter 1 reviews basic mathematical techniques. The section on vector calculus is particularly important to solve exercises that require the use of the transport equations.

First-year students may benefit from reading a general qualitative book such as Refs. [64, 60, 46] before delving into the details of specific areas of environmental physics. References [27, 28] contain entertaining and instructive solutions to selected problems using simplified quantitative models—for an advanced reference on environmental modeling, see Ref. [72]. Suggested readings are given at the beginning of each chapter or section. A recommendation for students just beginning in science and to whom many of these exercises are addressed: the problems should be solved using symbols for the physical quantities considered and the numerical values should only be inserted at the end of the mathematical calculations. It is strongly recommended to insert the corresponding units together with the numerical values of the various quantities, and to pay attention to the number of significant digits.

I have tried as much as possible to eliminate errors from the book, but I shall be grateful to readers informing of any errors that they may notice.

Lennoxville, Québec

March 2006

VALERIO FARAONI

Acknowledgments

I am grateful to all the colleagues who contributed useful suggestions, to the students who voluntarily (or most of the time involuntarily) tested the exercises of this book, and to my family for their patience during the compilation of this book. Finally, I wish to thank editor Dr. Gert-Jan Geraeds from Springer for friendly advice and assistance throughout the writing process.

Chapter 1

MATHEMATICAL METHODS

A part of the secret of analysis is the art of using notations well.
—Gottfried Wilhelm Leibnitz

Environmental problems are often posed in the context of data collection and statistics, and extensive discussion of the social, economic, and legal aspects of environmental science is certainly required. However, it is not sufficient to talk about environmental problems or to collect data and make statistics. To begin analyzing and finding solutions to problems in environmental physics requires a precise formulation in mathematical terms, and the methods of mathematical physics are widely used. Many problems—even if well-posed mathematically—are too difficult to solve because of the complexity resulting from interdisciplinarity and because of their intrinsic nonlinearity. As a result, simplified models are often employed.

Mathematical modeling is an art in which one needs to capture the essential features of the phenomenon under study, yet keep the model sufficiently simple so that it is useful. Complications and details can be added later by modifying a model that has provided physical insight, and observational data and statistics are required in order to formulate the necessary boundary and initial conditions. One ends up using approximations, which are usually found on the basis of physical intuition rather than mathematical convenience, although sometimes the temptation to kill complicated terms in the equations has led to meaningful approximations. The assumptions of the model, however, should not oversimplify—the old adagio applies: no model is better than its assumptions.

Environmental science takes advantage of virtually every mathematical tool developed—here we review the basic mathematical concepts used in the solution of the exercises of this book.

1.1   Complex numbers

Complex numbers are used to describe physical systems ruled by linear differential equations, to represent physical quantities with Fourier series and Fourier integrals, to compute definite integrals of functions of a single variable, in quantum mechanics, fluid dynamics, and in many other applications.

1 **(A)** Solve the complex algebraic equation

$$x + iy + 2 + 3i = 1 - 2i.$$

Solution

This equation can be rewritten as

$$x + iy = -1 - 5i$$

and, by equating the real (respectively, imaginary) part of the left-hand side to the real (respectively, imaginary) part of the right-hand side, we obtain the complex solution $z = -1 - 5i$.

2 **(A)** Solve the complex algebraic equation

$$z^2 - i = 0.$$

Solution

One can rewrite this equation using the polar form of $i = \cos(\pi/2) + i\sin(\pi/2) = e^{i\pi/2}$ as

$$z^2 = i = e^{i(\pi/2 + 2n\pi)} \qquad (n = 0, 1, 2, 3, \dots),$$

which has the two distinct solutions obtained for $n = 0, 1$

$$z_{1,2} = e^{i(\pi/4 + n\pi)} = \left[\cos\left(\frac{\pi}{4} + n\pi\right) + i\sin\left(\frac{\pi}{4} + n\pi\right)\right]$$

$$= \pm\frac{\sqrt{2}}{2}(1 + i).$$

3 **(A)** What regions of the complex plane correspond to the following?
a) $|z| < 1$

b) $\mathrm{Re}(z) > 3$
c) $\mathrm{Im}(z) > 2$
d) $|z + 5| \leq 1$
e) $-1 \leq \mathrm{Im}(z) \leq 1$
f) $2 < |z| < 3$

Solution
Let $z = x + iy$, where $x = \mathrm{Re}(z)$ and $y = \mathrm{Im}(z)$ are real. Then:
a) represents a circle of unit radius centered on the origin $z = 0$ and excluding the circumference of radius $r \equiv \sqrt{x^2 + y^2} = 1$
b) represents the half-plane $x > 3$ with arbitrary y
c) represents the half-plane $y > 2$ with x arbitrary
d) represents the circle of unit radius centered on $z = -5$ [or $(x, y) = (-5, 0)$] and including the circumference of unit radius
e) represents the horizontal strip $-1 \leq y \leq 1$ with arbitrary x
f) represents the annulus comprised between the circles of radii 2 and 3 and centered on the origin $z = 0$.

4 **(A)** Express the complex number $z = 1 + i\sqrt{3}$ in polar form.

Solution
The polar form is

$$z = \rho\, e^{i(\theta + 2n\pi)} \qquad (n = 0, 1, 2, 3, \ldots),$$

where $\rho = |z| = \sqrt{1^2 + \left(\sqrt{3}\right)^2} = 2$ and $\theta = \mathrm{tg}^{-1}(\sqrt{3}/1) = \pi/3$; hence

$$z = 2\, e^{i(\pi/3 + 2n\pi)} \qquad (n = 0, 1, 2, 3, \ldots).$$

The argument of z obtained for $n = 0$ is called the *principal argument* of z.

5 **(A)** When studying oscillations of a physical system described by ordinary differential equations, is it always legitimate to represent an oscillating quantity A using a complex exponential as $A = A_0 \exp(i\omega t)$, and to take the real part of A at the end of the calculations as the physical result? If $x(t)$ and $y(t)$ are oscillating quantities represented by complex exponentials, is $\mathrm{Re}(xy) = \mathrm{Re}(x) \cdot \mathrm{Re}(y)$?

Solution
No: the above representation is legitimate only when the oscillating quantity A obeys *linear* differential equations. Often a system described by a set of nonlinear equations may be described by the

linearized version of the full equations under the assumption of small motions or small oscillations (e.g., a simple pendulum), which may constitute a physically meaningful approximation.

If $x(t) = x_0 \, e^{i\omega_1 t}$ and $y(t) = y_0 \, e^{i\omega_2 t}$, then

$$x(t) \, y(t) = x_0 y_0 \, e^{i(\omega_1 + \omega_2)t};$$

however,

$$\operatorname{Re}(xy) = x_0 y_0 \cos\left[(\omega_1 + \omega_2)\, t\right]$$

$$= x_0 y_0 \left[\cos(\omega_1 t) \cos(\omega_2 t) - \sin(\omega_1 t) \sin(\omega_2 t)\right]$$

$$\neq \operatorname{Re}(x) \cdot \operatorname{Re}(y) = x_0 y_0 \left[\cos(\omega_1 t) \cos(\omega_2 t)\right].$$

6 **(A)** Prove that a phase factor $e^{i\theta}$, where θ is real, has unit modulus.

Solution
We have

$$|z| \equiv e^{i\theta} = |\cos\theta + i\sin\theta| = \left(\cos^2\theta + \sin^2\theta\right)^{1/2} = 1.$$

7 **(A)** Prove that

a) $\operatorname{Re}(z) = \frac{z + z^*}{2}$

b) $\operatorname{Im}(z) = \frac{z - z^*}{2i}$

c) $z^2 = (z^*)^2$ only if z is purely real or purely imaginary.

Solution
Let $z = x + iy$, where $x = \operatorname{Re}(z)$ and $y = \operatorname{Im}(z)$ are real. Then we have

$$\frac{z + z^*}{2} = \frac{(x + iy) + (x - iy)}{2} = x,$$

$$\frac{z - z^*}{2i} = \frac{(x + iy) - (x - iy)}{2i} = y;$$

the equation $z^2 = (z^*)^2$ is equivalent to

$$(x + iy)^2 = (x - iy)^2,$$

or

$$x^2 - y^2 + 2ixy = x^2 - y^2 - 2ixy.$$

Equating the real part of the left-hand side to the real part of the right-hand side and doing the same for the imaginary parts yields $xy = 0$, with solutions $x = 0$, or $y = 0$, or both x and y vanishing.

8 **(A)** Show that there are exactly n distinct roots of a complex number $z \neq 0$.

Solution
Write z in its polar form

$$z = \rho e^{i(\theta + 2k\pi)},$$

where $k = 0, 1, 2, 3, \ldots$ Then

$$z^{1/n} = \rho^{1/n} e^{i\left(\frac{\theta}{n} + \frac{2k}{n}\pi\right)}.$$

The n distinct roots of z are obtained from this formula by letting k assume the n values

$$k = 0, 1, 2, \ldots, (n-1).$$

9 **(B)** Use complex exponentials to derive the trigonometric identities

$$\sin(2\theta) = 2\sin\theta\cos\theta,$$

$$\cos(2\theta) = \cos^2\theta - \sin^2\theta.$$

Solution
The de Moivre formula

$$e^{i\theta} = \cos\theta + i\sin\theta$$

squared yields

$$e^{2i\theta} = \left(e^{i\theta}\right)^2 = (\cos\theta + i\sin\theta)^2 = \cos^2\theta - \sin^2\theta + 2i\sin\theta\cos\theta.$$

On the other hand,

$$e^{2i\theta} = \cos(2\theta) + i\sin(2\theta);$$

by comparing the two expressions of $e^{2i\theta}$ one deduces that

$$\cos(2\theta) + i\sin(2\theta) = \cos^2\theta - \sin^2\theta + 2i\sin\theta\cos\theta;$$

and by equating the real and the imaginary parts of the two sides of this equation, we obtain

$$\sin(2\theta) = 2\sin\theta\cos\theta,$$

$$\cos(2\theta) = \cos^2\theta - \sin^2\theta.$$

1.2   Differentiation and integration of functions of a single variable

Basic calculus begins by studying functions of a single variable: the most basic operations are taking limits, differentiation, and integration. In the mathematical modeling of a physical process or system one begins by choosing an independent variable and by letting other variables be functions of it—for example, in the problem of motion of a point particle the independent variable can be time and the particle coordinates are dependent variables.

1.2.1   Differentiation

1 **(B)** Compute the first and second derivatives of the function $f(x) = x\,e^x \ln x + x^3 \sin x$.

Solution
The function $f(x)$ is defined on $(0, +\infty)$ and has derivatives of all orders on this interval. The first derivative is

$$f'(x) = e^x \ln x + x\,e^x \ln x + e^x + 3x^2 \sin x + x^3 \cos x,$$

while the second derivative is

$$f''(x) = e^x \ln x + \frac{e^x}{x} + e^x \ln x + x\,e^x \ln x + x\,\frac{e^x}{x} + e^x + 6x \sin x$$

$$+ 3x^2 \cos x + 3x^2 \cos x - x^3 \sin x$$

$$= (x+2)\,e^x \ln x + \left(\frac{1}{x} + 2\right) e^x + x\left(6 - x^2\right) \sin x + 6x^2 \cos x.$$

2 **(B)** Compute the derivative df/dx, where

$$f(x) = \sqrt{\cos\left(\sin^2 x\right)}.$$

Solution
We have

$$\frac{df}{dx} = \frac{d\left(\cos\left(\sin^2 x\right)\right)/dx}{2\sqrt{\cos\left(\sin^2 x\right)}} = \frac{-\sin\left(\sin^2 x\right) d\left(\sin^2 x\right)/dx}{2\sqrt{\cos\left(\sin^2 x\right)}}$$

$$= \frac{-\sin x \cos x \sin\left(\sin^2 x\right)}{\sqrt{\cos\left(\sin^2 x\right)}}.$$

3 **(B)** Compute the derivative df/dx, where

$$f(x) = x^2 e^{-2x} \frac{3 e^{-2x} - 1}{(3 e^{-2x} + 1)^2}.$$

Solution
We have

$$\frac{df}{dx} = 2x e^{-2x} \frac{3 e^{-2x} - 1}{(3 e^{-2x} + 1)^2} - 2x^2 e^{-2x} \frac{3 e^{-2x} - 1}{(3 e^{-2x} + 1)^2}$$

$$+ 6x^2 e^{-4x} \left[\frac{- (3 e^{-2x} + 1) + 2 (3 e^{-2x} - 1)}{(3 e^{-2x} + 1)^3} \right]$$

$$= 2x e^{-2x} \left[\frac{(3 e^{-2x} - 1)}{(3 e^{-2x} + 1)^2} (1 - x) + 9x e^{-2x} \frac{(e^{-2x} - 1)}{(3 e^{-2x} + 1)^3} \right]$$

$$= \frac{2x e^{-2x}}{(3 e^{-2x} + 1)^3} \left(9 e^{-4x} - 9x e^{-2x} + x - 1 \right).$$

4 **(B)** Compute the derivative of the function

$$f(x) = x \ln \left(3x^4 + 2x^2 + |x| + 1 \right).$$

Solution
We apply the Leibnitz rule $(fg)' = f'g + fg'$ and the chain rule $\frac{d(f(g(x)))}{dx} dx = \frac{df}{dg} \frac{dg}{dx}$ obtaining, for $x \neq 0$,

$$\frac{df}{dx} = \ln \left(3x^4 + 2x^2 + |x| + 1 \right) + \frac{x \left(12x^3 + 4x + \frac{|x|}{x} \right)}{(3x^4 + 2x^2 + |x| + 1)}$$

$$= \ln \left(3x^4 + 2x^2 + |x| + 1 \right) + \frac{12x^4 + 4x^2 + |x|}{3x^4 + 2x^2 + |x| + 1}.$$

This result is obtained for $x \neq 0$, but the function $f(x)$ is defined at $x = 0$ and since the two limits

$$\lim_{x \to 0^-} \frac{df}{dx} = \lim_{x \to 0^-} \left[\ln \left(3x^4 + 2x^2 + |x| + 1 \right) + \frac{12x^4 + 4x^2 + |x|}{3x^4 + 2x^2 + |x| + 1} \right]$$

$$= 0,$$

$$\lim_{x \to 0^+} \frac{df}{dx} = \lim_{x \to 0^+} \left[\ln \left(3x^4 + 2x^2 + |x| + 1 \right) + \frac{12x^4 + 4x^2 + |x|}{3x^4 + 2x^2 + |x| + 1} \right]$$

$$= 0,$$

exist and are equal, we conclude that the derivative of $f(x)$ at $x = 0$ exists and is zero.

5 **(B)** Prove that

$$\arcsin x + \arccos x = \frac{\pi}{2}, \qquad -1 \le x \le 1,$$

$$\operatorname{arctg} x + \operatorname{arccotg} x = \frac{\pi}{2}, \qquad -\infty < x < +\infty.$$

Solution
Differentiate $\arcsin x + \arccos x$ in the interval $(-1, 1)$:

$$\frac{d}{dx} \left(\arcsin x + \arccos x \right) = \frac{1}{\sqrt{1 - x^2}} - \frac{1}{\sqrt{1 - x^2}} = 0,$$

hence $\arcsin x + \arccos x = $ const. in $(-1, 1)$ and, by continuity, also in $[-1, 1]$. At $x = 1$ we have $\arcsin 1 + \arccos 1 = \pi/2$, which fixes the value of the constant. Hence $\arcsin x + \arccos x = \pi/2$ in $[-1, 1]$.

Let us consider the function $\operatorname{arctg} x + \operatorname{arccotg} x$ on the real axis: differentiating in this interval we obtain

$$\frac{d}{dx} \left(\operatorname{arctg} x + \operatorname{arccotg} x \right) = \frac{1}{1 + x^2} - \frac{1}{1 + x^2} = 0$$

and hence $\operatorname{arctg} x + \operatorname{arccotg} x$ is constant. Since at $x = 1$

$$\operatorname{arctg} 1 + \operatorname{arccotg} 1 = \frac{\pi}{4} + \frac{\pi}{4} = \frac{\pi}{2},$$

the value of the constant is fixed and $\operatorname{arctg} x + \operatorname{arccotg} x = \pi/2$ over the entire real axis.

6 **(B)** Prove the identities

$$\operatorname{arctg} x + \operatorname{arctg} \left(\frac{1}{x} \right) = \frac{\pi}{2} \qquad (x > 0),$$

$$\operatorname{arctg} x + \operatorname{arctg} \left(\frac{1}{x} \right) = -\frac{\pi}{2} \qquad (x < 0).$$

Solution

The function $\mathrm{arctg}\, x + \mathrm{arctg}\left(\frac{1}{x}\right)$ is singular at $x = 0$ and therefore we have to consider separately the two semi-infinite intervals $x < 0$ and $x > 0$. Differentiation yields

$$\frac{d}{dx}\left[\mathrm{arctg}\, x + \mathrm{arctg}\left(\frac{1}{x}\right)\right] = \frac{1}{1 + x^2} + \frac{1}{1 + 1/x^2}\left(\frac{-1}{x^2}\right) = 0$$

for any $x \neq 0$. Hence the function $\mathrm{arctg}\, x + \mathrm{arctg}\left(\frac{1}{x}\right)$ is constant, but the value of the constant is different in the two disconnected intervals $x < 0$ and $x > 0$. In fact, for $x = -1$ it is

$$\mathrm{arctg}(-1) + \mathrm{arctg}(-1) = -\frac{\pi}{4} - \frac{\pi}{4} = -\frac{\pi}{2},$$

while for $x = 1$ it is

$$\mathrm{arctg}\, 1 + \mathrm{arctg}\, 1 = \frac{\pi}{4} + \frac{\pi}{4} = \frac{\pi}{2},$$

which fixes the values of the constants.

7 **(B)** Determine whether there exist values of α and β such that the curves representing the two functions

$$f(x) = \frac{3\alpha}{4} x^4 + \beta x^2,$$

$$g(x) = x^2 + 3,$$

have parallel tangents at some point, and find the values of x for which this happens.

Solution

The functions f and g are continuous with all their derivatives of any order on $(-\infty, +\infty)$. The points with the desired property are those where the first derivatives of f and g are equal, i.e., where

$$3\alpha x^3 + 2\beta x = 2x$$

or

$$x\left[3\alpha x^2 + 2\left(\beta - 1\right)\right] = 0.$$

The point $x = 0$ has the desired property for any value of α and β: the tangent to both curves representing $f(x)$ and $g(x)$ is horizontal here.

If $\alpha = 0$, one finds immediately that setting $\beta = 1$ all points x have the desired property.

If $\alpha \neq 0$, then the points x with the desired property satisfy the equation

$$x^2 = -\frac{2}{3\alpha}(\beta - 1);$$

this equation has solutions for $\alpha < 0$ and $\beta \geq 1$, or for $\alpha > 0$ and $\beta \leq 1$; the points x with the desired property are

$$x = \pm\sqrt{\frac{2}{3}\left|\frac{\beta - 1}{\alpha}\right|}.$$

To summarize, the values of α and β that allow for the desired property are
any (α, β) and $x = 0$,
$(\alpha, \beta) = (0, 1)$ and any x,
(α, β) with $\alpha < 0$ and $\beta \geq 1$, and $x = \pm\sqrt{\frac{2}{3}\left|\frac{\beta-1}{\alpha}\right|}$,
(α, β) with $\alpha > 0$ and $\beta \leq 1$, and $x = \pm\sqrt{\frac{2}{3}\left|\frac{\beta-1}{\alpha}\right|}$.

1.2.2   Integration
1 **(B)** Compute the indefinite integral

$$\int dx\,\left(x\,e^x + 3x^2\right).$$

Solution
Because of the linearity of the integral, we have

$$\int dx\,\left(x\,e^x + 3x^2\right) = \int dx\,x\,e^x + 3\int dx\,x^2.$$

The first integral on the right-hand side is evaluated by parts, obtaining

$$\int dx\,x\,e^x = x\,e^x - \int dx\,e^x = (x - 1)\,e^x.$$

As a check, one can take the derivative of this last term,

$$\frac{d}{dx}\left[(x - 1)\,e^x\right] = e^x + (x - 1)\,e^x = x\,e^x,$$

which assures us of the correctness of this first integral. The second integral is elementary,

$$\int dx\, x^2 = \frac{x^3}{3},$$

and therefore we have

$$\int dx\,\left(x\,e^x + 3x^2\right) = x\,e^x + x^3 + \text{constant}.$$

2 **(B)** Compute the definite integrals

$$I_1 = \int_{-\infty}^{+\infty} dx\, f(x)\, x,$$

$$I_2 = \int_{-\infty}^{+\infty} dx\, f(x)\, x^3,$$

$$I_3 = \int_{-\infty}^{+\infty} dx\, g(x)\, x^2,$$

$$I_4 = \int_{-\infty}^{+\infty} dx\, g(x)\, x^8,$$

where the functions $f(x)$ and $g(x)$ are defined and regular over the entire real axis and are, respectively, even and odd, i.e., $f(-x) = f(x)$ and $g(-x) = -g(x)$ for any real value of x.

Solution
We have
$$I_1 = I_2 = I_3 = I_4 = 0$$
because in all these cases the integrand is an odd function of x and the integrals are computed over an interval symmetric with respect to $x = 0$ (the entire real axis). The contribution to the integral coming from regions with $x < 0$ cancels the corresponding contribution, with opposite sign, from symmetric regions with $x > 0$.

3 **(B)** Compute the integral

$$\int_{1}^{+\infty} dx\, \frac{1}{x\,(x+1)}.$$

Solution
We decompose the fraction in the integrand as follows:

$$\frac{1}{x\,(x+1)} - \frac{A}{x} + \frac{B}{x+1},$$

where the constants A and B are determined by writing the two terms on the right-hand side with common denominator

$$\frac{A}{x} + \frac{B}{x+1} = \frac{(A+B)\,x + A}{x\,(x+1)}$$

and setting this equal to $1/x(x+1)$, which yields

$$A + B = 0,$$

$$A = 1,$$

or $(A, B) = (1, -1)$. Therefore,

$$
\begin{aligned}
\int_1^{+\infty} dx\, \frac{1}{x\,(x+1)} &= \int_1^{+\infty} \frac{dx}{x} - \int_1^{+\infty} \frac{dx}{x+1} \\[2mm]
&= \left[\ln x - \ln\,(x+1)\right]_1^{+\infty} \\[2mm]
&= \lim_{M \to +\infty} \left[\ln\left(\frac{M}{M+1} \right) - \ln \frac{1}{2} \right] \\[2mm]
&= \left[\lim_{M \to +\infty} \left(\frac{M}{M+1} \right) \right] + \ln 2 = \ln 2.
\end{aligned}
$$

4 (B) Consider a river modeled as a straight channel of width a with irregular depth. Using horizontal x- and y- axes pointing in the direction of the flow and in the transversal direction, respectively, the depth profile across the river is given by the function

$$
h(y) = \begin{cases} -h_0 \sin\left[\pi\,\left(1 - \frac{y}{a}\right)\right] & \text{if } 0 \le y < a, \\ 0 & \text{if } y < 0 \text{ or } y \ge a, \end{cases}
$$

where h_0 is a constant with the dimensions of a length. Compute the cross-sectional area of the river.

Solution
The area of a cross section of the river is

$$
\begin{aligned}
A &= \int_0^a dy\, |h(y)| = -h_0 \int_0^a dy \sin\left[\pi \left(\frac{y}{a} - 1\right)\right] \\[2mm]
&= h_0\, \frac{a}{\pi}\, \cos\left[\pi \left(\frac{y}{a} - 1\right)\right]\Big|_0^a = \frac{2 h_0 a}{\pi} \simeq 0.6367 h_0 a.
\end{aligned}
$$

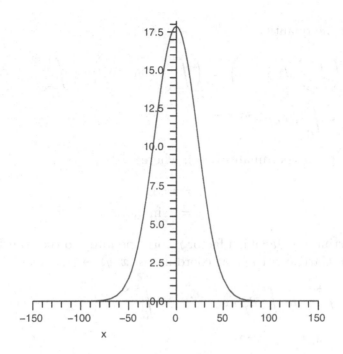

Figure 1.1. The normalized Gaussian (1.2).

5 **(B)** Compute the integral

$$I = \int_{-\infty}^{+\infty} dx\, e^{-\alpha x^2},$$ (1.1)

which represents the area of the region of plane delimited by the x-axis and by the graph of a Gaussian[1] (Fig. 1.1). Normalize the Gaussian in such a way that

$$f(x) \equiv N\, e^{-\alpha x^2},$$ (1.2)

where N is a constant, satisfies

$$\int_{-\infty}^{+\infty} dx\, f(x) = 1.$$

---

[1]The Gaussian function is widely used in statistics and in many models (*Gaussian plume models*) describing the spreading of pollutants in water or in the atmosphere.

Solution

Consider the quantity

$$I^2 = \left(\int_{-\infty}^{+\infty} dx\, e^{-\alpha x^2} \right)^2 = \left(\int_{-\infty}^{+\infty} dx\, e^{-\alpha x^2} \right) \cdot \left(\int_{-\infty}^{+\infty} dy\, e^{-\alpha y^2} \right)$$

$$= \int \int_{R^2} dx\, dy\, e^{-\alpha(x^2+y^2)}.$$

By using polar coordinates (r, φ), where

$$x = r\cos\varphi,$$
$$y = r\sin\varphi,$$

and inserting the Jacobian factor r corresponding to the transformation from Cartesian to polar coordinates $(x, y) \to (r, \varphi)$, we obtain

$$I^2 = \int_0^{+\infty} dr \int_0^{\pi} d\varphi\, r\, e^{-\alpha r^2} = 2\pi \int_0^{+\infty} dr \left(-\frac{1}{2\alpha} \right) \frac{d}{dr} \left(e^{-\alpha r^2} \right)$$

$$= -\frac{\pi}{\alpha} \left[e^{-\alpha r^2} \right]_0^{+\infty} = \frac{\pi}{\alpha},$$

and therefore

$$I = \sqrt{\frac{\pi}{\alpha}}.$$

In order to find a normalization factor N such that $\int_{-\infty}^{+\infty} dx\, f(x) = 1$, one needs to impose the condition

$$\int_{-\infty}^{+\infty} dx\, f(x) = 1,$$

and hence $N = 1/I$. With the choice $N = \sqrt{\alpha/\pi}$, the normalized Gaussian

$$f(x) = \sqrt{\frac{\alpha}{\pi}}\, e^{-\alpha x^2}$$

satisfies

$$\int_{-\infty}^{+\infty} dx\, N e^{-\alpha x^2} = 1.$$

1.2.3   Maxima, minima, and graphs

Calculus allows one to compute maxima, minima, and inflection points, to study the behavior of functions of a single variable and to construct their graphs. In physics, these tools are used to find states of equilibrium, study stability, or optimize choices.

1 **(B)** Study the graph of the function $f(x) = x^2 \ln |x|$.

Solution
The function is defined on $(-\infty, 0) \cup (0, +\infty)$ and is continuous with all its derivatives of any order there. The function is even, i.e., $f(x) = f(-x)$ for all values of x in the intervals on which f is defined. We also notice that $f(x) > 0$ for $|x| > 1$, that $f(x) < 0$ in the intervals $-1 < x < 0$ and $0 < x < 1$, and $f(\pm 1) = 0$. The points $x = \pm 1$ are the only zeros of f.

Let us compute the limits of $f(x)$; as $x \to 0$ we have

$$\lim_{x \to 0} f(x) = \lim_{x \to 0} \frac{\ln |x|}{1/x^2} = \lim_{x \to 0} \frac{\frac{1}{|x|}\frac{|x|}{x}}{\frac{-2}{x^3}} = \lim_{x \to 0} \frac{-x^2}{2} = 0,$$

by using de l'Hôpital rule. The function $f(x)$ can be redefined so that

$$\tilde{f}(x) \equiv \begin{cases} f(x) & \text{if } x \neq 0, \\ 0 & \text{if } x = 0 \end{cases}$$

is continuous at $x = 0$. The other limits of $f(x)$ are

$$\lim_{x \to \pm\infty} x^2 \ln |x| = +\infty.$$

The first derivative of the function f is

$$f'(x) = x\left(1 + 2\ln |x|\right),$$

and its sign is determined by studying the sign of $1 + 2\ln |x|$, which is positive for $|x| > e^{-1/2}$, negative for $-e^{-1/2} < x < e^{-1/2}$, and zero at $\pm e^{-1/2}$. Therefore:
$f'(x) < 0$ and f is strictly decreasing if $x < -1/\sqrt{e}$ and $0 < x < 1/\sqrt{e}$;
$f'(\pm 1/\sqrt{e}) = 0$ and f has horizontal tangent there;
$f'(x) > 0$ and f is strictly increasing if $-1/\sqrt{e} < x < 0$ and $x > 1/\sqrt{e}$.
This information, plus what we know about the continuity of f, is sufficient to establish that $f(x)$ has local and absolute minima at $x = \pm 1/\sqrt{e}$, and the minimum is $f(\pm 1/\sqrt{e}) = -1/2e$. The graph of the function is reported in Fig. 1.2.

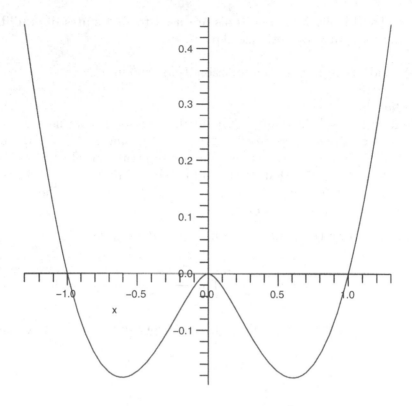

Figure 1.2. The graph of $f(x) = x^2 \ln |x|$.

2 **(B)** It can be shown [12, 27, 4] that the work delivered by a heat engine is

$$f(x) = xa\left(T_H - \frac{T_C}{1-x}\right),$$

where x is the efficiency of the engine and T_H and T_C (with $0 < T_C < T_H$) are the absolute temperatures of the hot and cold reservoir, respectively, while a is a positive constant. Find the efficiency that maximizes the work delivered in the interval[2] $[0, 1 - T_C/T_H]$.

Solution
We look for a maximum of the function $f(x)$ in the efficiency interval $[0, 1 - T_C/T_H]$. The function f is continuous with all its derivatives

---

[2]Thermodynamics imposes the fundamental upper limit on the efficiency $0 \le x \le x_c$, where $x_c \equiv 1 - T_C/T_H < 1$ is the *Carnot factor*.

in this interval and its first derivative is

$$\frac{df}{dx} = a\left(T_H - \frac{T_c}{1-x}\right) - \frac{aT_C x}{(1-x)^2} = \frac{aT_H}{(1-x)^2}\left[(1-x)^2 - \frac{T_C}{T_H}\right].$$

One has $df/dx > 0$ and f strictly increasing if $(1-x)^2 > T_C/T_H$, $df/dx = 0$ (horizontal tangent) if $(1-x)^2 = T_C/T_H$, while $df/dx < 0$ and f is strictly decreasing if $(1-x)^2 < T_C/T_H$. The inequality $(1-x)^2 > T_C/T_H$ corresponds to

$$x < 1 - \sqrt{\frac{T_C}{T_H}} < 1 - \frac{T_C}{T_H},$$

and the above results are sufficient to conclude that $f(x)$ has a local maximum at $x_* = 1 - \sqrt{T_C/T_H}$, which has the value

$$f_{\max} = f\left(1 - \sqrt{\frac{T_C}{T_H}}\right) = x_* a T_H\left(1 - \frac{1}{1-x_*}\frac{T_C}{T_H}\right)$$

$$= a\left(1 - \sqrt{\frac{T_C}{T_H}}\right)^2.$$

Since $f_{\max} > f(0) = f(1 - T_C/T_H) = 0$ and f is continuous on $[0, 1 - T_C/T_H]$, the local maximum is also an absolute maximum.

3 **(B)** Study the graph of the function $f(x) = x\,|x|\,e^x$.

Solution
The function is defined on $(-\infty, +\infty)$ and is continuous in this interval. All its derivatives exist and are continuous on $(-\infty, 0) \cup (0, +\infty)$. The limits of the function at the boundaries of this interval are

$$\lim_{x \to +\infty} f(x) = +\infty,$$

$$\lim_{x \to -\infty} f(x) = 0.$$

The function is negative for $x < 0$, vanishes only at $x = 0$, and is positive for $x > 0$. The first derivative of $f(x)$ for $x \neq 0$ is

$$f'(x) = |x|\,e^x\,(x+2).$$

Since both limits

$$\lim_{x \to 0^-} f'(x) = 0,$$

$$\lim_{x \to 0^+} f'(x) = 0$$

exist and are finite and equal, the first derivative of $f(x)$ exists also at $x = 0$ and has zero value.

The study of the sign of $f'(x)$ allows one to conclude that
$f'(x) < 0$ for $x < -2$, where f is strictly decreasing;
$f'(-2) = 0$, where the graph of f has horizontal tangent;
$f'(x) > 0$ for $x > -2$, where $f(x)$ is strictly increasing.
The function f has a local minimum at $x = -2$, which is $f(-2) = -4/e^2$. This minimum is also an absolute minimum.

The second derivative $f''(x)$ is defined on the set $(-\infty, 0) \cup (0, +\infty)$ and has the value

$$f''(x) = |x|\, e^x \left(\frac{2}{x} + x + 4 \right):$$

it is not defined at $x = 0$. By studying the sign of $f''(x)$ one concludes that
$f''(x) < 0$ for $x < -(2 + \sqrt{2})$ and for $-2 + \sqrt{2} < x < 0$; the graph of the function has concavity facing downward in these intervals.
$f''(-2 \pm \sqrt{2}) = 0$ and the graph of $f(x)$ changes concavity at $x = -2 \pm \sqrt{2}$.
$f''(x) > 0$ for $-(2 + \sqrt{2}) < x < -2 + \sqrt{2}$ and for $x > 0$, where the curve representing $f(x)$ has upward-facing concavity. Therefore, the graph of $f(x)$ is as follows: the x-axis is a horizontal asymptote as $x \to -\infty$. Beginning from $x \to -\infty$, the function is negative with downward-facing concavity, decreases until it reaches its absolute minimum at $x = -2$ (changing concavity at $x = -2-\sqrt{2}$ before it reaches its minimum), then it starts increasing and is always strictly increasing for $x > -2$ (it changes concavity again at $x = -2 + \sqrt{2}$ past its minimum point). It reaches its zero at $x = 0$, where the second derivative has a jump discontinuity (from -2 as $x \to 0^-$ to $+2$ as $x \to 0^+$), and diverges as $x^2\, e^x$ as $x \to +\infty$. The graph is reported in Fig. 1.3.

4 **(B)** Study the function $f(x) = x\, e^{\lambda x}$ as the real parameter λ varies, and sketch its graph.

Solution
The function is continuous with all its derivatives of any order on $(-\infty, +\infty)$. The sign of $f(x)$ is easy to study—we have, for any real value of the parameter λ:
$f(x) > 0$ if $x > 0$;
$f(x) = 0$ only at $x = 0$;
$f(x) < 0$ if $x < 0$.

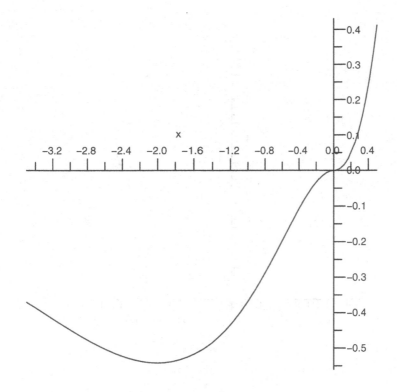

Figure 1.3. The graph of $f(x) = x\,|x|\,e^x$.

The first and second derivatives of f are

$$f'(x) = (\lambda x + 1)\,e^{\lambda x},$$

$$f''(x) = \lambda\,e^{\lambda x}\,(\lambda x + 2),$$

respectively. We now consider the possible values of λ separately. If $\lambda > 0$, the limits of $f(x)$ are

$$\lim_{x \to +\infty} x\,e^{\lambda x} = +\infty,$$

$$\lim_{x \to -\infty} x\,e^{\lambda x} = 0.$$

The sign of the first derivative of f is as follows:
$f'(x) < 0$ for $x < -1/\lambda$, where f is strictly decreasing;
$f'(-1/\lambda) = 0$ (f has horizontal tangent);
$f'(x) > 0$ for $x > -1/\lambda$, where f is strictly increasing.

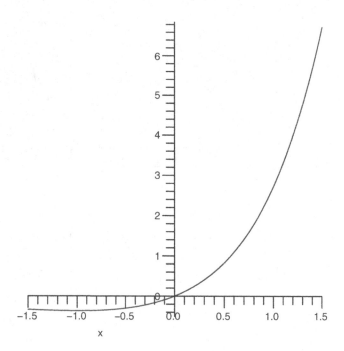

Figure 1.4. The graph of $f(x) = x\,e^{\lambda x}$ for $\lambda = 1$.

Hence, $f(x)$ has a local minimum at $x = -1/\lambda$, which has the value $f(-1/\lambda) = -(\lambda e)^{-1}$. Given what we know on the limits of f and the fact that f is continuous everywhere, we can state that the local minimum is also an absolute minimum. We can also study the sign of the second derivative of f, concluding that

$f''(x) < 0$ for $x < -2/\lambda$, and the concavity of the graph of f is facing downward;

$f''(-2/\lambda) = 0$, where there is a change of concavity;

$f''(x) > 0$ for $x > -2/\lambda$, and the concavity of the graph of f is facing upward.

As $\lambda \to 0^+$, the minimum point and the inflection point get closer and closer to zero—they coincide for $\lambda = 0$.

For $\lambda = 0$ the function reduces to $f(x) = x$, whose graph is the straight line passing through the origin and making a 45° angle with both axes. This is the limiting case of the situations for $\lambda > 0$ or

$\lambda < 0$ as the parameter λ becomes smaller and smaller.

For $\lambda < 0$ the limits of f are

$$\lim_{x \to +\infty} x\,e^{\lambda x} = 0,$$

$$\lim_{x \to -\infty} x\,e^{\lambda x} = -\infty.$$

The sign of the first derivative is as follows:
$f'(x) > 0$ for $x < 1/|\lambda|$, where $f(x)$ is strictly increasing;
$f'(1/|\lambda|) = 0$, where f has horizontal tangent and an inflection point;
$f'(x) < 0$ for $x > 1/|\lambda|$, where $f(x)$ is strictly decreasing.

Therefore, $f(x)$ has an absolute maximum at $x = 1/|\lambda|$, with value $f(1/|\lambda|) = -1/(\lambda e) > 0$. The second derivative $f''(x)$ has sign given by the following:
$f''(x) < 0$ for $x < 2/|\lambda|$, where the graph of f has downward-facing concavity;
$f''(2/|\lambda|) = 0$, where there is a change of concavity;
$f''(x) > 0$ for $x > 2/|\lambda|$, where the graph of f has upward-facing concavity.
As $\lambda \to 0^-$, the maximum point and the inflection point get closer and closer to zero—they coincide for $\lambda = 0$.

The graphs of $f(x)$ for $\lambda = +1$ and $\lambda = -1$ are given in Fig. 1.4 and Fig. 1.5, where the qualitative variation as λ increases from negative to positive values is evident.

5 **(B)** Study the function $f(x) = x^2(1 + \lambda x)$ as the real parameter λ varies, and sketch its graph.

Solution
The function is continuous with all its derivatives of any order on $(-\infty, +\infty)$; in addition, $f(0) = 0$. For large values of $|x|$ the functions is asymptotic to the cubic λx^3. We consider the possible values of λ separately.

For $\lambda < 0$ we have $f'(x) = x(2 - 3|\lambda|x)$ and $f(x) \approx -|\lambda|x^3$ as $x \to +\infty$, with

$$\lim_{x \to \pm\infty} f(x) = \mp\infty.$$

The sign of the first derivative is obtained by separately studying the signs of the terms x and $(2 - 3|\lambda|x)$ and then putting the results

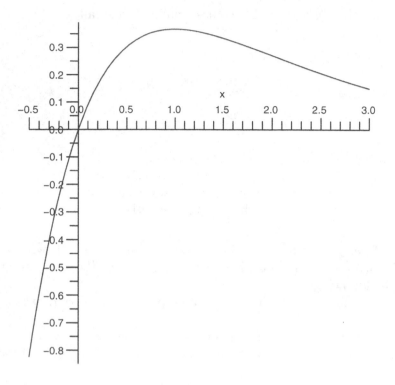

Figure 1.5. The graph of $f(x) = x\,e^{\lambda x}$ for $\lambda = -1$.

together, obtaining

$f'(x) < 0$ for $x < 0$, where f is strictly decreasing;

$f'(0) = 0$, where f has horizontal tangent;

$f'(x) > 0$, for $0 < x < 2/\left(3\,|\lambda|\right)$ where f is strictly increasing;

$f'\left(2/3\,|\lambda|\right) = 0$, where f has horizontal tangent again;

$f'(x) < 0$ for $x > 2/3\,|\lambda|$, where f is strictly decreasing.

Since f is continuous, this is sufficient to establish that there is a local minimum of f at $x = 0$ [which is $f(0) = 0$] and a local maximum at $x = 2/3\,|\lambda|$, with value $f\left(2/3\,|\lambda|\right) = 4/\left(27\lambda^2\right)$.

For $\lambda = 0$ the function reduces to the parabola $f(x) = x^2$ with an absolute minimum $f(0) = 0$ at $x = 0$ and $f(x) > 0$ for $x \neq 0$

For $\lambda > 0$ the function is asymptotic to λx^3 as $x \to +\infty$ and

$$\lim_{x \to +\infty} = +\infty,$$

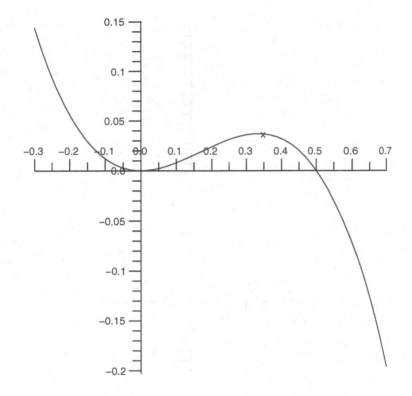

Figure 1.6. The graph of $f(x) = x^2 (1 + \lambda x)$ for $\lambda = -2$.

$$\lim_{x \to -\infty} = -\infty.$$

The sign of the first derivative is

$f'(x) > 0$ for $x < -2/(3\lambda)$ and for $x > 0$, where $f(x)$ is strictly increasing;

$f'(0) = f'(-2/(3\lambda)) = 0$, where the graph of f has horizontal tangent;

$f'(x) < 0$ for $-2/(3\lambda) < x < 0$, where $f(x)$ is strictly decreasing.

Since f is continuous everywhere, it has a local maximum

$$f_{max} = f\left(\frac{-2}{3\lambda}\right) = \frac{4}{27\lambda^2}$$

at $x = -2/(3\lambda)$ and a local minimum $f(0) = 0$ at $x = 0$.

The graphs of $f(x)$ for $\lambda = -2$ and $\lambda = 2$ are given in Fig. 1.6 and Fig. 1.7, respectively.

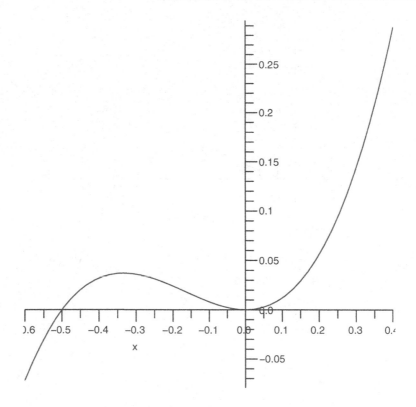

Figure 1.7. The graph of $f(x) = x^2 (1 + \lambda x)$ for $\lambda = 2$.

1.3   Ordinary differential equations

Ordinary differential equations (ODEs) describe physical systems with a finite number of degrees of freedom. Often the solution of partial differential equations, which describe systems with an infinite number of degrees of freedom, can be reduced to the problem of solving a set of ODEs. Many so called zero-dimensional models or box models in earth sciences or environmental sciences neglect the spatial variation of the relevant quantities and retain only their temporal variation: the result is a simplified model of environmental processes based on ODEs that admit analytical solutions. These models provide valuable physical insight and are the starting point for more sophisticated models based on partial differential equations (PDEs).

1.3.1   Solution methods

Here the basic solution methods for first- and second-order ODEs are reviewed.

1 **(B)** Solve the first-order ODE

$$y' = x^2 y^2.$$

Solution

If the solution $y(x)$ is not identically zero, we can divide both sides of the ODE by y^2, obtaining

$$\frac{y'}{y^2} = \frac{d}{dx}\left(-\frac{1}{y}\right) = x^2,$$

which is immediately integrated:

$$\int_{y_0}^{y} d\left(\frac{1}{z}\right) = -\int_{x_0}^{x} ds\, s^2.$$

The general solution of the ODE is therefore

$$\frac{1}{y} - \frac{1}{y_0} = -\frac{x^3}{3} + \frac{x_0^3}{3}$$

or

$$y = \frac{3}{C - x^3},$$

where C is an arbitrary integration constant.

Since we have divided by y^2 assuming that $y(x)$ is not identically zero, we now have to check the case $y \equiv 0$: by inspection one concludes immediately that this is also a solution.

2 **(B)** Solve implicitly the first-order ODE

$$y' = \frac{x + 2}{3y^2 + 1}.$$

Solution

This equation is separable and can be written as

$$(3y^2 + 1)\frac{dy}{dx} = x + 2$$

and integrated

$$\int dy\,(3y^2 + 1) = \int dx\,(x + 2)$$

to yield the solution in implicit form

$$\left(y^2 + 1\right) y = x^2 + 2x + C,$$

where C is an arbitrary integration constant.

3 **(B)** Find the general solution of the first-order ODE

$$xy' + y = x^2.$$

Solution
For $x \neq 0$ the ODE is equivalent to

$$y' + \frac{y}{x} = x.$$

The general solution is the sum of the general solution of the associated homogeneous equation $y' + y/x = 0$ and of a particular solution y_p of the nonhomogeneous equation. The associated homogeneous equation can be written as

$$\frac{y'}{y} = -\frac{1}{x},$$

or

$$\frac{d}{dx}\left(\ln\left|\frac{y}{y_0}\right|\right) = -\frac{d}{dx}\left(\ln\left|\frac{x}{x_0}\right|\right),$$

where x_0 and y_0 are arbitrary constant. Therefore,

$$|y| = \frac{|y_0|\,|x_0|}{|x|},$$

which can be written as

$$y(x) = \frac{C}{x},$$

with C an arbitrary constant. For a particular solution of the inhomogeneous ODE, we try the form $y_p(x) = Ax^2$, with A a constant. Substitution into the ODE yields $A = 1/3$, and therefore the general solution of the given ODE for $x \neq 0$ is

$$y(x) = \frac{C}{x} + \frac{x^2}{3}.$$

4 **(B)** A physical system is described by the nonlinear ODE

$$\frac{dx}{dt} = \alpha\, x^\beta,$$

where α is a positive constant and β is a parameter that can be changed in steps.

a) Assume that initially $\beta = -1$ and solve the ODE. On what interval(s) is the solution defined?

b) Repeat for $\beta = 0$.

c) Repeat for $\beta = 1$.

d) Repeat for $\beta = 2$.

What is the effect of progressively varying β on the growth rate of the solution $x(t)$? Interpret your result in terms of feedback on the solution.

Solution

The ODE can be written as

$$x^{-\beta}\frac{dx}{dt} = \alpha$$

and integrated, obtaining

$$\frac{1}{1-\beta}x^{1-\beta} = \alpha(t - t_0)$$

if $\beta \neq 1$, where t_0 is an integration constant.

a) If $\beta = -1$, this solution applies and $x(t) = \pm\sqrt{2\alpha(t - t_0)}$, which is defined on $[t_0, +\infty)$.

b) If $\beta = 0$, the solution still applies and is linear, $x(t) = \alpha(t - t_0)$, defined on $(-\infty, +\infty)$.

c) For $\beta = 1$ the previous solution does not apply. The ODE is now $dx/dt = \alpha x$, which has the solution $x(t) = C\,e^{\alpha t}$, where C is an integration constant and is defined on $(-\infty, +\infty)$.

d) For $\beta = 2$ the solution still applies, and

$$x(t) = \frac{1}{\alpha(t_0 - t)}$$

has two separate branches defined on $(-\infty, t_0)$ or $(t_0, +\infty)$.

The effect of increasing the parameter β is to increase the growth rate of the solution $x(t)$, which is proportional to x^β. If $\beta < 0$, the growth rate decreases although $x(t)$ still increases: for $\beta = -1$ we find $x(t) \propto \sqrt{t - t_0}$. For $\beta = 0$ the solution grows faster (linearly) and is defined everywhere, for $\beta = 1$ it grows exponentially fast. In this case, as x grows, dx/dt increases at the same rate, which implies a larger x, which implies a larger dx/dt, and so on. There is a positive feedback. The solution grows fast, but it keeps growing for an infinite time. When β is increased to values larger than unity, the solution $x(t) \propto 1/(t - t_0)^{\frac{1}{|\beta - 1|}}$, the derivative $dx/dt \propto x^\beta$ has such a large positive feedback and grows so fast that it cannot keep growing for an infinite time: the system "explodes" in a finite time at the vertical asymptote $t = t_0$.

5 **(B)** Solve the nonlinear ordinary differential equation

$$\frac{du}{dt} + cu^2 + a = 0 \qquad (1.3)$$

(*Riccati equation* [32]), where $c \neq 0$ and $a \neq 0$ are constants.

Solution
If $c = 0$, it is easy to see that $u(t) = -at + u_0$, where u_0 is a constant (this includes the particular case $u = \text{const.}$ when $a = 0$). If both c and a are different from zero, the solution is found by introducing the auxiliary variable w defined by

$$u = \frac{1}{c} \frac{w'}{w}$$

(where a prime denotes differentiation with respect to the variable t), which changes Eq. (1.3) into the linear equation

$$w'' + acw = 0,$$

the solution of which is elementary. If $ca > 0$, we have the harmonic oscillator equation with general solution

$$w(t) = A\cos(\omega t) + B\sin(\omega t), \qquad \omega = \sqrt{ca}.$$

If $ca < 0$, the solution is instead the sum of two exponentials

$$w(t) = C\,e^{\omega t} + D\,e^{-\omega t}, \qquad \omega = \sqrt{|ca|}.$$

A, B, C, and D are integration constants. By returning to the original variable u, we obtain

$$u(t) = \frac{\omega}{c} \frac{-A\sin(\omega t) + B\cos(\omega t)}{A\cos(\omega t) + B\sin(\omega t)}$$

if $ca > 0$, and

$$u(t) = \frac{\omega}{c}\, \frac{C\,e^{\omega t} - D\,e^{-\omega t}}{C\,e^{\omega t} + D\,e^{-\omega t}}$$

if $ca < 0$.

6 **(B)** Find the general solution of the ODE $\quad y'' + 3y = 0$.

Solution
This is a linear second-order ODE with constant coefficients (the harmonic oscillator equation), the solution of which is elementary. The complementary equation is

$$r^2 + 3 = 0,$$

which has complex conjugate roots $r_\pm = \pm i\sqrt{3}$. A set of linearly independent solutions on $(-\infty, +\infty)$ is $\left\{ e^{\pm i\sqrt{3}\,x} \right\}$. Instead of this set one can use the equivalent linearly independent set

$$\left\{ \sin\left(\sqrt{3}\,x\right), \cos\left(\sqrt{3}\,x\right) \right\},$$

obtaining the general solution

$$y(x) = C_1 \cos\left(\sqrt{3}\,x\right) + C_2 \sin\left(\sqrt{3}\,x\right)$$

on $(-\infty, +\infty)$, where $C_{1,2}$ are arbitrary integration constants.

7 **(B)** Find the solution of the initial-value problem

$$\begin{cases} y'' - 5y = 0, \\ y(0) = 2, \\ y'(0) = 1. \end{cases}$$

Solution
The theorems on ODEs guarantee existence and uniqueness of the solution for this initial-value problem. The second-order linear ODE with constant coefficients is elementary, and the complementary equation is

$$r^2 - 5 = 0,$$

with real roots $r_\pm = \pm\sqrt{5}$. The general solution of the ODE on $(-\infty, +\infty)$ is therefore

$$y(x) = C_1\, e^{\sqrt{5}\,x} + C_2\, e^{-\sqrt{5}\,x},$$

where $C_{1,2}$ are integration constants to be determined by the initial conditions.

Imposing $y(0) = 2$ yields $C_1 + C_2 = 2$, while the second initial condition yields $\sqrt{5}\,(C_1 - C_2) = 1$. The linear system

$$\begin{cases} C_1 + C_2 = 2, \\ C_1 - C_2 = \frac{1}{\sqrt{5}}, \end{cases}$$

has the solution $C_{1,2} = 1 \pm 1/\left(2\sqrt{5}\right)$ and the unique solution of the initial-value problem on $(-\infty, +\infty)$ is

$$\begin{aligned} y(x) &= \left(1 + \frac{1}{2\sqrt{5}}\right) e^{\sqrt{5}\,x} + \left(1 - \frac{1}{2\sqrt{5}}\right) e^{-\sqrt{5}\,x} \\ &= 2\cosh\left(\sqrt{5}\,x\right) + \frac{1}{\sqrt{5}} \sinh\left(\sqrt{5}\,x\right). \end{aligned}$$

8 (B) Find the general solution of the ODE $x^2 y'' + 3xy' - y = 0$.

Solution
This second-order ODE is recognized to be an *Euler–Cauchy* (or *equidimensional* [3]) equation, which has the general form

$$x^2 y'' + \alpha x y' + \beta y = 0,$$

with α and β constants. The solution is defined on $(-\infty, 0) \cup (0, +\infty)$ and $x = 0$ is a (regular) singular point.

We look for power-law solutions $y(x) = x^r$; substitution of this form into the ODE yields the algebraic equation for the power r

$$r^2 + 2r - 1 = 0,$$

with real distinct roots $r = -1 \pm \sqrt{2}$. The solutions $y_1(x) = x^{\sqrt{2}-1}$ and $y_2(x) = x^{-(1+\sqrt{2})}$ are linearly independent and the general solution of the ODE on $(-\infty, 0) \cup (0, +\infty)$ is therefore

$$y(x) = C_1\, |x|^{\sqrt{2}-1} + C_2\, |x|^{-(\sqrt{2}+1)}.$$

9 (B) Solve the linear ODE $x^2 y'' + 3xy' + y = 0$.

---

[3]The name *equidimensional* comes from the fact that if x is replaced by kx, where k is a dimensional constant, the left-hand side does not change dimensions.

Solution

This second-order ODE is recognized to be an *Euler–Cauchy* or *equidimensional* equation, with solutions defined on $(-\infty, 0) \cup (0, +\infty)$. The point $x = 0$ is a (regular) singular point.

By looking for power-law solutions $y(x) = x^r$ and substituting into the ODE, one finds the algebraic equation for the power r:

$$r^2 + 2r + 1 = (r+1)^2 = 0,$$

with real coincident roots $r = -1$. Therefore, $y_1(x) = 1/x$ is a solution. A second, linearly independent, solution is $y_2(x) = x^r \ln x = \ln x / x$. The general solution of the ODE on $(-\infty, 0) \cup (0, +\infty)$ is therefore

$$y(x) = \frac{1}{|x|} \left(C_1 + C_2 \ln |x| \right),$$

where $C_{1,2}$ are arbitrary integration constants.

10 **(B)** Solve the two-point boundary-value problem in $[0, 1]$

$$\begin{cases} y'' + \pi^2 y = 0, \\[2mm] y(0) = 1, \\[2mm] y(1) = 0. \end{cases}$$

Solution

This second-order ODE is the harmonic oscillator equation with general solution on $(-\infty, +\infty)$

$$y(x) = C_1 \cos(\pi x) + C_2 \sin(\pi x),$$

where $C_{1,2}$ are arbitrary integration constants. The boundary condition at $x = 0$ implies that $C_1 = 0$, while the second boundary condition at $x = 1$ requires $C_2 \sin \pi = 0$, which is always satisfied for any value of C_2. Therefore, the solution of the boundary-value problem is

$$y(x) = C_2 \sin(\pi x),$$

where C_2 can assume any value different from zero [if $C_2 = 0$, then the constant function $y \equiv 0$ does not satisfy the boundary condition $y(0) = 1$].

11 **(B)** Solve the eigenvalue problem

$$\begin{cases} y'' + \lambda y = 0, \\ y(0) = 0, \\ y(\pi) = 0, \end{cases}$$

i.e., find all the real values of λ (*eigenvalues*) for which the problem has nontrivial solutions (*eigenfunctions*), and these solutions.

Solution
If $\lambda = 0$, the ODE is simply $y'' = 0$, which has the general linear solution $y(x) = \alpha x + \beta$, where α and β are integration constants. The boundary condition at $x = 0$ requires $\beta = 0$, while the boundary condition at $x = \pi$ is only satisfied if $\alpha = 0$, yielding the trivial solution. There are no nontrivial solutions for $\lambda = 0$.

If $\lambda > 0$, the ODE is the harmonic oscillator equation with general solution

$$y(x) = C_1 \cos\left(\sqrt{\lambda}\,x\right) + C_2 \sin\left(\sqrt{\lambda}\,x\right),$$

with $C_{1,2}$ integration constants. The boundary condition at $x = 0$ implies that $C_1 = 0$, while to satisfy the boundary condition at $x = \pi$ it must be $C_2 \sin\left(\sqrt{\lambda}\,\pi\right) = 0$. Since C_2 cannot be zero (otherwise we are left only with the trivial solution) it must be $\sin\left(\sqrt{\lambda}\,\pi\right) = 0$, or $\sqrt{\lambda}\,\pi = n\pi$ and $\lambda = n^2$, $n = 1, 2, 3, \ldots$ Note that the value $n = 0$ is not acceptable because it yields the trivial solution.

If $\lambda < 0$, the ODE has the general solution

$$y(x) = C_1 e^{\sqrt{|\lambda|}\,x} + C_2 e^{-\sqrt{|\lambda|}\,x},$$

with $C_{1,2}$ integration constants. By imposing the boundary conditions, we obtain

$$C_1 + C_2 = 0,$$

$$C_1 e^{\sqrt{|\lambda|}\,\pi} + C_2 e^{-\sqrt{|\lambda|}\,\pi} = 0.$$

Substitution of $C_2 = -C_1$ from the first equation into the second one yields $\sinh\left(\sqrt{|\lambda|}\,\pi\right) = 0$, which has no solutions for $\lambda < 0$ (the function $\sinh x$ intersects the x-axis only at $x = 0$).

To summarize, the eigenvalues of the problem are $\lambda_n = n^2$ with $n = 1, 2, 3, \ldots$, and the eigenfunctions are $y_n(x) = \sin(nx)$.

1.3.2   Qualitative analysis

Very often one cannot provide analytical solutions of nonlinear ODEs, but it is still possible to perform a rigorous analysis of the qualitative behavior of the solutions in phase space. The concepts of stability, instability, phase space, and attractor are very relevant in the environmental sciences. Very often one is not interested in transient solutions that decay relatively rapidly as much as in final states of equilibrium—then it is important to be able to decide whether these states are stable or unstable. Here stability is reviewed through exercises on population dynamics, a field of interest to both the biologist and the environmental scientist.

1 **(B)** a) The population, defined as the number of live individuals $P(t)$, of a certain species in the presence of unlimited food and in the absence of predators or competing species, is described by the *Malthus model*

$$\frac{dP}{dt} = aP,$$

where the birth rate a is constant. Find the future evolution of the species population.

b) A better model takes into account the fact that, as the population grows, its members begin competing between themselves for food or other resources, or get poisoned by their own waste products, and the growth cannot continue indefinitely at the same rate. The *Verhulst model* describes the rate of change of the population with the famous *logistic equation*[4]

$$\frac{dP}{dt} = aP\left(1 - \alpha P\right),$$

where a and α are positive constants. Solve for the evolution of the population $P(t)$.

Solution
a) This elementary ODE is solved by writing

$$\frac{1}{P}\frac{dP}{dt} = \frac{d}{dt}\left(\ln\frac{P}{P_0}\right) = a,$$

---

[4]The logistic equation $x' - ax\,(1 - x)$ exhibits chaos associated with its nonlinear character when the parameter a is varied: there is a bifurcation as the value of a increases.

which is immediately integrated, yielding the solution

$$P(t) = P_0 \, e^{at},$$

where $P_0 = P(0)$. The growth of the population is exponential and unbounded in the Malthusian model.

b) To solve the Verhulst model, we write the logistic equation as

$$\frac{1}{P(1 - \alpha P)} \frac{dP}{dt} = a$$

and decompose the fraction on the left-hand side as follows:

$$\frac{A}{P} + \frac{B}{1 - \alpha P} = \frac{1}{P(1 - \alpha P)}.$$

The appropriate values of A and B that make this decomposition possible are easily found to be $A = 1$ and $B = \alpha$, hence

$$\frac{1}{P} \frac{dP}{dt} + \frac{\alpha}{1 - \alpha P} \frac{dP}{dt} = a.$$

Integration gives

$$\int_{P_0}^{P} \frac{dP'}{P'} - \int_{(\alpha P_0 - 1)}^{(\alpha P - 1)} \frac{d(\alpha P' - 1)}{\alpha P' - 1} = at,$$

and

$$\ln\left(\frac{P}{P_0}\right) - \ln\left|\frac{\alpha P - 1}{\alpha P_0 - 1}\right| = at.$$

By using the properties of logarithms and taking the exponential of both sides, we obtain

$$\frac{P(\alpha P_0 - 1)}{P_0(\alpha P - 1)} = e^{\alpha t}$$

and finally, after straightforward algebraic manipulations,

$$P(t) = \frac{P_0 \, e^{at}}{1 - \alpha P_0 + \alpha P_0 \, e^{at}} = \frac{P_0}{\alpha P_0 + (1 - \alpha P_0) \, e^{-at}},$$

where again, the integration constant has the meaning $P_0 = P(0)$.

The late time (formally, as $t \to +\infty$) state of the population is $P \approx 1/\alpha$ irrespective of the initial condition P_0—the population does not grow indefinitely as in the Malthus model, but it asymptotically reaches a constant value. If one begins with a very small population, i.e., if the initial condition satisfies $P_0 \ll \alpha^{-1}$, the approximation

$$P(t) = \frac{P_0\, e^{at}}{1 - \alpha P_0 + \alpha P_0\,(1 + \ldots)} \approx P_0\, e^{at} \qquad (t \to 0)$$

holds and the Malthus solution is recovered. The effects of the limiting factor α are felt when the population becomes larger.

The asymptotic state of equilibrium $P = 1/\alpha$ is an exact solution of the logistic equation, as seen by inspection, and it can be found by setting dP/dt in the search for steady-state equilibrium solutions. It is a stable solution and an attractor in the (t, P) plane. In fact, if the initial population is $P_0 > \alpha^{-1}$, the solution will always be larger than $1/\alpha$; otherwise the curve corresponding to $P(t)$ would cross the straight line representing $P = \alpha^{-1}$ in the (t, P) plane: this is forbidden by the uniqueness theorems of the solutions of ODEs. (Similarly, if $P_0 < \alpha^{-1}$, then it is $P(t) < \alpha^{-1}$ for all times t.) Therefore, if $P_0 > \alpha^{-1}$, it is $dP/dt = aP\,(1 - \alpha P) < 0$ and $P(t)$ is a monotonically decreasing function. It cannot go to minus infinity because otherwise it would cross the line $P = \alpha^{-1}$; hence it must converge asymptotically to its lower bound α^{-1} with $dP/dt \to 0$ as $t \to +\infty$ (similar conclusions hold if $P_0 < \alpha^{-1}$). Therefore, $P = \alpha^{-1}$ is a stable solution and an attractor in the phase space.

2 **(B)** The dynamics of a population $P(t)$ of animals broken into spatially separate subpopulations (a *metapopulation*) can be described by the ODE [43, 25]

$$\frac{dP}{dt} = cP\,(1 - P) - mP,$$

where c and m are positive constants describing the intrinsic growth rate and the mortality rate, respectively. Analyze qualitatively the solutions of the ODE and predict the future of the population.

Solution
If the initial condition $P(0)$ is in a small neighborhood of the origin in the (t, P) plane, i.e., if $P(0) \ll 1$, we can neglect the quadratic term in the ODE, which reduces to the elementary asymptotic equation

$$\frac{dP}{dt} \approx (c - m)\, P \qquad (t \to 0, P(0) \ll 1),$$

which has the solution

$$P(t) = P(0)\, e^{(c-m)t}.$$

If $c > m$ (birth rate larger than the mortality rate), this asymptotic solution grows exponentially fast and the approximation $P \ll 1$ breaks down. If $c < m$ instead, the species declines exponentially fast and soon becomes extinct ($P(t) \to 0$). Since in this case $P(t) < P(0) \ll 1$, the approximation that led to the asymptotic equation holds better and better and this solution remains valid at all times $t > 0$. However, it is limited to initial conditions such that $P(0) \ll 1$.

Let us consider now the case of arbitrary initial conditions: if $c = m$, the ODE can be rewritten as

$$-\frac{1}{P^2}\frac{dP}{dt} = \frac{d}{dt}\left(\frac{1}{P}\right) = c,$$

which is integrated to yield

$$P(t) = \frac{1}{c\,(t - t_0)}$$

with t_0 an integration constant. In terms of the initial condition $P_0 \equiv P(0)$, one finds

$$P(t) = \frac{P_0}{1 + P_0 c t}.$$

It should be noted, however, that this exact solution corresponds to an unrealistic fine-tuning in the parameters c and m.

For arbitrary initial conditions, let us look for equilibrium solutions with $dP/dt = 0$. These are obtained by setting $[(c - m) - cP]\,P = 0$, which yields (discarding the trivial solution)

$$P = P_* \equiv 1 - \frac{m}{c},$$

and it only exists for $c \geq m$ because P is forced to be nonnegative.

Let us study the stability of this equilibrium solution for $c > m$. Consider a generic solution $P(t)$, which satisfies the ODE rewritten as

$$\frac{dP}{dt} = c\,(P_* - P)\,P :$$

because c and m are positive, the sign of dP/dt is the same as the sign of $P_* - P$ and, due to the uniqueness theorems for the solutions of ODEs, the curve representing the solution $P(t)$ cannot cross the straight line $P \equiv P_*$; hence it is always $P > P_*$ or $P < P_*$.

If $P > P_*$, then $dP/dt < 0$, and the solution $P(t)$ is always strictly decreasing. Therefore, it will asymptotically approach its lower bound

P_* with $dP/dt \to 0$ as $t \to +\infty$. If instead $P < P_*$, then $dP/dt > 0$, and the solution $P(t)$ always increases approaching its upper bound P_* asymptotically. Therefore, $P = P_*$ is a stable solution and an attractor in the (t, P) plane.

For $c < m$ there are no equilibrium solutions,

$$\frac{dP}{dt} = (c - m) P - cP^2$$

is always negative, and any solution $P(t)$ is always strictly decreasing approaching zero asymptotically: the species becomes extinct.

1.4 Functions of two or more variables

As soon as the dependence of physical quantities on space—in addition to time—is introduced, one faces functions of more than one variable. Many ideas from the theory of functions of a single variable are generalized, with obvious complications arising from the increased number of dimensions.

1.4.1 Differentiation

1 **(B)** Find an expression for the gradient $\vec{\nabla} f(r)$, where $f(r)$ is a regular function of the radius $r \equiv |\vec{x}| = \sqrt{x^2 + y^2 + z^2}$.

Solution
We have

$$\vec{\nabla} f(r) = f' \vec{\nabla} r = f' \left(\frac{\partial r}{\partial x}, \frac{\partial r}{\partial y}, \frac{\partial r}{\partial z} \right)$$

$$= f' \left(\frac{x}{r}, \frac{y}{r}, \frac{z}{r} \right) = f' \frac{\vec{x}}{r}, \tag{1.4}$$

where $f' \equiv df/dr$.

2 **(B)** Compute $\vec{\nabla} f$ and the directional derivative $\partial f / \partial s$, where

$$f(x, y) = x \sin \left(x^2 + y^2 \right)$$

and $\vec{s} = \frac{1}{\sqrt{5}} (1, 2)$.

Solution
The partial derivatives of f are

$$\frac{\partial f}{\partial x} = \sin \left(x^2 + y^2 \right) + 2x^2 \cos \left(x^2 + y^2 \right),$$

$$\frac{\partial f}{\partial y} = 2xy \cos \left(x^2 + y^2 \right);$$

hence the gradient of f is

$$\vec{\nabla} f = \left(\frac{\partial f}{\partial x}, \frac{\partial f}{\partial y} \right)$$

$$= \left(\sin \left(x^2 + y^2 \right) + 2x^2 \cos \left(x^2 + y^2 \right), 2xy \cos \left(x^2 + y^2 \right) \right).$$

The directional derivative of f along $\vec{s}$ is

$$\frac{\partial f}{\partial s} \equiv \vec{s} \cdot \vec{\nabla} f = \left(\frac{1}{\sqrt{5}}, \frac{2}{\sqrt{5}} \right)$$

$$\cdot \left(\sin \left(x^2 + y^2 \right) + 2x^2 \cos \left(x^2 + y^2 \right), 2xy \cos \left(x^2 + y^2 \right) \right)$$

$$= \frac{1}{\sqrt{5}} \left[\sin \left(x^2 + y^2 \right) + 2x \left(x + 2y \right) \cos \left(x^2 + y^2 \right) \right].$$

3 (B) Find the directional derivative $\partial f / \partial u$ of the polynomial

$$f(x, y, z) = x^2 y + 3xyz$$

in the direction identified by the unit vector $\vec{u} = (l, m, n)$, where $l^2 + m^2 + n^2 = 1$.

Solution
The directional derivative is

$$\frac{\partial f}{\partial u} \equiv \vec{u} \cdot \vec{\nabla} f = (l, m, n) \cdot \left(2xy + 3yz, x^2 + 3xz, 3xy \right)$$

$$= ly \left(2x + 3z \right) + mx \left(x + 3z \right) + 3nxy.$$

4 (B) Compute the Laplacian of the function

$$f(x, y, z) = \sin x \sin y \cos z.$$

Solution

The first partial derivatives of f are

$$\frac{\partial f}{\partial x} = \cos x \sin y \cos z,$$

$$\frac{\partial f}{\partial y} = \sin x \cos y \cos z,$$

$$\frac{\partial f}{\partial z} = -\sin x \sin y \sin z,$$

and the second partial derivatives appearing in the Laplacian are

$$\frac{\partial^2 f}{\partial x^2} = \frac{\partial^2 f}{\partial y^2} = \frac{\partial^2 f}{\partial z^2} = -\sin x \sin y \cos z.$$

The Laplacian of f is therefore

$$\nabla^2 f \equiv \frac{\partial^2 f}{\partial x^2} + \frac{\partial^2 f}{\partial y^2} + \frac{\partial^2 f}{\partial z^2} = -3 \sin x \sin y \cos z.$$

5 **(B)** Prove *Euler's theorem*: if $f(x, y, z)$ is a regular function of the three variables $x, y,$ and z that satisfies the property

$$f(\lambda x, \lambda y, \lambda z) = \lambda f(x, y, z) \tag{1.5}$$

for any real number λ and for any triple (x, y, z), then

$$f(x, y, z) = x \frac{\partial f}{\partial x} + y \frac{\partial f}{\partial y} + z \frac{\partial f}{\partial z}. \tag{1.6}$$

Solution

To prove the theorem, differentiate Eq. (1.5) with respect to λ, obtaining

$$\frac{\partial f}{\partial(\lambda x)} x + \frac{\partial f}{\partial(\lambda y)} y + \frac{\partial f}{\partial(\lambda z)} z = f(x, y, z).$$

This equation holds true for any real $\lambda \neq 0$, in particular for $\lambda = 1$. By setting $\lambda = 1$, the desired property (1.6) follows.

6 **(B)** Prove that the expression $f(x, y) = 0$, where $f(x, y) = x \sin y + y$ defines implicitly a function $y(x)$ in a neighborhood of the point $(0, 0)$. Is this function unique? Motivate your answer and compute the derivative dy/dx.

Solution
The function f is continuous with continuous first derivatives on the entire (x, y) plane, $f(0, 0) = 0$, and $f_y(0, 0) = 1 \neq 0$. The hypotheses of the implicit function theorem are satisfied and the theorem guarantees that an implicit function $y(x)$ is uniquely defined and is continuous in a neighborhood $[-\epsilon, \epsilon]$ of $x = 0$, and is continuously differentiable in $(-\epsilon, \epsilon)$. The first partial derivatives of f are

$$f_x = \sin y, \qquad f_y = x \cos y + 1,$$

and the implicit function $y(x)$ has derivative

$$\frac{dy}{dx} = \frac{-f_x\left(x, y(x)\right)}{f_y\left(x, y(x)\right)} = -\frac{\sin\left(y(x)\right)}{x \cos\left(y(x)\right) + 1}.$$

7 (B) Prove that the surfaces of constant f and g, where

$$f(x) \;\;=\;\; x^2 - y^2,$$

$$g(x) \;\;=\;\; xy + C$$

(with C a constant), are orthogonal to each other.

Solution
The functions f and g are defined and differentiable at any point (x, y) in the plane. The surfaces of constant f and g are orthogonal if and only if the gradients $\vec{\nabla}f$ and $\vec{\nabla}g$ are mutually perpendicular. We have

$$\vec{\nabla}f \;\;\equiv\;\; \left(\frac{\partial f}{\partial x}, \frac{\partial f}{\partial y}\right) = (2x, -2y),$$

$$\vec{\nabla}g \;\;\equiv\;\; \left(\frac{\partial g}{\partial x}, \frac{\partial g}{\partial y}\right) = (y, x),$$

and

$$\vec{\nabla}f \cdot \vec{\nabla}g = 2xy - 2yx = 0.$$

The two surfaces are mutually orthogonal at every point (x, y) in the plane.

1.4.2   Integration

1 (B) Compute the integral

$$I = \int\int\int_Q dx dy dz \, xy^2 z^3 \, e^x,$$

where Q is the cube $Q = [0, 1] \times [0, 1] \times [0, 1]$ in the three-dimensional space (x, y, z).

Solution
The integrand is the product of three factors that depend only on x, y, and z, and therefore the integral splits into the product of three integrals,

$$I = \int\int\int_Q dx dy dz \, xy^2 z^3 \, e^x = \int_0^1 dx \, x \, e^x \cdot \int_0^1 dy y^2 \cdot \int_0^1 dz \, z^3.$$

We compute the first integral by parts, obtaining

$$\int_0^1 dx\, x\, e^x = [x\, e^x] - \int_0^1 dx\, e^x = [(x-1)\, e^x]_0^1 = 1.$$

Then

$$I = 1 \cdot \left[\frac{y^3}{3}\right]_0^1 \cdot \left[\frac{z^4}{4}\right]_0^1 = \frac{1}{12}.$$

2 **(B)** Compute the integral

$$I = \int\int_T dx dy\, y\, \sin x,$$

where T is the triangle delimited by the x-axis and the straight lines of equations $x = \pi/2$ and $y = x$.

Solution
It is convenient to integrate first with respect to y and then with respect to x:

$$\int\int_T dx dy\, y\, \sin x = \int_0^{\pi/2} dx \int_0^x dy\, y\, \sin x = \int_0^{\pi/2} dx\, \sin x \left[\frac{y^2}{2}\right]_0^x$$

$$= \frac{1}{2}\int_0^{\pi/2} dx\, x^2 \sin x.$$

The last integral is computed by parts,

$$\int dx\, x^2 \sin x = -x^2 \cos x + \int dx\, 2x \cos x$$

$$= -x^2 \cos x + 2\left[x \sin x - \int dx\, \sin x\right]$$

$$= -x^2 \cos x + 2x \sin x + 2 \cos x,$$

and therefore

$$I = \frac{1}{2}\left[-x^2 \cos x + 2x \sin x + 2 \cos x\right]_0^{\pi/2} = \frac{\pi}{2} - 1.$$

3 **(B)** Compute the integral

$$\int\int_S dx dy\, x^2 y,$$

where S is the region of the plane delimited by the two parabolas of equations $y = x^2$ and $y = 3x^2$, and the lines $x = \pm 1$.

Solution
It is convenient to integrate first with respect to the variable y and then with respect to the variable x:

$$
\iint_S dxdy\, x^2 y = \int_{-1}^{+1} dx\, x^2 \int_{x^2}^{3x^2} dy\, y = \int_{-1}^{+1} dx\, x^2 \left[\frac{y^2}{2}\right]_{x^2}^{3x^2}
$$

$$
= \frac{1}{2}\int_{-1}^{+1} dx\, x^2 \left[9x^4 - x^4\right] = 4\int_{-1}^{+1} dx\, x^6
$$

$$
= \left[\frac{4x^7}{7}\right]_{-1}^{+1} = \frac{8}{7}.
$$

4 **(B)** Compute the integral

$$
\iint_S dxdy\, \frac{1}{(x^2 + y^2)^2},
$$

where S is the region of the plane between the circles of radii 2 and 3 centered on the origin.

Solution
It is convenient to use polar coordinates (r, φ); then the region S is simply given by $(r, \varphi) \in [2, 3] \times [0, 2\pi]$ and the integrand is r^{-4}. By inserting the Jacobian factor r, we obtain

$$
\iint_S dxdy\, \frac{1}{(x^2 + y^2)^2} = \int_2^3 dr \int_0^{2\pi} d\varphi\, \frac{r}{r^4} = \int_0^{2\pi} d\varphi \cdot \int_2^3 dr\, r^{-3}
$$

$$
= 2\pi \left[-\frac{1}{2} r^{-2}\right]_2^3 = -\pi \left(\frac{1}{3^2} - \frac{1}{2^2}\right) = \frac{5\pi}{36}.
$$

5 **(B)** Compute the area of the circle C of radius R by using double integrals. Compute the length of the circumference γ using line integrals.

Solution
Let the circle C be centered on the origin $(0, 0,)$; its area is computed using polar coordinates (r, φ) with Jacobian r,

$$
A \equiv \iint_C dxdy = \int_C \int dr d\varphi\, r = \int_0^R dr\, r \int_0^{2\pi} d\varphi
$$

$$= 2\pi \int_0^R dr\, r = 2\pi \left[\frac{r^2}{2}\right]_0^R = \pi R^2,$$

the well-known result.

The circumference γ has the parametric representation

$$x(t) = R\cos t,$$

$$y(t) = R\sin t,$$

with $0 \le t \le 2\pi$. The length of this curve is

$$l \equiv \int_\gamma dl = \int_0^{2\pi} dt\, \left\|\vec{T}(t)\right\|,$$

where $\vec{T}(t)$ is the tangent vector to γ. We have

$$\vec{T} = \left(\frac{dx}{dt}, \frac{dy}{dt}\right) = (-R\sin t, R\cos t)$$

and

$$l = \int_0^{2\pi} dt\, \sqrt{R^2\sin^2 t + R^2\cos^2 t} = R\int_0^{2\pi} dt = 2\pi R,$$

another well-known result.

6 **(B)** Compute the area and the volume of the sphere of radius R by using double and triple integrals, respectively.

Solution
Let S be the sphere of radius R centered on the origin $(0,0,0)$; its area and volume are computed using polar coordinates (r,θ,φ), with Jacobian $r^2\sin\theta$. The area of the sphere is

$$A \equiv \iint_{S^2} R^2 d\theta d\varphi \sin\theta = R^2 \int_0^\pi d\theta \sin\theta \int_0^{2\pi} d\varphi$$

$$= 2\pi R^2 \left[-\cos\theta\right]_0^\pi = 4\pi R^2,$$

a well-known result. The volume of the sphere is

$$V \equiv \iiint_S dx dy dz = \iiint_S dr\, d\theta\, d\varphi\, r^2 \sin\theta$$

$$= \int_0^R dr\, r^2 \int_0^\pi d\theta \sin\theta \int_0^{2\pi} d\varphi$$

$$= 2\pi \left[\frac{r^3}{3}\right]_0^R [-\cos\theta]_0^\pi = \frac{4\pi R^3}{3},$$

another well-known result.

7 **(B)** Compute the integral $\int_\gamma dl\, xy$, where the curve γ is the segment of the circle of unit radius centered on the origin and lying in the first quadrant ($x > 0$ and $y > 0$).

Solution
A parametric representation of the curve γ is

$$x(t) = \cos t,$$

$$y(t) = \sin t,$$

where $0 \le t \le \pi/2$. The tangent to the circle is the vector

$$\vec{T} = \left(\frac{dx}{dt}, \frac{dy}{dt}\right) = (-\sin t, \cos t),$$

and the integral is

$$\int_\gamma dl\, xy \equiv \int_0^{\pi/2} dt\, x(t) y(t) \left\| \vec{T}(t) \right\|$$

$$= \int_0^{\pi/2} dt\, \sin t \cos t \sqrt{\sin^2 t + \cos^2 t}$$

$$= \int_0^{\pi/2} dt\, \sin t \cos t = \left[\frac{\sin^2 t}{2}\right]_0^{\pi/2} = \frac{1}{2}.$$

8 **(B)** Compute the integral

$$\int_\gamma dl\, (x^2 + y^2) \ln (x^2 + y^2),$$

where the curve γ is the segment of the spiral of equation $r = e^\varphi$ comprised between $r = 0$ and $r = 10$.

Solution
It is convenient to use polar coordinates (r, φ); then a parametric representation of the spiral is its equation $r = e^{\varphi}$ using φ as a parameter. We are interested in the portion of the spiral between $r = 0$ and $r = 10$, which correspond to $\varphi \to -\infty$ and $\varphi = \ln 10$, respectively. In polar coordinates

$$x(\varphi) = r \cos \varphi,$$

$$y(\varphi) = r \sin \varphi,$$

we have, along the spiral,

$$x = e^{\varphi} \cos \varphi,$$

$$y = e^{\varphi} \sin \varphi,$$

and the tangent to γ is the vector

$$\vec{T}(\varphi) = \left(\frac{dx}{d\varphi}, \frac{dy}{d\varphi} \right) = e^{\varphi} \left(\cos \varphi - \sin \varphi, \sin \varphi + \cos \varphi \right),$$

and, setting

$$f(x, y) \equiv (x^2 + y^2) \ln (x^2 + y^2) = e^{2\varphi} 2\varphi e^{\varphi} \sqrt{2},$$

the required integral

$$\int_{\gamma} dl \, f(x, y) \equiv \int_{\gamma} d\varphi \, f(x(\varphi), y(\varphi)) \, \|\vec{T}(\varphi)\|$$

$$= \int_{-\infty}^{\ln 10} d\varphi \, \sqrt{2} \, e^{\varphi} 2\varphi \, e^{2\varphi}$$

$$= 2\sqrt{2} \int_{-\infty}^{\ln 10} d\varphi \, \varphi \, e^{3\varphi} = \frac{2\sqrt{2}}{3} \left[\frac{e^{3\varphi}}{3} \left(\varphi - \frac{1}{3} \right) \right]_{-\infty}^{\ln 10}$$

$$= \frac{2\sqrt{2}}{3} \frac{10^3}{3} \left(\ln 10 - \frac{1}{3} \right) \simeq 618.9.$$

1.5   Vector calculus

Vector calculus is covered in many excellent mathematics textbooks. Alternatively, more physically minded students who are not inclined toward such books may find it more practical to refer to textbooks on mechanics or electromagnetism that usually devote an entire introductory chapter to vector calculus (e.g., Refs. [59, 22]).

1 **(C, B)** Explain the difference between scalars and vectors and provide examples of scalar and vectorial physical quantities. What is a scalar field? What is a vector field?

Solution 1 (level C)
Physically, a *scalar* quantity is completely characterized by its magnitude while a *vector* quantity is characterized by its magnitude and direction. Examples of scalars are mass, temperature, time, while position with respect to a fixed origin, velocity, acceleration, force, electric and magnetic field are vector quantities.

A *scalar field* is a scalar function $f(\vec{x})$ of the position $\vec{x} = (x, y, z)$, while a *vector field* is a vector quantity that depends on position, $\vec{a} = \vec{a}(\vec{x})$.

Solution 2 (level B)
A mathematically more precise definition of scalars and vectors can be given by using the transformation properties of their components under a coordinate transformation

$$x^i \longrightarrow x'^i = x'^i(x^j). \tag{1.7}$$

A *scalar* s is unchanged by coordinate transformation, i.e., $s' = s$, while a *covariant vector* $\vec{a}$ with components a^i transforms according to

$$a^i \longrightarrow a'^i = \sum_j \frac{\partial x'^i}{\partial x^j} a^j.$$

A *1-form* or *contravariant vector* with components ω_i instead transforms according to

$$\omega_i \longrightarrow \omega'_i = \sum_l \frac{\partial x^l}{\partial x'^i} \omega_l.$$

Both vectors and 1-forms can be used to represent vectorial quantities.

2 **(A)** Find the vector product of the vectors

$$\vec{a} = (3, 0, 2), \qquad \vec{b} = (1, 5, 0).$$

Solution
The vector product of $\vec{a}$ and $\vec{b}$ is given by the pseudodeterminant

$$\vec{a} \times \vec{b} = \begin{vmatrix} \vec{i} & \vec{j} & \vec{k} \\ 3 & 0 & 2 \\ 1 & 5 & 0 \end{vmatrix} = -10\,\vec{i} + 2\,\vec{j} + 15\,\vec{k} = (-10, 2, 15),$$

where $\vec{i}, \vec{j}$, and $\vec{k}$ denote the unit vectors in the x, y, and z directions, respectively.

3 **(A)** Show that the following vectors are perpendicular to each other:

$$\vec{a} = (1, 2, 5), \qquad \vec{b} = (2, -2, 2/5).$$

Solution
The scalar product of $\vec{a}$ and $\vec{b}$ is

$$\vec{a} \cdot \vec{b} = (1, 2, 5) \cdot \begin{pmatrix} 2 \\ -2 \\ 2/5 \end{pmatrix} = 2 - 4 + 2 = 0,$$

and therefore $\vec{a}$ and $\vec{b}$ are perpendicular.

4 **(B)** Compute $\vec{a} \times \vec{a} \times \vec{a} \times \vec{x}$, where $\vec{a}$ is a constant vector and $\vec{x}$ is the position vector.

Solution
Let $\vec{a} = (a_x, a_y, a_z)$ and $\vec{x} = (x, y, z)$ in Cartesian coordinates; then

$$\vec{a} \times \vec{x} = \begin{vmatrix} \vec{e}_x & \vec{e}_y & \vec{e}_z \\ a_x & a_y & a_z \\ x & y & z \end{vmatrix}$$

$$= \vec{e}_x \left(a_y z - a_z y \right) - \vec{e}_y \left(a_x z - a_z x \right) + \vec{e}_z \left(a_x y - a_y x \right)$$

and

$$\vec{a} \times \vec{a} \times \vec{x} = \begin{vmatrix} \vec{e}_x & \vec{e}_y & \vec{e}_z \\ a_x & a_y & a_z \\ (a_y z - a_z y) & (a_z x - a_x z) & (a_x y - a_y x) \end{vmatrix}$$

$$= \quad \vec{e}_x \left[a_y \left(a_x y - a_y x \right) - a_z \left(a_z x - a_x z \right) \right]$$

$$- \vec{e}_y \left[a_x \left(a_x y - a_y x \right) - a_z \left(a_y z - a_z y \right) \right]$$

$$+ \vec{e}_z \left[a_x \left(a_z x - a_x z \right) - a_y \left(a_y z - a_z y \right) \right]$$

$$= \quad \vec{e}_x \left(a_x a_y y - a_y^2 x - a_z^2 x + a_x a_z z \right)$$

$$+ \vec{e}_y \left(a_y a_z z - a_z^2 y - a_x^2 y + a_x a_y x \right)$$

$$+ \vec{e}_z \left(a_x a_z x - a_x^2 z - a_y^2 z + a_y a_z y \right)$$

$$\equiv \quad A_x \vec{e}_x + A_y \vec{e}_y + A_z \vec{e}_z$$

and finally

$$\vec{a} \times \vec{a} \times \vec{a} \times \vec{x} = \begin{vmatrix} \vec{e}_x & \vec{e}_y & \vec{e}_z \\ a_x & a_y & a_z \\ A_x & A_y & A_z \end{vmatrix}$$

$$= \quad \vec{e}_x \left(-a_x^2 a_y z - a_y^3 z + a_y^2 a_z y - a_y a_z^2 z + a_z^3 y + a_x^2 a_z y \right)$$

$$- \vec{e}_y \left(a_x^2 a_z x - a_x^3 z - a_x a_y^2 z + a_y^2 a_z x + a_z^3 x - a_x a_z^2 z \right)$$

$$+ \vec{e}_z \left(-a_x a_z^2 y - a_x^3 y + a_x^2 a_y z - a_x a_y^2 y + a_y^3 x + a_y a_z^2 x \right) .$$

5 **(B)** Compute the divergences of the vector fields

$$\vec{a} \quad = \quad \left(x^2 + yz \right) \vec{i} + \left(y^2 + 3xz \right) \vec{j} + \left(z^2 - xy \right) \vec{k},$$

$$\vec{b} \quad = \quad \left(x^2 + xz \right) \vec{i} + \left(y^2 + xyz \right) \vec{j} + \left(3zy^2 \right) \vec{k}.$$

Solution
The divergence of $\vec{a}$ is

$$\vec{\nabla} \cdot \vec{a} \equiv \frac{\partial a_x}{\partial x} + \frac{\partial a_y}{\partial y} + \frac{\partial a_z}{\partial z} = 2(x + y + z),$$

while the divergence of $\vec{b}$ is

$$\vec{\nabla} \cdot \vec{b} \equiv \frac{\partial b_x}{\partial x} + \frac{\partial b_y}{\partial y} + \frac{\partial b_z}{\partial z}$$

$$= (2x + z) + (2y + xz) + 3y^2$$

$$= 2(x + y) + 3y^2 + z(1 + x).$$

6 (B) Calculate the Laplacian of

$$f(x, y, x) = x^2 \sin y + \sin x \sin z.$$

Solution

The first partial derivatives of f are

$$\frac{\partial f}{\partial x} = 2x \sin y + \cos x \sin z,$$

$$\frac{\partial f}{\partial y} = x^2 \cos y,$$

$$\frac{\partial f}{\partial z} = \sin x \cos z,$$

while the second derivatives needed to form the Laplacian are

$$\frac{\partial^2 f}{\partial x^2} = 2 \sin y - \sin x \sin z,$$

$$\frac{\partial^2 f}{\partial y^2} = -x^2 \sin y,$$

$$\frac{\partial^2 f}{\partial z^2} = -\sin x \sin z.$$

The Laplacian of f is

$$\nabla^2 f = \frac{\partial^2 f}{\partial x^2} + \frac{\partial^2 f}{\partial y^2} + \frac{\partial^2 f}{\partial z^2} = (2 - x^2) \sin y - 2 \sin x \sin z.$$

7 (B) Prove that the divergence of a gradient is equal to the Laplacian operator, or

$$\vec{\nabla} \cdot \left(\vec{\nabla} f \right) = \nabla^2 f$$

for any regular function f.

Solution
It is sufficient to verify the identity in Cartesian coordinates and then it will be true in any coordinate system because this quantity is a scalar. We have

$$\vec{\nabla} \cdot \left(\vec{\nabla} f\right) = \frac{\partial}{\partial x} \left[\left(\vec{\nabla} f\right)_x\right] + \frac{\partial}{\partial y} \left[\left(\vec{\nabla} f\right)_y\right] + \frac{\partial}{\partial z} \left[\left(\vec{\nabla} f\right)_z\right]$$

$$= \frac{\partial}{\partial x} \left(\frac{\partial f}{\partial x}\right) + \frac{\partial}{\partial y} \left(\frac{\partial f}{\partial y}\right) + \frac{\partial}{\partial z} \left(\frac{\partial f}{\partial z}\right) = \frac{\partial^2 f}{\partial x^2} + \frac{\partial^2 f}{\partial y^2} + \frac{\partial^2 f}{\partial z^2} \equiv \nabla^2 f.$$

8 **(B)** Prove that

$$\nabla^2 (fg) = f\nabla^2 g + 2\vec{\nabla} f \cdot \vec{\nabla} g + g\nabla^2 f.$$

Solution
We have

$$\begin{aligned}
\nabla^2 (fg) &= \vec{\nabla} \cdot \left[\vec{\nabla} (fg)\right] \\
&= \vec{\nabla} \cdot \left[\left(\vec{\nabla} f\right) g + f \left(\vec{\nabla} g\right)\right] \\
&= \vec{\nabla} \cdot \left[\left(\vec{\nabla} f\right) g\right] + \vec{\nabla} \cdot \left(f\vec{\nabla} g\right) \\
&= \left[\vec{\nabla} \cdot \vec{\nabla} f\right] g + \left(\vec{\nabla} f\right) \cdot \left(\vec{\nabla} g\right) + \left(\vec{\nabla} f\right) \cdot \left(\vec{\nabla} g\right) \\
&\quad + f\left[\vec{\nabla} \cdot \left(\vec{\nabla} g\right)\right] \equiv (\nabla^2 f) g + 2\vec{\nabla} f \cdot \vec{\nabla} g + f\nabla^2 g.
\end{aligned}$$

9 **(B)** Find the expressions of $\vec{\nabla} \cdot \vec{x}$ and of $\vec{\nabla} \times \vec{x}$, where $\vec{x} = (x, y, z)$ is the position vector in Cartesian coordinates. Can the position vector be expressed as the gradient of a scalar function?

Solution
The divergence of the position vector is

$$\vec{\nabla} \cdot \vec{x} = \vec{\nabla} \cdot (x, y, z) = \frac{\partial x}{\partial x} + \frac{\partial y}{\partial y} + \frac{\partial z}{\partial z} = 1 + 1 + 1 = 3,$$

while the curl of the position vector is given by

$$\vec{\nabla} \times \vec{x} = \begin{vmatrix} \vec{i} & \vec{j} & \vec{k} \\ \partial_x & \partial_y & \partial_z \\ x & y & z \end{vmatrix}$$

$$= \vec{i} \left(\partial_y z - \partial_z y\right) - \vec{j} \left(\partial_x z - \partial_z x\right) + \vec{k} \left(\partial_x y - \partial_y x\right) = 0.$$

The vanishing of the curl tells us that one can express the position vector as the gradient of a scalar; in fact, it is straightforward to verify[5] that $\vec{x} = \vec{\nabla} \left(r^2/2\right)$, where $r = \sqrt{x^2 + y^2 + z^2} = |\vec{x}|$.

10 **(B)** Find the curl of the vector field

$$\vec{a} \left(\vec{x}\right) = \vec{i} \left(x\,e^y\right) + \vec{j} \left(xy \ln z\right) + \vec{k} \left(xyz\,e^z\right).$$

Solution
In the calculation of the curl of $\vec{a}$ it is convenient to use the pseudo-determinant

$$\vec{\nabla} \times \vec{a} = \begin{vmatrix} \vec{i} & \vec{j} & \vec{k} \\ \partial_x & \partial_y & \partial_z \\ a_x & a_y & a_z \end{vmatrix}$$

$$= \vec{i} \left(\partial_y a_z - \partial_z a_y\right) - \vec{j} \left(\partial_x a_z - \partial_z a_x\right) + \vec{k} \left(\partial_x a_y - \partial_y a_x\right)$$

$$= x \left(z\,e^z - \frac{y}{z}\right) \vec{i} - yz\,e^z \vec{j} + \left(y \ln z - x\,e^y\right) \vec{k}.$$

11 **(B)** The notation $\vec{\nabla} \times \vec{a}$ for the curl of the vector field $\vec{a}$ may suggest that $\vec{\nabla} \times \vec{a}$ is orthogonal to the vector field $\vec{a}$. Is this true?

Solution
In order to decide whether or not this is true, we examine the scalar product $\vec{a} \cdot \left(\vec{\nabla} \times \vec{a}\right)$ and see if it is zero (which would mean that $\vec{a}$

---

[5]This is trivial in one dimensions, in which $x = d\left(x^2/2\right)/dx$.

and $\vec{\nabla} \times \vec{a}$ are orthogonal). We have

$$\vec{\nabla} \times \vec{a} = \begin{vmatrix} \vec{e}_x & \vec{e}_y & \vec{e}_z \\ \partial_x & \partial_y & \partial_z \\ a_x & a_y & a_z \end{vmatrix}$$

$$= \vec{e}_x \left(\partial_y a_z - \partial_z a_y \right) - \vec{e}_y \left(\partial_x a_z - \partial_z a_x \right) + \vec{e}_z \left(\partial_x a_y - \partial_y a_x \right),$$

and the required scalar product is

$$\vec{a} \cdot \left(\vec{\nabla} \times \vec{a} \right) = a_x \left(\partial_y a_z - \partial_z a_y \right) - a_y \left(\partial_x a_z - \partial_z a_x \right) + a_z \left(\partial_x a_y - \partial_y a_x \right).$$

This scalar product does not vanish in general, and therefore the suggested property of $\vec{\nabla} \times \vec{a}$ is not true.

12 **(B)** Prove that, for any vector $\vec{a}$,

$$\left(\vec{a} \cdot \vec{\nabla} \right) \vec{x} = \vec{a}.$$

Solution
We have

$$\left(\vec{a} \cdot \vec{\nabla} \right) \vec{x} \equiv \left(a_x \frac{\partial \vec{x}}{\partial x} + a_y \frac{\partial \vec{x}}{\partial y} + a_z \frac{\partial \vec{x}}{\partial z} \right)$$

$$= a_x \left(1, 0, 0 \right) + a_y \left(0, 1, 0 \right) + a_z \left(0, 0, 1 \right) = \vec{a}.$$

13 **(B)** Prove that $\vec{\nabla} \times \vec{a} = 0$ if $\vec{a}$ is the gradient of a scalar field.

Solution
Let $\vec{a} = \vec{\nabla} f$, where $f(\vec{x})$ is a regular function; then

$$\vec{\nabla} \times \vec{\nabla} f = \begin{vmatrix} \vec{i} & \vec{j} & \vec{k} \\ \partial_x & \partial_y & \partial_z \\ \partial_x f & \partial_y f & \partial_z f \end{vmatrix}$$

$$= \vec{i} \left(\partial_y \partial_z f - \partial_z \partial_y f \right) - \vec{j} \left(\partial_x \partial_z f - \partial_z \partial_x f \right)$$

$$+ \vec{k} \left(\partial_x \partial_y f - \partial_y \partial_x f \right) = 0,$$

due to the fact that mixed second derivatives commute for a regular function,

$$\frac{\partial^2 f}{\partial x^i \partial x^j} = \frac{\partial^2 f}{\partial x^j \partial x^i} \qquad (i, j = 1, 2, 3).$$

14 **(B)** Prove that $\vec{\nabla} \cdot \vec{a} = 0$ if $\vec{a} = \vec{\nabla} \times \vec{b}$, where $\vec{b}$ is a regular vector field.

Solution

We have

$$\vec{a} = \vec{\nabla} \times \vec{b} = \begin{vmatrix} \vec{i} & \vec{j} & \vec{k} \\ \partial_x & \partial_y & \partial_z \\ b_x & b_y & b_z \end{vmatrix}$$

$$= \vec{i} \left(\partial_y b_z - \partial_z b_y \right) - \vec{j} \left(\partial_x b_z - \partial_z b_x \right) + \vec{k} \left(\partial_x b_y - \partial_y b_x \right)$$

and

$$\vec{\nabla} \cdot \vec{a} = \partial_x a_x + \partial_y a_y + \partial_z a_z$$

$$= \partial_x \partial_y b_z - \partial_x \partial_z b_y - \partial_y \partial_x b_z + \partial_y \partial_z b_x + \partial_z \partial_x b_y - \partial_z \partial_y b_x = 0$$

because mixed second derivatives commute for a regular field, i.e., $\partial_i \partial_j b_s = \partial_j \partial_i b_s$ for $i, j, s = 1, 2, 3$.

15 **(B)** Compute the Laplacian of the vector field

$$\vec{a} = \left(x^3, -3xz, 2x^2 y^2 z \right) \equiv \left(a_x, a_y, a_z \right).$$

Solution

By definition, the Laplacian of the vector field $\vec{a}$ is

$$\nabla^2 \vec{a} \equiv \left(\nabla^2 a_x, \nabla^2 a_y, \nabla^2 a_z \right).$$

The first partial derivatives of the vector components are

$$\frac{\partial a_x}{\partial x} = 3x^2, \quad \frac{\partial a_x}{\partial y} = 0, \quad \frac{\partial a_x}{\partial z} = 0,$$

$$\frac{\partial a_y}{\partial x} = -3z, \quad \frac{\partial a_y}{\partial y} = 0, \quad \frac{\partial a_y}{\partial z} = -3x,$$

$$\frac{\partial a_z}{\partial x} = 4xy^2 z, \quad \frac{\partial a_z}{\partial y} = 4x^2 yz, \quad \frac{\partial a_z}{\partial z} = 2x^2 y^2.$$

The second derivatives needed to construct the Laplacian of each component are

$$\frac{\partial^2 a_x}{\partial x^2} = 6x, \quad \frac{\partial^2 a_x}{\partial y^2} = 0, \quad \frac{\partial^2 a_x}{\partial z^2} = 0,$$

$$\frac{\partial^2 a_y}{\partial x^2} = 0, \quad \frac{\partial^2 a_y}{\partial y^2} = 0, \quad \frac{\partial^2 a_y}{\partial z^2} = 0,$$

$$\frac{\partial^2 a_z}{\partial x^2} = 4y^2 z, \quad \frac{\partial^2 a_z}{\partial y^2} = 4x^2 z, \quad \frac{\partial^2 a_z}{\partial z^2} = 0,$$

and the Laplacian of $\vec{a}$ is

$$\nabla^2 \vec{a} = \left(\nabla^2 a_x, \nabla^2 a_y, \nabla^2 a_z \right) = \left(6x, 0, 4y^2 z + 4x^2 z \right).$$

16 **(B)** Prove that

$$\vec{\nabla} \times \left(\vec{\nabla} \times \vec{a} \right) = \vec{\nabla} \left(\vec{\nabla} \cdot \vec{a} \right) - \nabla^2 \vec{a} \qquad (1.8)$$

in Cartesian coordinates.[6]

Solution
Let $\vec{b}$ denote the curl

$$\vec{b} \equiv \vec{\nabla} \times \vec{a} = \begin{vmatrix} \vec{i} & \vec{j} & \vec{k} \\ \partial_x & \partial_y & \partial_z \\ a_x & a_y & a_z \end{vmatrix}$$

$$= \vec{i} \left(\partial_y a_z - \partial_z a_y \right) - \vec{j} \left(\partial_x a_z - \partial_z a_x \right) + \vec{k} \left(\partial_x a_y - \partial_y a_x \right).$$

The curl of $\vec{b}$ is

$$\vec{\nabla} \times \vec{b} = \begin{vmatrix} \vec{i} & \vec{j} & \vec{k} \\ \partial_x & \partial_y & \partial_z \\ b_x & b_y & b_z \end{vmatrix} = \vec{i} \left(\partial_y b_z - \partial_z b_y \right)$$

$$- \vec{j} \left(\partial_x b_z - \partial_z b_x \right) + \vec{k} \left(\partial_x b_y - \partial_y b_x \right)$$

---

[6] Note that Eq. (1.8) is valid only in Cartesian coordinates.

$$= \vec{i}\left(\partial_y\partial_x a_y - \partial_{yy}^2 a_x + \partial_z\partial_x a_z - \partial_{zz}^2 a_x\right)$$

$$-\vec{j}\left(\partial_{xx}^2 a_y - \partial_x\partial_y a_x - \partial_z\partial_y a_z + \partial_{zz}^2 a_y\right)$$

$$+\vec{k}\left(-\partial_{xx}^2 a_z + \partial_x\partial_z a_x - \partial_{yy}^2 a_z + \partial_y\partial_z a_y\right)$$

$$= \vec{i}\left\{\partial_x\left(\partial_y a_y + \partial_z a_z\right) - \left(\partial_{yy}^2 + \partial_{zz}^2\right)a_x\right\}$$

$$+\vec{j}\left\{\partial_y\left(\partial_x a_x + \partial_z a_z\right) - \left(\partial_{xx}^2 + \partial_{zz}^2\right)a_y\right\}$$

$$+\vec{k}\left\{\partial_z\left(\partial_x a_x + \partial_y a_y\right) - \left(\partial_{xx}^2 + \partial_{yy}^2\right)a_z\right\}$$

$$= \vec{i}\left\{\partial_x\left(\vec{\nabla}\cdot\vec{a}\right) - \nabla^2 a_x\right\} + \vec{j}\left\{\partial_y\left(\vec{\nabla}\cdot\vec{a}\right) - \nabla^2 a_y\right\}$$

$$+\vec{k}\left\{\partial_z\left(\vec{\nabla}\cdot\vec{a}\right) - \nabla^2 a_z\right\}$$

$$= \vec{\nabla}\left(\vec{\nabla}\cdot\vec{a}\right) - \nabla^2\vec{a}.$$

17 (B) Compute the curl

$$\vec{\nabla}\times\left(\frac{\vec{e}_r}{r^2}\right).$$

Solution
It is convenient to use spherical coordinates (r, θ, φ) with $\vec{e}_r, \vec{e}_\theta$, and $\vec{e}_\varphi$ denoting the associated unit vectors; then, for a generic n (and in particular for $n = -2$, which is the case given)

$$\vec{A} \equiv r^n\,\vec{e}_r = (r^n, 0, 0).$$

By using the expression of the curl operator in spherical coordinates (see Appendix C), we obtain

$$\vec{\nabla}\times\vec{A} = \frac{1}{r\sin\theta}\left[\frac{\partial}{\partial\theta}\left(\sin\theta\,A_\varphi\right) - \frac{\partial A_\theta}{\partial\varphi}\right]\vec{e}_r$$

$$+\frac{1}{r}\left[\frac{1}{\sin\theta}\frac{\partial A_r}{\partial\varphi} - \frac{\partial}{\partial r}\left(r A_\varphi\right)\right]\vec{e}_\theta$$

$$+\frac{1}{r}\left[\frac{\partial}{\partial r}\left(r A_\theta\right) - \frac{\partial A_r}{\partial\theta}\right]\vec{e}_\varphi$$

$$= \frac{1}{r\sin\theta} \frac{\partial (r^n)}{\partial\varphi} \vec{e}_\theta - \frac{1}{r}\frac{\partial (r^n)}{\partial\theta} \vec{e}_\varphi = 0.$$

18 **(B)** Prove Green's theorem

$$\int\int\int_V d^3\vec{x}\,(f\nabla^2 g - g\nabla^2 f) = \int\int_S \left(f\vec{\nabla}g - g\vec{\nabla}f\right)\cdot d\vec{S} \quad (1.9)$$

by using the Gauss theorem, where V is a finite volume delimited by the closed surface S.

Solution
By applying the divergence operator to the vector $\left(f\vec{\nabla}g - g\vec{\nabla}f\right)$, we obtain

$$\vec{\nabla}\cdot\left(f\vec{\nabla}g - g\vec{\nabla}f\right) = \vec{\nabla}f\cdot\vec{\nabla}g + f\nabla^2 g - \vec{\nabla}g\cdot\vec{\nabla}f - g\nabla^2 f = f\nabla^2 g - g\nabla^2 f.$$
$$(1.10)$$

By integrating the left-hand side of Eq. (1.10) over the volume V and applying the Gauss theorem, we obtain

$$\int\int\int_V d^3\vec{x}\,\vec{\nabla}\cdot\left(f\vec{\nabla}g - g\vec{\nabla}f\right) = \int\int_S \left(f\vec{\nabla}g - g\vec{\nabla}f\right)\cdot d\vec{S}. \quad (1.11)$$

The integration of Eq. (1.10) over the same volume yields

$$\int\int\int_V d^3\vec{x}\,\vec{\nabla}\cdot\left(f\vec{\nabla}g - g\vec{\nabla}f\right) = \int\int\int_V d^3\vec{x}\,(f\nabla^2 g - g\nabla^2 f);$$
$$(1.12)$$

a comparison of Eqs. (1.11) and (1.12) then yields Green's theorem (1.9).

1.6   Partial differential equations

While ordinary differential equations (ODEs) describe physical systems with a finite number of degrees of freedom (such as, in mechanics, point particles and rigid bodies), distributed systems with an infinite number of degrees of freedom are more realistic, and they are described by partial differential equations (PDEs). Often the solution of a PDE can be reduced to the problem of solving a set of ODEs, e.g., by the method of separation of variables.

1 **(B, C)** Provide physical examples of elliptic, parabolic, and hyperbolic partial differential equations.

Solution

An example of elliptic PDE is the Laplace equation, which in Cartesian coordinates assumes the form

$$\nabla^2 f = \frac{\partial^2 f}{\partial x^2} + \frac{\partial^2 f}{\partial y^2} + \frac{\partial^2 f}{\partial z^2} = 0. \tag{1.13}$$

The Laplace equation is satisfied, e.g., by the Newtonian gravitational potential outside mass distributions, by the electrostatic potential outside charge distributions, by the temperature, the concentration of a diffusing chemical, or by the hydraulic potential in stationary situations.

An example of parabolic PDE is

$$\frac{\partial f}{\partial t} = \alpha \nabla^2 f, \tag{1.14}$$

where α is a constant. If $f(t, \vec{x})$ represents the temperature, Eq. (1.14) describes heat transfer by conduction in a homogeneous medium (see Chapter 5). If instead f describes the concentration of a pollutant, Eq. (1.14) describes its spreading by diffusion processes (see Chapter 8).

An example of hyperbolic PDE is the d'Alembert or wave equation

$$\Box f \equiv \frac{\partial^2 f}{\partial x^2} + \frac{\partial^2 f}{\partial y^2} + \frac{\partial^2 f}{\partial z^2} - \frac{1}{v^2} \frac{\partial^2 f}{\partial t^2} = 0,$$

where v is a constant representing the speed of propagation of the wave $f(t, x, y, z)$ in a homogeneous medium.

2 **(B, C)** Oceanic circulation, weather, and climate are examples of fluid-dynamical systems. The basic equations of fluid dynamics are the Navier–Stokes equations

$$\rho \frac{dv^i}{dt} = -\frac{\partial P}{\partial x^i} + 2 \sum_{j=1}^{3} \frac{\partial}{\partial x^j} (\eta \sigma_{ij}) + f_i, \tag{1.15}$$

where $\vec{v}$, ρ, and η are, respectively, the velocity field, the density, and the dynamic viscosity coefficient of the fluid. P is the pressure, f_i is the volume density of external forces, and σ_{ij} is the stress tensor. Oceanic circulation, weather, and climate are known to exhibit chaotic phenomena in the form of *turbulence*—this is the reason why it is impossible to obtain an accurate weather forecast on a scale of weeks or months—and chaos occurs in systems obeying nonlinear and dissipative equations. Point out the nonlinearity and the dissipative features of the Navier–Stokes Eqs. (1.15).

Solution
Dissipation is present through the term $2\partial\left(\eta\sigma_{ij}\right)/\partial x^j$ representing the action of viscosity. Viscosity describes internal friction in the fluid and friction between the fluid and its boundaries, both sources of energy dissipation. Nonlinearity is evident in the velocity term

$$\frac{dv^i(t,\vec{x})}{dt} = \frac{\partial v^i}{\partial t} + \sum_{j=1}^{3}\frac{\partial v^i}{\partial x^j}\frac{dx^j}{dt} = \frac{\partial v^i}{\partial t} + \sum_{j=1}^{3}\frac{\partial v^i}{\partial x^j}v^j = \frac{\partial v^i}{\partial t} + \left(\vec{v}\cdot\vec{\nabla}\vec{v}\right)^i,$$

where the advective term

$$\left(\vec{v}\cdot\vec{\nabla}\vec{v}\right)^i = \sum_{k=1}^{3}v^k\frac{\partial v^i}{\partial x^k}$$

is nonlinear.

3 **(B)** Is $f(x,y) = \alpha\left(x^2 - y^2\right) + \beta xy$, where α and β are constants, a harmonic function?

Solution
By definition a harmonic function satisfies the Laplace equation

$$\nabla^2 f = 0.$$

The first derivatives of f are

$$\frac{\partial f}{\partial x} = 2\alpha x + 2\beta y, \qquad \frac{\partial f}{\partial y} = -2\alpha y + \beta x,$$

and the second derivatives needed to form the Laplacian are

$$\frac{\partial^2 f}{\partial x^2} = 2\alpha, \qquad \frac{\partial^2 f}{\partial y^2} = -2\alpha.$$

The Laplacian of f is

$$\nabla^2 f \equiv \frac{\partial^2 f}{\partial x^2} + \frac{\partial^2 f}{\partial y^2} = 0,$$

and therefore f is a harmonic function.

4 **(B)** Solve the two-dimensional boundary-value problem

$$
\begin{cases}
\frac{\partial^2 u}{\partial x^2} + \frac{\partial^2 u}{\partial y^2} = 0, \\[4pt]
u\,(0, y) = 0, \\[4pt]
u\,(a, y) = f(y), \\[4pt]
u\,(x, 0) = 0, \\[4pt]
u\,(x, b) = 0,
\end{cases}
$$

for $(x, y) \in [0, a] \times [0, b]$, where $f(y)$ is a regular function and, for consistency, $f(0) = f(b) = 0$.

Solution
We proceed by separation of variables looking for solutions of the form

$$
u\,(x, y) = X(x)Y(y).
$$

Substitution into the Laplace equation and division by $u = XY$ yield

$$
\frac{1}{X}\frac{d^2 X}{dx^2} + \frac{1}{Y}\frac{d^2 Y}{dy^2} = 0.
$$

The first term on the left-hand side depends only on x, the second term depends only on y, and this equation can only be satisfied if both terms are constant and opposite to each other. We set the first term equal to λ, $d^2 X/X dx^2 = \lambda$, and the second term equal to $-\lambda$, $d^2 Y/Y dy^2 = -\lambda$, obtaining the two ODEs

$$
\frac{d^2 X}{dx^2} - \lambda X = 0,
$$

$$
\frac{d^2 Y}{dy^2} + \lambda Y = 0,
$$

with the boundary conditions $X(0) = 0$ and $Y(0) = Y(b) = 0$.

If $\lambda < 0$, we set $\lambda \equiv -\mu^2$ (where $\mu > 0$) for convenience and the general solution of the equation for $Y(y)$ is

$$
Y(y) = C_1 \, e^{\mu y} + C_2 \, e^{-\mu y},
$$

with $C_{1,2}$ integration constants. The boundary condition $Y(0) = 0$ yields $C_1 + C_2 = 0$, while the boundary condition $Y(b) = 0$ yields

$C_1 e^{\mu b} + C_2 e^{-\mu b} = 0$. By substituting $C_2 = -C_1$ from the first equation into the second equation, one obtains $C_1 \sinh(\mu b) = 0$. Since $\sinh(\mu b) \neq 0$ for $\lambda < 0$, it must be $C_1 = -C_2 = 0$, yielding only the trivial solution, which is not acceptable because it does not satisfy the boundary condition $u(a, y) = f(y)$. There are no solutions for $\lambda < 0$.

If $\lambda = 0$, the equations for X and Y are $d^2X/dx^2 = 0$ and $d^2Y/dy^2 = 0$, with linear solution $Y(y) = \alpha y + \beta$, with α and β constants. The boundary condition $Y(0) = 0$ implies that $\beta = 0$, while the boundary condition $Y(b) = 0$ requires that also $\alpha = 0$. Again, this leaves only the trivial solution, which is not acceptable. There are no solutions for $\lambda = 0$.

If $\lambda > 0$, we set $\lambda = \mu^2$ for convenience and the general solutions of the equations for X and Y are

$$X(x) = C_1 e^{\mu x} + C_2 e^{-\mu x},$$

$$Y(y) = D_1 \cos(\mu y) + D_2 \sin(\mu y),$$

with $C_{1,2}, D_{1,2}$ integration constants. The boundary condition $X(0) = 0$ yields $C_1 + C_2 = 0$ and

$$X(x) = C \sinh(\mu x),$$

while the boundary conditions $Y(0) = Y(b) = 0$ yield $D_1 = 0$ and $D_2 \sin(\mu b) = 0$, which implies $\mu b = n\pi$ with $n = 1, 2, 3 \ldots$ or the discrete values

$$\lambda_n = \mu_n^2 = \left(\frac{n\pi}{b}\right)^2 \qquad (n = 1, 2, 3 \ldots)$$

for the separation constant. The fundamental solutions of the problem are

$$u_n(t, x) = X_n(x)Y_n(y) = \sinh\left(\frac{n\pi x}{b}\right) \sin\left(\frac{n\pi y}{b}\right).$$

According to the superposition principle expressing the linearity of the Laplace equation, the general solution of the problem is the series

$$u(x, y) = \sum_{n=1}^{+\infty} c_n \sinh\left(\frac{n\pi x}{b}\right) \sin\left(\frac{n\pi y}{b}\right).$$

The coefficients c_n are determined by imposing the last boundary condition $u(a, y) = f(y)$, i.e.,

$$f(y) = \sum_{n=1}^{+\infty} c_n \sinh\left(\frac{n\pi a}{b}\right) \sin\left(\frac{n\pi y}{b}\right).$$

This equation expresses the fact that the series is the Fourier series of $f(y)$ on $[0, b]$ and therefore the coefficients c_n are the Fourier coefficients

$$c_n = \frac{2}{b \sinh(n\pi a/b)} \int_0^b dy\, f(y) \sin\left(\frac{n\pi y}{b}\right) \qquad (n = 1, 2, 3, \ldots).$$

5 **(B)** Consider the Laplace equation in the n-dimensional space with coordinates $\left(x^1, x^2, \ldots, x^n\right)$

$$\nabla^2 u = \frac{\partial^2 u}{\partial\left(x^1\right)^2} + \frac{\partial^2 u}{\partial\left(x^2\right)^2} + \ldots + \frac{\partial^2 u}{\partial\left(x^n\right)^2} = 0,$$

and find all its solutions (*harmonic functions*) that depend only on $r \equiv \sum_{k=1}^n \left(x^k\right)^2$.

Solution

Let $u = u(r)$ and $\nabla^2 u = 0$. The first derivatives of u are

$$\frac{\partial u}{\partial x^i} = \frac{du}{dr}\frac{\partial r}{\partial x^i} \equiv u'\frac{x^i}{r}.$$

The second derivatives of u needed to form the Laplacian $\nabla^2 u$ are

$$\frac{\partial^2 u}{\partial\left(x^i\right)^2} = u''\left(\frac{x^i}{r}\right)^2 + \frac{u'}{r} - \frac{u'x^i}{r^2}\frac{\partial r}{\partial x^i} =$$

$$= u''\left(\frac{x^i}{r}\right)^2 + \frac{u'}{r} - \frac{u'\left(x^i\right)^2}{r^3}.$$

The Laplacian of u is therefore, using $\sum_{i=1}^n \left(x^i\right)^2 = r^2$,

$$\nabla^2 u = \sum_{i=1}^n \left[\frac{u''\left(x^i\right)^2}{r^2} + \frac{u'}{r} - \frac{u'\left(x^i\right)^2}{r^3}\right] = u'' + (n-1)\frac{u'}{r} = 0.$$

This ODE can easily be integrated by writing

$$\frac{u''}{u'} + \frac{n-1}{r} = 0$$

or

$$\left(\ln u'\right)' + (n-1)\left(\ln r\right)' = 0,$$

which, using the properties of logarithms, becomes

$$u' = \frac{C}{r^{n-1}},$$

where C is an integration constant. This equation can be further integrated. If $n \neq 2$, we have

$$u(r) = C\,r^{2-n} + D \qquad (n \neq 2),$$

while for $n = 2$ we have

$$u(r) = C\,\ln r + D \qquad (n = 2),$$

where C and D are arbitrary integration constants.

6 **(B)** Solve the one-dimensional initial-boundary-value problem

$$\begin{cases} \dfrac{\partial u}{\partial t} = a\,\dfrac{\partial^2 u}{\partial x^2}, \\[2mm] \dfrac{\partial u}{\partial x}\,(t,0) = 0, \\[2mm] \dfrac{\partial u}{\partial x}\,(t,l) = 0, \\[2mm] u\,(0,x) = f(x), \end{cases}$$

for $x \in [0,l]$ and $t \geq 0$, where $f(x)$ is a regular function on $[0,l]$. This is a heat conduction problem in a rod with insulated ends—in fact, u is the temperature and the flux density $q'' = -k\partial u/\partial x$ is zero at both ends.

Solution
We proceed by separation of variables looking for solutions of the form

$$u\,(t,x) = T(t)X(x).$$

By substituting this form into the one-dimensional heat equation, one obtains

$$X\,\frac{dT}{dt} = aT\,\frac{d^2 X}{dx^2}.$$

Division by $au = a\,TX$ yields

$$\frac{1}{aT}\,\frac{dT}{dt} = \frac{1}{X}\,\frac{d^2 X}{dx^2},$$

where the left-hand side depends only on t and the right-hand side depends only on x. This equation can only be satisfied if both sides are constant and have the same value (*separation constant*), which we will call $-\lambda$. Hence we obtain the two ODEs

$$\frac{dT}{dt} + \lambda a T = 0,$$

$$\frac{d^2 X}{dx^2} + \lambda X = 0.$$

The PDE is reduced to two simple ODEs, and the price to pay for this simplification is that we have two equations instead of one. The time equation is immediately integrated, yielding

$$T(t) = e^{-\lambda a t}.$$

We do not attribute a multiplicative integration constant to this solution because we retain arbitrary integration constants in $X(x)$, which multiplies $T(t)$ in the solution.

The boundary condition $u_x(t,0) = 0$ yields $T(t)dX/dx(0) = 0$ and, since $T(t)$ cannot vanish identically, it must be $dX/dx(0) = 0$. Similarly, the boundary condition at $x = l$, $u_x(t,l) = 0$, yields $dX/dx(l) = 0$. Let us consider all the possible real values of the separation constant λ.

If $\lambda < 0$, we set $\lambda \equiv -\mu^2$ (where $\mu > 0$) for convenience. The spatial equation is

$$\frac{d^2 X}{dx^2} - \mu^2 X = 0,$$

and it has the general solution in $[0, l]$

$$X(x) = C_1 e^{\mu x} + C_2 e^{-\mu x},$$

with $C_{1,2}$ integration constants. The boundary condition $dX/dx = 0$ at $x = 0, l$ implies that

$$\mu (C_1 - C_2) = 0,$$

$$\mu \left(C_1 e^{\mu l} - C_2 e^{-\mu l} \right) = 0.$$

This linear system of algebraic equations admits only the solution $(C_1, C_2) = (0,0)$; therefore, for $\lambda < 0$ there is only the trivial solution, which is not acceptable because it does not satisfy the initial condition $u(0,x) = f(x)$.

If $\lambda = 0$, the spatial equation is $d^2 X/dx^2 = 0$, which has the general linear solution $X(x) = \alpha x + \beta$, with α and β constants. The boundary condition $dX/dx(0) = 0$ implies that $\alpha = 0$ and the only acceptable solution for $\lambda = 0$ is $X =$ constant.

If $\lambda > 0$, we set $\lambda = \mu^2$ for convenience and the spatial equation is the harmonic oscillator equation

$$\frac{d^2 X}{dx^2} + \mu^2 X = 0$$

with general solution

$$X(x) = C_1 \cos(\mu x) + C_2 \sin(\mu x),$$

with $C_{1,2}$ integration constants. The boundary condition $dX/dx(0) = 0$ implies $C_2 = 0$, while the boundary condition $dX/dx(l) = 0$ yields $C_1 \sin(\mu l) = 0$. Since C_1 cannot be zero (otherwise we have only the trivial solution, which does not satisfy the initial conditions), it must be $\sin(\mu l) = 0$, or $\mu l = n\pi$ with $n = 1, 2, 3\ ...$ (it cannot be $n = 0$ for $\mu \neq 0$). The separation constant therefore can only assume the discrete values (eigenvalues of the problem)

$$\lambda_n = \mu_n^2 = \left(\frac{n\pi}{l}\right)^2 \qquad (n = 1, 2, 3\ ...),$$

and the corresponding eigenfunctions are

$$X_n(x) = \cos\left(\frac{n\pi x}{l}\right).$$

The fundamental solutions of the problem are

$$u_n(t, x) = T_n(t)X_n(x) = e^{-\frac{n^2\pi^2 at}{l^2}} \cos\left(\frac{n\pi x}{l}\right).$$

The heat equation is linear and satisfies the superposition principle, which allows one to obtain the general solution as the series

$$u(t, x) = \frac{c_0}{2} + \sum_{n=1}^{+\infty} c_n e^{-\frac{n^2\pi^2 at}{l^2}} \cos\left(\frac{n\pi x}{l}\right),$$

where the first term on the left-hand side represents a steady-state constant solution corresponding to $\lambda = 0$, and the second term is a transient that decays as time goes by. The solution is not complete though, because the coefficients c_n of the series have not yet been determined. This is achieved by imposing the initial condition $u(0, x) = f(x)$, which yields

$$f(x) = \frac{c_0}{2} + \sum_{n=1}^{+\infty} c_n \cos\left(\frac{n\pi x}{l}\right).$$

This equation expresses the fact that the coefficients c_n are the Fourier coefficients of the function $f(x)$ on $[0, l]$, or

$$c_n = \frac{2}{l} \int_0^l dx\, f(x) \cos\left(\frac{n\pi x}{l}\right) \qquad (n = 1, 2, 3, ...).$$

7 **(B)** Verify that the plane wave

$$f(t, \vec{x}) = f_0 \, e^{i(\vec{k}\cdot\vec{x}-\omega t)},$$

where $f_0, \vec{k}$, and ω are constants, satisfies the d'Alembert or wave equation

$$\nabla^2 f - \frac{1}{c^2}\frac{\partial^2 f}{\partial t^2} = 0$$

provided that $\omega = c|\vec{k}|$.

Solution
The first-order partial derivatives of f are

$$\frac{\partial f}{\partial x^i} = i k_i f_0 \, e^{i(\vec{k}\cdot\vec{x}-\omega t)} = i k_i f$$

or, in compact notation, $\vec{\nabla} f = i\vec{k}f$, while

$$\frac{\partial f}{\partial t} = -i\omega f_0 \, e^{i(\vec{k}\cdot\vec{x}-\omega t)} = -i\omega f,$$

and

$$\nabla^2 f - \frac{1}{c^2}\frac{\partial^2 f}{\partial t^2}$$

$$= \frac{\partial^2 f}{\partial x^2} + \frac{\partial^2 f}{\partial y^2} + \frac{\partial^2 f}{\partial z^2} - \frac{1}{c^2}\frac{\partial^2 f}{\partial t^2}$$

$$= \left(i\vec{k}\right)^2 f - \frac{1}{c^2}(i\omega)^2 f$$

$$= -\left(\vec{k}^2 - \frac{\omega^2}{c^2}\right) f.$$

The last expression vanishes if $\omega = c\left|\vec{k}\right|$, which is equivalent to the usual relation $c = \lambda\nu$ between the wavelength λ, the phase velocity c, and the frequency ν of the wave if one remembers that $k = 2\pi/\lambda$ and $\omega = 2\pi\nu$.

8 **(B)** Solve the one-dimensional initial-boundary-value problem

$$\begin{cases} \dfrac{\partial^2 u}{\partial x^2} - \dfrac{1}{v^2}\dfrac{\partial^2 u}{\partial t^2} = 0, \\[2mm] u\,(t,0) = 0, \qquad u\,(t,l) = 0, \\[2mm] u\,(0,x) = f\,(x), \\[2mm] \dfrac{\partial u}{\partial t}\,(0,x) = 0 \end{cases}$$

for $x \in [0,l]$ and $t \geq 0$, where $f(x)$ is a regular function on $[0,l]$ with $f(0) = f(l) = 0$.

Solution
We proceed by separation of variables assuming that

$$u\,(t,x) = T(t)X(x);$$

substitution into the wave equation and division by $u = TX$ yield

$$\frac{1}{X}\frac{d^2 X}{dx^2} = \frac{1}{v^2}\frac{d^2 T}{dt^2}.$$

Since the left-hand side depends only on x and the right-hand side depends only t, this equation can only be satisfied if both sides are constant and have the same value $-\lambda$ (*separation constant*). This yields the two ODEs

$$\frac{d^2 T}{dt^2} + \lambda v^2 T = 0,$$

$$\frac{d^2 X}{dx^2} + \lambda X = 0.$$

If $\lambda < 0$, we set $\lambda \equiv -\mu^2$ (with $\mu > 0$) for convenience. The spatial equation

$$\frac{d^2 X}{dx^2} - \mu^2 X = 0$$

has the general solution

$$X(x) = C_1\,e^{\mu x} + C_2\,e^{-\mu x},$$

with $C_{1,2}$ integration constants. The boundary conditions $u\,(t,0) = 0$ and $u\,(t,l) = 0$ yield $X(0) = 0$ and $X(l) = 0$, respectively, which translate into

$$C_1 + C_2 = 0,$$

$$C_1 e^{\mu l} + C_2 e^{-\mu l} = 0.$$

Substitution of the first equation into the second gives $C_1 \sinh(\mu l) = 0$, which, for $\mu \neq 0$, is only satisfied by $C_1 = -C_2 = 0$; there is only the trivial solution for $\lambda < 0$ and it is not acceptable because it does not satisfy the initial condition $u(0, x) = f(x)$.

If $\lambda = 0$, the spatial equation is $d^2 X / dx^2 = 0$, which has the general solution $X(x) = \alpha x + \beta$, with α and β constants. The boundary conditions at $x = 0, l$ yield $\beta = 0$ and $\alpha = 0$, respectively, and there are no nontrivial solutions for $\lambda = 0$.

If $\lambda > 0$, we set $\lambda = \mu^2$ for convenience. The spatial equation

$$\frac{d^2 X}{dx^2} + \mu^2 X = 0$$

has the general solution

$$X(x) = C_1 \cos(\mu x) + C_2 \sin(\mu x),$$

with $C_{1,2}$ integration constants. The boundary condition $X(0) = 0$ yields $C_1 = 0$, while the second boundary condition $X(l) = 0$ yields $C_2 \sin(\mu l) = 0$. Since C_2 cannot be zero (otherwise the only solution is the trivial solution, which does not satisfy the initial conditions), it must be $\sin(\mu l) = 0$, or $\mu_n = n\pi/l$ with $n = 1, 2, 3 \ldots$. The eigenvalues of the problem are therefore the discrete values

$$\lambda_n = \left(\frac{n\pi}{l}\right)^2 \qquad (n = 1, 2, 3 \ldots),$$

and the corresponding eigenfunctions are

$$y_n(x) = \sin\left(\frac{n\pi x}{l}\right).$$

Let us solve now the time equation

$$\frac{d^2 T}{dt^2} + \lambda_n v^2 T = 0;$$

the general solution is

$$T_n(t) = \alpha_n \cos\left(\sqrt{\lambda_n}\, vt\right) + \beta_n \sin\left(\sqrt{\lambda_n}\, vt\right).$$

The initial condition $u_t(0, x) \equiv 0$ yields $\beta_n = 0$ for all n, and the fundamental solutions of the problem are

$$u_n(t, x) = T_n(t) X_n(x) = \cos\left(\frac{n\pi vt}{l}\right) \sin\left(\frac{n\pi x}{l}\right).$$

The wave equation is linear and we can superpose the fundamental solutions to form the general solution

$$u(t, x) = \sum_{n=1}^{+\infty} \alpha_n \cos\left(\frac{n\pi vt}{l}\right) \sin\left(\frac{n\pi x}{l}\right).$$

The coefficients α_n are determined by imposing the initial condition $u(0, x) = f(x)$, which has not yet been used. This yields

$$f(x) = \sum_{n=1}^{+\infty} \alpha_n \sin\left(\frac{n\pi x}{l}\right),$$

i.e., the coefficients α_n are the Fourier coefficients of the function $f(x)$ on $[0, l]$, or

$$\alpha_n = \frac{2}{l} \int_0^l dx\, f(x) \sin\left(\frac{n\pi x}{l}\right) \qquad (n = 1, 2, 3 \ldots).$$

9 **(B)** Provide a physical interpretation of the fact that

$$f(\vec{x}) = \frac{1}{r} \equiv \frac{1}{\sqrt{x^2 + y^2 + z^2}}$$

is a (distributional) solution of the partial differential equation

$$\nabla^2 f = -4\pi\delta^{(3)}(x, y, z),$$

where $\delta^{(3)}(\vec{x}) = \delta(x)\delta(y)\delta(z)$ is the Dirac delta in three dimensions [17], i.e., a generalized function such that

$$\int\int\int_{\text{all space}} d^3\vec{x}\, \delta^{(3)}(\vec{x}) = 1.$$

Hint: Consider the Newtonian potential generated by a point mass M for $r > 0$.

Solution
To recognize that $1/r$ is a solution of this PDE (in the sense of distributions [17]) consider the Newtonian gravitational potential Φ due to a point mass M located at the origin of the coordinates. It is well

known that $\Phi = -GM/r$, where G is the gravitational constant and that, in general, Φ satisfies the Poisson equation

$$\nabla^2 \Phi = -4\pi G \rho\,(x, y, z)\,,$$

where $\rho(\vec{x})$ is the mass density. Since the mass M is concentrated at the point $\vec{x} = 0$, one can express the corresponding mass density as

$$\rho\,(x, y, z) = M\,\delta^{(3)}\,(x, y, z)$$

and then

$$\nabla^2 \Phi = -GM\,\nabla^2 \left(\frac{1}{r}\right) = 4\pi GM\,\delta^{(3)}\,(x, y, z)\,,$$

from which it follows that $1/r$ is a solution with the Dirac delta as a source

$$\nabla^2 \left(\frac{1}{r}\right) = -4\pi\,\delta^{(3)}\,(x, y, z)\,.$$

The total mass in space is obtained, as usual, by integrating the mass density ρ over all space and is given by

$$\int\int\int d^3\vec{x}\,\rho\,(x, y, z) = M\,\int\int\int d^3\vec{x}\,\delta^{(3)}\,(x, y, z) = M.$$

1.7 Tensors

Once the transition from scalars (naively, quantities with no indices) to vectors (naively, quantities with one index) is made, one can generalize to quantities with more than one index: *tensors*. Although they usually do not appear in elementary physics and mathematics courses, tensors are necessary to describe, for example, stresses in fluid dynamics and elasticity theory, the inertial properties of a rigid body, the transfer of momentum by an electromagnetic field, thermal conductivity in an anisotropic medium, or the hydraulic conductivity in an anisotropic aquifer in groundwater hydrology. An approachable reference is [65].

1 **(B, C)** What is a tensor with two contravariant indices? A tensor with two covariant indices? A tensor with two mixed indices? Provide an example of a two-index tensor.

Solution

A *tensor with two contravariant indices* is an object with components T^{ij} labeled by two upper indices, that transforms according to

$$T^{ij} \longrightarrow T'^{\,ij} = \sum_{l,m} \frac{\partial x'^i}{\partial x^l}\,\frac{\partial x'^j}{\partial x^m}\,T^{lm}$$

under the coordinate transformation $x^i \longrightarrow x'^i$. A *tensor with two covariant indices* is an object with components T_{ij} labeled by two lower indices, that transforms according to

$$T_{ij} \longrightarrow T'_{ij} = \sum_{l,m} \frac{\partial x^l}{\partial x'^i} \frac{\partial x^m}{\partial x'^j} T_{lm}$$

under the coordinate transformation $x^i \longrightarrow x'^i$.

A *tensor with two mixed indices* is an object with components $T_i{}^j$ labeled by one upper and one lower index, that transforms according to

$$T_i{}^j \longrightarrow T'_i{}^j = \sum_{l,m} \frac{\partial x^l}{\partial x'^i} \frac{\partial x'^j}{\partial x^m} T_l{}^m$$

under $x^i \longrightarrow x'^i$. The difference between covariant and contravariant indices is relevant when curvilinear coordinates are used instead of Cartesian ones, or when the metric is not Euclidean. Note that the order of the mixed indices is important.

A simple example of a two-index tensor is built out a vector field, e.g., the velocity field v^i in a fluid. The quantity $T^{ij} \equiv v^i v^j$ is a tensor with two contravariant indices and it is symmetric because $T^{ji} = T^{ij}$ for all values of i and j. Another example is the tensor $\partial v_i / \partial x^j$, often used as an approximation of the stress tensor in fluid mechanics.

2 **(B)** Show that if the components of a tensor vanish in a coordinate system, they vanish in any other coordinate system.

Solution
Let $T^{ab\cdots}{}_{cd\cdots} = 0$ in the coordinate system $\{x^i\}$. Then in any other coordinate system $\{x'^i\}$ it is

$$T'^{ab\cdots}{}_{cd\cdots} = \sum_{e,f,\dots=1}^{n} \sum_{g,h,\dots=1}^{n} \frac{\partial x'^a}{\partial x^e} \frac{\partial x'^b}{\partial x^f} \cdots \frac{\partial x^g}{\partial x'^c} \frac{\partial x^h}{\partial x'^d} \cdots T^{ef\cdots}{}_{gh\cdots} = 0$$

because $T^{ef\cdots}{}_{gh\cdots} = 0$.

3 **(B)** Prove that the identity tensor with components δ_j^i is invariant under coordinate transformations, i.e., that its components assume the same values in any coordinate system. By using this property, show that the trace

$$T \equiv \sum_{i=1}^{n} T^i{}_i$$

of a two-index tensor $T^i{}_j$ is invariant under coordinate transformations.

Solution
Under a coordinate transformation $x^i \longrightarrow x'^i$, the components of the identity tensor transform according to

$$\delta'^i{}_j = \sum_{k,l=1}^{n} \frac{\partial x'^i}{\partial x^k} \frac{\partial x^l}{\partial x'^j} \delta^k_l$$

$$= \sum_{k=1}^{n} \frac{\partial x'^i}{\partial x^k} \frac{\partial x^k}{\partial x'^j} =$$

(using the chain rule)

$$= \frac{\partial x'^i}{\partial x'^j} = \delta^i_j.$$

The trace of a two-index tensor transforms under a change of coordinates according to

$$T' \equiv \sum_{i=1}^{n} T'^i{}_i = \sum_{i,j=1}^{n} T'^i{}_j \delta'^j{}_i = \sum_{i,k,l=1}^{n} \frac{\partial x'^i}{\partial x^k} \frac{\partial x^l}{\partial x'^j} T^k{}_l \delta^j_i =$$

$$= \sum_{i,k,l=1}^{n} \frac{\partial x'^i}{\partial x^k} \frac{\partial x^l}{\partial x'^i} T^k{}_l = \sum_{i,k,l=1}^{n} \frac{\partial x^l}{\partial x'^i} \frac{\partial x'^i}{\partial x^k} T^k{}_l$$

$$= \sum_{k,l=1}^{n} \frac{\partial x^l}{\partial x^k} T^k{}_l = \sum_{k,l=1}^{n} \delta^k_l T^k{}_l = \sum_{i=1}^{n} T'^l{}_l \equiv T.$$

4 **(B)** Any two-index tensor T_{ij} can be decomposed into a symmetric part $T_{(ij)}$ and an antisymmetric part $T_{[ij]}$ as follows:

$$T_{ij} = \frac{T_{ij} + T_{ji}}{2} + \frac{T_{ij} - T_{ji}}{2} \equiv T_{(ij)} + T_{[ij]},$$

where

$$T_{(ij)} = \frac{T_{ij} + T_{ji}}{2},$$

$$T_{[ij]} = \frac{T_{ij} - T_{ji}}{2}.$$

The identity is trivial. Show that this decomposition is unique.

Solution
We need to show that, if $T_{ij} = p_{ij} + q_{ij}$ is another decomposition of T_{ij} into a symmetric and an antisymmetric part, with $p_{ij} = p_{ji}$ and $q_{ij} = -q_{ji}$, then p_{ij} and q_{ij} coincide with $T_{(ij)}$ and $T_{[ij]}$, respectively.

Let us use the symmetry and antisymmetry properties of p_{ij} and q_{ij} to write

$$T_{ij} \;=\; p_{ij} + q_{ij}, \tag{1.16}$$

$$T_{ji} \;=\; p_{ij} - q_{ij}. \tag{1.17}$$

Adding Eqs. (1.16) and (1.17) term to term yields

$$T_{ij} + T_{ji} = 2p_{ij},$$

or

$$p_{ij} = T_{(ij)},$$

while subtracting Eq. (1.17) from Eq. (1.16) term to term yields

$$T_{ij} - T_{ji} = 2q_{ij},$$

or

$$q_{ij} = T_{[ij]},$$

which is what we wanted to show.

5 **(B)** By using tensor algebra, prove that the scalar product $\vec{a} \cdot \vec{b}$ of two vectors $\vec{a}$ and $\vec{b}$ is a scalar under coordinate transformations. Repeat the proof for the contraction

$$\sum_{i,j=1}^{n} A_{ij} B^{ij}$$

of the two-index tensors A_{ij} and B^{ij}.

Solution
For two vectors $\vec{a}$ and $\vec{b}$ in an n-dimensional space, we have

$$\vec{a} \cdot \vec{b} = \sum_{i=1}^{n} a_i b^i$$

in the coordinate system $\{x^i\}$. Under the coordinate transformation $x^i \longrightarrow x'^i$, the scalar product transforms as

$$\vec{a}' \cdot \vec{b}' \equiv \sum_{i=1}^{n} a'_i b'^i = \sum_{i,k,l=1}^{n} \frac{\partial x^l}{\partial x'^i} a_l \frac{\partial x'^i}{\partial x^k} b^k$$

$$= \sum_{i,k,l=1}^{n} \left(\frac{\partial x^l}{\partial x'^i} \frac{\partial x'^i}{\partial x^k} \right) a_l b^k = \sum_{i,k,l=1}^{n} \frac{\partial x^l}{\partial x^k} a_l b^k$$

$$= \sum_{i,k,l=1}^{n} \delta^l_k \, a_l b^k = \sum_{i,k,l=1}^{n} a_l b^l \equiv \vec{a} \cdot \vec{b}.$$

For the two-index tensors A_{ij} and B_{ij}, we have

$$\sum_{i,j=1}^{n} A'_{ij} B'^{ij} = \sum_{i,j,k,l,r,s=1}^{n} \frac{\partial x^k}{\partial x'^i} \frac{\partial x^l}{\partial x'^j} A_{kl} \frac{\partial x'^i}{\partial x^r} \frac{\partial x'^j}{\partial x^s} B^{rs}$$

$$= \sum_{i,j,k,l,r,s=1}^{n} \left(\frac{\partial x^k}{\partial x'^i} \frac{\partial x'^i}{\partial x^r} \right) \left(\frac{\partial x^l}{\partial x'^j} \frac{\partial x'^j}{\partial x^s} \right) A_{kl} \, B^{rs}$$

$$= \sum_{k,l,r,s=1}^{n} \delta^k_r \, \delta^l_s \, A_{kl} B^{rs} = \sum_{k,l=1}^{n} A_{kl} B^{kl},$$

i.e., the quantity $\sum_{i,j=1}^{n} A_{ij} B^{ij}$ does not change under coordinate transformations.

1.8   Dimensional analysis

Dimensional analysis is a subsidiary tool to avoid a more detailed analysis of physical problems—it shortcuts to a result that is usually correct within one order of magnitude. Although the method of dimensional analysis cannot provide an exact result (if it does, it is a coincidence), it is nevertheless very useful to estimate physical quantities of interest in a physical problem before attacking it with the full arsenal of mathematical weapons available, or else resorting to experiment. The method of dimensional analysis is also useful as a check of mathematical formulas and very often reveals errors by showing the dimensional inconsistency of a formula. In fact, the equations used must be dimensionally homogeneous, i.e., the left-hand side and the right-hand side must have the same dimensions, and only quantities with the same dimensions can be added or subtracted in a mathematical expression.[7] Similarly, the argument of trigonometric, logarithmic, or exponential functions must be dimensionless.

Dimensional analysis can range from a simple back-of-the-envelope calculation to the more sophisticated analyses used in environmental engineering (see, e.g., Ref. [4]). In certain situations where a complete mathematical solution of an environmental physics or engineering problem is not possible, dimensional analysis proves to be extremely valuable.

Most first-year physics textbooks have a section on dimensional analysis that can be used for reference; Ref. [36] is entirely devoted to this subject.

1 **(B)** Derive an approximate formula for the period of a simple pendulum using only the method of dimensional analysis and neglecting friction.

Solution
The period P of the pendulum has the dimensions of time. If we neglect friction the physical quantities that are in principle of possible relevance to determine the pendulum's period are its length l, the mass m of the bob, and the acceleration of gravity g. Assume that $P = l^\alpha m^\beta g^\gamma$: then

$$[T] = [L^\alpha] \left[M^\beta\right] \left[L^\gamma T^{-2\gamma}\right] = \left[L^{\alpha+\gamma} T^{-2\gamma} M^\beta\right],$$

and it must be

$$\alpha + \gamma = 0,$$

$$-2\gamma = 1,$$

$$\beta = 0.$$

This linear system has solution

$$\alpha = \frac{1}{2}, \qquad \beta = 0, \qquad \gamma = -\frac{1}{2};$$

hence, the desired approximate formula for the period of the pendulum is

$$P = \sqrt{\frac{l}{g}}.$$

---

[7]This *principle of dimensional homogeneity* was stated explicitly by Jean Baptiste Joseph Fourier in his work *La Théorie Analitique de la Chaleur* in 1822.

The mass of the bob does not appear in this formula: this is due to the universality of free fall, i.e., to the fact that the acceleration of gravity is the same for all bodies placed in a given gravitational field, independently of their mass. Even if we forget about this experimental fact discovered long ago by Galileo, the method of dimensional analysis allows us to recover it. For comparison, the exact formula for the period of the simple pendulum (in the approximation of small oscillations) is

$$P = 2\pi\sqrt{\frac{l}{g}}.$$

2 **(B)** You are writing a physics exam and you don't remember the formula for the period of the compound pendulum, which you need in order to solve a problem. Is it $P = 2\pi\sqrt{mgl/I}$ or $P = 2\pi\sqrt{I/(mgl)}$? (Here I, m, and l are the moment of inertia, mass, and distance between the center of gravity and the rotation axis of the pendulum, while g is the acceleration of gravity.) Use the method of dimensions to find out.

Solution
The dimensions of I and g are $[I] = [ML^2]$ and $[g] = [LT^{-2}]$, so

$$\left[\sqrt{\frac{mgl}{I}}\right] = \left[\left(\frac{MLT^{-2}L}{ML^2}\right)^{1/2}\right] = [T^{-1}];$$

the expression $\sqrt{mgl/I}$ has the dimensions of the inverse of a time and is therefore incorrect. On the other hand,

$$\left[\sqrt{\frac{I}{mgl}}\right] = [T],$$

and $P = 2\pi\sqrt{I/mgl}$ is the correct expression for the period of the compound pendulum.

3 **(B)** Verify the dimensional correctness of Einstein's famous formula $E = mc^2$, where m is the mass of a particle, E is its (rest) energy, and c is the speed of light in vacuum.

Solution
Energy is a quantity homogeneous to work, which is dimensionally a force multiplied by a length; hence,

$$[E] = [F][L] = [L^2T^{-2}M];$$

while
$$[m\,c^2] = [M]\,[L^2T^{-2}]\,;$$
hence $[E] = [mc^2]$.

4 **(B)** Suppose that you adopt length, time, and force $(L, T,$ and $F)$ as
fundamental quantities instead of length, time, and mass $(L, T,$ and
$M)$. What would be the dimensions of mass, moment of inertia, and
mass density in this new hypothetical system of units?

Solution
Because force = mass × acceleration, in this hypothetical system of
units the dimensions of mass would be

$$[M] = \left[\frac{F}{a}\right] = \left[FL^{-1}T^2\right].$$

A moment of inertia I would have dimensions

$$[I] = \left[ML^2\right] = \left[FL^{-1}T^2\right]\left[L^2\right] = \left[FLT^2\right],$$

while mass density ρ would have dimensions

$$[\rho] = \left[ML^{-3}\right] = \left[FL^{-1}T^2\right]\left[L^{-3}\right] = \left[FL^{-4}T^2\right].$$

5 **(B)** In gravitational physics there is a system of units in which the
gravitational constant G and the speed of light c are dimensionless,
and assume the value unity. This procedure avoids the need to write
G and c many times. Show that in these units length, time, and mass
all have the same dimensions. Find the conversion factor from mass
to length once the dimensions and numerical values of G and c are
restored in SI units.

Solution
In units in which $G = c = 1$, we have

$$[c] = \left[\frac{L}{T}\right] = [0]\,,$$

which implies that $[L] = [T]$. Further, the dimensions of the gravitational constant G are $[G] = \left[L^3T^{-2}M^{-1}\right]$, which yields

$$[0] = \left[LM^{-1}\right]$$

and $[M] = [L] = [T]$.

When G and c are restored in SI units we have

$$
\begin{aligned}
[L] &= \left[G^\alpha c^\beta M \right] = \left[M^{-\alpha} L^{3\alpha} T^{-2\alpha} \right] \left[L^\beta T^{-\beta} \right] [M] \\
&= \left[L^{3\alpha+\beta} T^{-2\alpha-\beta} M^{-\alpha+1} \right],
\end{aligned}
$$

which yields the system

$$
-\alpha + 1 = 0,
$$

$$
3\alpha + \beta = 1,
$$

$$
-2\alpha - \beta = 0,
$$

with solution

$$
\alpha = 1, \qquad \beta = -2,
$$

and therefore GM/c^2 has the dimensions of a length and G/c^2 is the conversion factor from mass to length.

6 **(B)** It is believed that the classical description of gravity must break down at a very small length scale (the *Planck length* introduced by Max Planck), at which gravity and quantum mechanics should merge into an as-yet unknown theory of quantum gravity. Using only the fundamental constants of gravity (the Newton constant G), electromagnetism (the speed of light in vacuum c), and quantum mechanics (the reduced Planck constant $\hbar$), derive expressions and numerical values for the Planck scales of length, time, mass, and energy.

Solution
The fundamental constants G, c, and $\hbar$ have dimensions

$$
[G] = \left[L^3 M^{-1} T^{-2} \right],
$$

$$
[c] = \left[LT^{-1} \right],
$$

$$
[\hbar] = \left[L^2 M T^{-1} \right].
$$

One forms the Planck length $l_p = G^\alpha c^\beta \hbar^\gamma$ with

$$
\begin{aligned}
[l_p] &= \left[L^{3\alpha} M^{-\alpha} T^{-2\alpha} \right] \left[L^\beta T^{-\beta} \right] \left[L^{2\gamma} M^\gamma T^{-\gamma} \right] \\
&= \left[L^{3\alpha+\beta+2\gamma} M^{-\alpha+\gamma} T^{-2\alpha-\beta-\gamma} \right],
\end{aligned}
$$

and therefore it must be

$$3\alpha + \beta + 2\gamma \;= 1,$$

$$-\alpha + \gamma \;= 0,$$

$$-2\alpha - \beta - \gamma \;= 0.$$

This system has the solution

$$\alpha = \gamma = \frac{1}{2}, \qquad \beta = -\frac{3}{2};$$

hence the Planck length is

$$l_p \;=\; \sqrt{\frac{G\hbar}{c^3}} = \sqrt{\frac{(6.67 \cdot 10^{-11}\,\mathrm{N} \cdot \mathrm{m}^2 \cdot \mathrm{kg}^{-2})\,(1.05 \cdot 10^{-34}\,\mathrm{J} \cdot \mathrm{s})}{(3 \cdot 10^8\,\mathrm{m/s})^3}}$$

$$=\; 1.6 \cdot 10^{-35}\,\mathrm{m},$$

while the Planck time is simply

$$t_p = \frac{l_p}{c} = \sqrt{\frac{G\hbar}{c^5}} = 5.3 \cdot 10^{-44}\,\mathrm{s}.$$

The Planck mass is given by $m_p = G^\alpha c^\beta \hbar^\gamma$, which now yields

$$[m_p] \;=\; \left[L^{3\alpha} M^{-\alpha} T^{-2\alpha}\right] \left[L^\beta T^{-\beta}\right] \left[L^{2\gamma} M^\gamma T^{-\gamma}\right]$$

$$=\; \left[L^{3\alpha+\beta+2\gamma} M^{-\alpha+\gamma} T^{-2\alpha-\beta-\gamma}\right]$$

and

$$3\alpha + \beta + 2\gamma = 0,$$

$$-\alpha + \gamma 7 = 1,$$

$$-2\alpha - \beta - \gamma \;= 0,$$

with the solution

$$\alpha = -\frac{1}{2}, \qquad \beta = \gamma = \frac{1}{2};$$

hence, the Planck mass is

$$m_p = \sqrt{\frac{c\hbar}{G}} = \frac{c^2}{G} l_p = 2.2 \cdot 10^{-8}\,\text{kg},$$

and the Planck energy scale is found by noting that

$$E_p = m_p c^2 = \sqrt{\frac{c^5 \hbar}{G}} = l_p \frac{c^4}{G} = 1.9 \cdot 10^9\,\text{J}.$$

7 **(B, C)** Give a reason why the argument of trigonometric, exponential, logarithmic, or hyperbolic functions must always be dimensionless — consider, for example, the function $\sin x$.

Solution
All these functions can be expanded in power series—for example, for $\sin x$,

$$\sin x = \sum_{n=0}^{+\infty} (-1)^n \frac{x^{2n+1}}{(2n+1)!} = x - \frac{x^3}{3!} + \frac{x^5}{5!} + \dots.$$

If the argument x of the sine had dimensions, $\sin x$ should *simultaneously* have the dimensions of x and x^3 and x^5, ..., which is clearly impossible.

Chapter 2

PLANET EARTH IN SPACE

When we contemplate the whole globe as one great dewdrop, stripped and dotted with continents and islands, flying through space with other stars all singing and shining together as one, the whole universe appears as an infinite storm of beauty.

—John Muir, *Travels in Alaska*

Physics has identified four fundamental forces: gravity, the electromagnetic interaction, the strong nuclear force, and the weak interaction. Although the gravitational force is the weakest of the four, it dominates on the scale of the Earth, the planets, stars, stellar systems, and on larger scales. Gravity rules the dynamics of planets, stars, galaxies, galaxy clusters and superclusters, and of the universe itself. Newton's law of gravity dictates the shape of stars and planets, which is modified only slightly by rotation.

The particular location of the Earth near a rather average star called the Sun creates environmental conditions that are just right for life. Probably many—if not most—of the stars in the Milky Way have planets and maybe some of them host forms of life. However, the presence of life on the Earth should not be taken for granted, as several concomitant factors contribute to make it possible. First, the distance of the Earth from the Sun is such that temperature extremes found on planets closer or further away from the Sun are avoided. This property is also due to the presence of the atmosphere, which mitigates temperature fluctuations. The Earth's atmosphere can be retained because the mass and size of our planet provide sufficient gravitational attraction to keep the molecules of atmospheric gases that surround us moving at the speeds corresponding to the temperatures typically found in the atmosphere. The presence of liquid water in the oceans and on land and in the biomass is made possible by the surface temperature of the Earth being in the range between 0°C and 100°C. The atmosphere screens life from the harmful γ-rays, X-rays, and ultraviolet radiation that are common in outer space.

The rotation of the Earth around the Sun and the spinning of the Earth about its axis are responsible for the existence of the seasons, the day and night cycle, the different insolation of different parts of the planet, and for the various local climates. Differential absorption of solar radiation and the consequent temperature gradients maintain global and local atmospheric circulation with transport of mass, energy, and momentum around the globe.

Photosynthesis by plants and algae provides a means of converting the electromagnetic energy from the Sun into chemical energy available to the food chain. Almost all the energy stored and used in physical, chemical, and biological processes on the planet comes from the Sun.

2.1   Astronomy

The Earth is a planet orbiting a star in a stellar system comprising other planets and celestial bodies: this basic fact determines essential features such as the energy received from the Sun, which sustains life, the diurnal and seasonal cycles, the reinforcing of the tides due to planetary conjunctions, the rotation of the Earth affecting winds and oceanic currents, or the fall of meteorites of various sizes on the planet.

Any standard first-year physics textbook will suffice as a reference for the material in this section.

1 **(A)** Compute the angular and linear velocity of the Earth around the Sun. Approximate the orbit with a circle of radius $r = 1$ A.U.$=$ $1.50 \cdot 10^{11}$ m.

Solution
The orbital period is $T = 1$ year; hence, the angular velocity is

$$\omega = \frac{2\pi}{T} = \frac{2\pi}{365 \cdot 24 \cdot 3600 \, \text{s}} = 1.99 \cdot 10^{-7} \, \frac{\text{rad}}{\text{s}}.$$

The linear speed of the Earth in its orbit is

$$v = \omega r = \left(1.99 \cdot 10^{-7} \, \frac{\text{rad}}{\text{s}} \right) \left(1.50 \cdot 10^{11} \, \text{m} \right) = 29.9 \, \frac{\text{km}}{\text{s}}.$$

2 **(A)** Derive Kepler's third law of planetary motion from Newton's theory of gravity under the assumption of circular orbits.

Solution
Assuming circular orbits, the centripetal force acting on any of the
planets is

$$F = ma_c = m\omega^2 r,$$

where m is the mass of the planet, ω its angular velocity, and r
the orbital radius. Newton's law of gravitation yields the attractive
gravitational force between the planet and the Sun,

$$F = \frac{GM_\odot m}{r^2},$$

where $M_\odot$ is the mass of the Sun. By comparing the last two equations the mass of the individual planet drops out and the rest of the
calculation does not depend on any individual planet. We obtain

$$\omega^2 r = \frac{GM_\odot}{r^2},$$

or

$$\omega^2 r^3 = GM_\odot,$$

where the right-hand side is a constant that has the same value for all
planets. Kepler's third law (a phenomenological law) is thus derived
from Newton's law of gravitation (a fundamental one).

3 **(A, B)** Compute the escape speed the minimum speed that a body
must have in order to escape to infinity and forever leave the gravitational attraction of the planet—from the surface of a planet of mass
M and radius R. Does the escape speed depend on the mass of the
body? Neglect the friction of the body with the atmosphere.

Solution 1 (level A)
The gravitational potential energy of a body of mass m and radial
coordinate r in the gravitational field of the planet is

$$E_g = -\frac{GMm}{r},$$

and it vanishes at infinity ($r \to +\infty$). The escape speed is computed
by assuming that the initial kinetic energy $mv^2/2$ of the body leaving
the surface of the planet in radial (vertical) motion be just sufficient
to overcome the gravitational attraction—the body arrives at infinity
with asymptotically vanishing speed. The initial energy must equal
the final energy, i.e.,

$$\frac{1}{2}mv^2 - \frac{GMm}{R} = 0.$$

The mass m of the body cancels out and therefore the escape speed does not depend on it—this is consistent with the equivalence principle (universality of free fall [73]). The escape speed is

$$v_e = \left(\frac{2GM}{R}\right)^{1/2}.$$

Solution 2 (level B)
In order to leave the planet, the initial kinetic energy of the body of mass m on the planet's surface must equal the work done against the gravitational force $-GMm/r^2$ and required to take the mass m to $r = +\infty$,

$$\frac{1}{2}mv^2 = \int_R^{+\infty} dr\, \frac{GMm}{r^2} = \left[-\frac{GMm}{r}\right]_R^{+\infty} = \frac{GMm}{R}.$$

Hence, one finds

$$v_e = \left(\frac{2GM}{R}\right)^{1/2}.$$

The mass m of the body cancels out and therefore the escape speed does not depend on it—this is consistent with the equivalence principle (universality of free fall—see Ref. [73]). If friction can be neglected, the escape speed is the same for a molecule and for a satellite.

4 **(A)** Arizona's Meteor Crater was created 50,000 years ago by a metallic meteorite with a mass estimated to be $5.00 \cdot 10^8$ kg. On impact the meteorite vaporized. Assuming for simplicity that the meteorite was composed of iron, compare the minimum kinetic energy of the meteorite and the energy necessary to melt and vaporize it. The latent heats of fusion and vaporization of iron are, respectively, $2.89 \cdot 10^5$ J/kg and $63.4 \cdot 10^5$ J/kg, the melting point and boiling point are 1535°C and 3000°C, and the specific heat is $c = 0.107$ Kcal/kg.

Solution
Consider a meteorite that starts from the radial distance $r = \infty$ with zero kinetic energy and zero gravitational potential energy, hence zero total energy. Assuming conservation of energy and neglecting friction with the atmosphere (which is appropriate for a large meteorite), it arrives on the surface of the Earth ($r = R$) with zero total energy

$$E = \frac{1}{2}mv^2 - \frac{GM}{r} = 0$$

from which one deduces its speed $v = \sqrt{2GM/r}$, which is the escape speed. If the initial kinetic energy of the meteorite is larger than zero, its speed on arrival is larger than the escape speed from the planet $v_{esc} = \sqrt{2GM/R} = 11.2\,\mathrm{km/s}$. The kinetic energy of the meteorite that generated Meteor Crater was

$$T \geq \frac{1}{2} m v_{esc}^2 = \frac{1}{2}\,(5.00 \cdot 10^8\,\mathrm{kg}) \cdot \left(1.12 \cdot 10^4\,\frac{\mathrm{m}}{\mathrm{s}}\right)^2 = 3.14 \cdot 10^{16}\,\mathrm{J}.$$

The energy necessary to melt and vaporize the mass of the meteorite (assumed to be iron) is the sum of latent heat $L_f\, m$, the heat necessary to raise its temperature from the melting to the boiling point $Q = cm\Delta T$, and the latent heat[1] $L_v m$,

$$
\begin{aligned}
E &= (L_f + c\Delta T + L_v)\,m \\[2mm]
&= \left[2.89 \cdot 10^5\,\mathrm{J \cdot kg^{-1}} + 0.107 \cdot \left(4186\,\frac{\mathrm{J}}{\mathrm{kg \cdot (^\circ C)}}\right)\right. \\[2mm]
&\quad \times (3000^\circ\mathrm{C} - 1535^\circ\mathrm{C}) \\[2mm]
&\quad \left. + \left(63.4 \cdot 10^5\,\mathrm{J \cdot kg^{-1}}\right)\right] (5.00 \cdot 10^8\,\mathrm{kg}) \\[2mm]
&= 3.64 \cdot 10^{15}\,\mathrm{J}.
\end{aligned}
$$

The minimum kinetic energy of the meteorite was certainly sufficient to vaporize it. The rest of its kinetic energy went into mechanical work to create the crater, melting the rocks on the site (fused silica have been found in the crater), and in the generation of earthquake and sound waves.

5 **(B)** The probability of a 10 km-sized asteroid colliding with the Earth is estimated to one event every 10^8 years. Many people believe that such an impact caused the extinction of dinosaurs approximately 65 million years ago. Compute the minimum speed and kinetic energy of a 10 km-sized iron asteroid falling on the Earth in a head-on collision. For simplicity, assume that the asteroid starts from rest very far away and describe it as a homogeneous sphere. Compare your result with the energy of a 10 megaton hydrogen bomb. The density of iron is $\rho = 7.8 \cdot 10^3\,\mathrm{kg/m^3}$ and 1 ton (of TNT) is equivalent to

---

[1]We make the simplifying assumption that the vaporized iron is not heated further.

$4.2 \cdot 10^9$ J.

Solution
Assume that the asteroid starts from rest at infinity in a head-on collision. For a large asteroid only a small fraction of the kinetic energy is dissipated into friction with the atmosphere and one can assume that the total energy E of the asteroid is conserved. At $r = \infty$ the total energy is

$$E = \frac{1}{2} mv^2 - \frac{GMm}{r} = 0$$

and, upon impact with the surface of the Earth at $r = R$, the asteroid has the same energy. The previous equation yields, at the Earth radius R,

$$v = \sqrt{\frac{2GM}{R}} = 11.2 \frac{\text{km}}{\text{s}}.$$

The speed of the asteroid equals the escape speed from the surface of the Earth. The mass of the asteroid (assumed to be a homogeneous sphere of radius $r = 5$ km and uniform mass density equal to the density of iron) is

$$m = \frac{4\pi r^3}{3} \rho = \frac{4\pi}{3} \left(5 \cdot 10^3\,\text{m}\right)^3 \left(7.8 \cdot 10^3 \frac{\text{kg}}{\text{m}^3}\right) = 4.1 \cdot 10^{15}\,\text{kg}.$$

The minimum kinetic energy of the asteroid is then

$$\frac{1}{2} mv^2 = \frac{1}{2} \left(4.1 \cdot 10^{15}\,\text{kg}\right) \left(1.2 \cdot 10^3 \frac{\text{m}}{\text{s}}\right)^2 = 3 \cdot 10^{21}\,\text{J}.$$

This energy is equivalent to

$$\frac{3 \cdot 10^{21}\,\text{J}}{4.2 \cdot 10^{16}\,\text{J}} = 7 \cdot 10^4$$

hydrogen bombs: the impact of such an asteroid with the Earth would be catastrophic.

6 **(A)** Assuming that the Sun radiates like a blackbody at 5800 K, what is the value of the solar constant on the surface of Neptune, 30.2 astronomical units (A.U.) away? 1 A.U. $= 1.5 \cdot 10^{11}$ m, and the radius of the Sun is $R_\odot = 7 \cdot 10^8$ m.

Solution
The solar constant on the surface of Neptune is the energy received per unit time and per unit of normal area,

$$S_{\text{Nept}} = \frac{L_\odot}{4\pi d^2} = \frac{\left(\sigma T_\odot^4\right) \cdot \left(4\pi R_\odot^2\right)}{4\pi d^2} = \sigma T_\odot^4 \left(\frac{R_\odot}{d}\right)^2,$$

where $L_\odot$ is the solar luminosity (energy emitted per unit time), d is the average distance between Neptune and the Sun, and σ is the Stefan–Boltzmann constant. Numerically, we have

$$S_{\text{Nept}} = \left(5.67 \cdot 10^{-8} \, \frac{\text{W}}{\text{m}^2 \cdot \text{K}^4}\right) \cdot (5800 \, \text{K})^4 \cdot \left(\frac{7 \cdot 10^8 \, \text{m}}{30.2 \cdot 1.5 \cdot 10^{11} \, \text{m}}\right)^2$$

$$= 1.5 \, \frac{\text{W}}{\text{m}^2}.$$

2.2 Planet Earth

The physical and chemical parameters of planet Earth—such as mass, radius, angular momentum, average density, chemical composition, and solar constant—differentiate it from the other planets in the solar system and determine its essential features. These include the strength of gravity, the surface temperature, the existence and chemical composition of the atmosphere, the atmospheric pressure, geological activity, and so forth. In turn, these features are responsible for the existence of liquid oceans, deserts, forested areas, icecaps, and mountains, for the global climate, the atmospheric and oceanic circulation, the thermal inertia of the oceans, and for the hydrologic, rock, and other cycles.

1 **(A, C)** As a first approximation the Earth is considered as a sphere of average radius $R = 6370$ km. This neglects a bulge at the equator, flattening at the poles, and local irregularities. What fraction of the Earth's radius is the tallest mountain? The depth of the deepest ocean trench? The average ocean depth? Does your result mean that one can treat the Earth as a perfect sphere when planning the operation of a satellite?

Solution
The average radius of the Earth is $R = 6370$ km; the tallest mountain is Mt. Everest at 8848 m, for which

$$\frac{h}{R} = \frac{8848 \, \text{m}}{6.37 \cdot 10^6 \, \text{m}} \simeq 1.4 \cdot 10^{-3}.$$

The deepest trench is the Marianas trench, 11 km deep, for which

$$\frac{h}{R} = \frac{11 \, \text{km}}{6370 \, \text{km}} \simeq 1.7 \cdot 10^{-3}.$$

The average ocean depth is 3.8 km and

$$\frac{h}{R} = \frac{3.8 \, \text{km}}{6370 \, \text{km}} \approx 6 \cdot 10^{-4}.$$

It is reasonable to neglect these local irregularities in a first approximation. However, the gravitational field of the Earth is not spherically symmetric because local overdensities and underdensities determine gravitational anomalies and deviations from the monopole field of a perfect sphere that are important for satellite motion.

2 **(A)** a) What is the distance along the surface of the Earth between two points on the same meridian separated by one degree of latitude?

b) What is the distance between two points at the same latitude $\lambda = 50°$S separated by one degree of longitude φ?

Solution
a) This is the distance between two points on a the perimeter of a great circle[2] (meridian) and is given by $L = \lambda R$, where R is the Earth radius and λ the latitude angle separating the two points, expressed in radians. Numerically,

$$L = \lambda R = \left(\pi \frac{1°}{180°} \right) (6.37 \cdot 10^6 \, \text{m}) = 1.11 \cdot 10^5 \, \text{m} = 111 \, \text{km}.$$

b) Two points at the same latitude λ lie on a parallel, which is a circle of radius $R \cos \lambda$ (not a great circle): if they are a longitudinal angle φ apart, their linear distance L' along the Earth's surface is $\varphi R \cos \lambda$, where φ must be expressed in radians. Numerically, for the case assigned,

$$
\begin{aligned}
L' &= \varphi R \cos \lambda = \left(\pi \frac{1°}{180°} \right) (6.37 \cdot 10^6 \, \text{m}) \cos (50°) \\
&= 7.15 \cdot 10^4 \, \text{m} = 71.5 \, \text{km}.
\end{aligned}
$$

3 **(A)** At a point of latitude λ the angle θ between the radial direction and the solar rays is given by

$$\cos \theta = \sin \lambda \sin \delta + \cos \lambda \cos \delta \cos (\omega t),$$

where the *declination* δ is the latitude at which the Sun is at the zenith at noon (δ depends on the day of the year) and $\omega = 2\pi/P$, with $P = 24$ hours (the phase is chosen so that $t = 0$ at noon). Find the latitudes $\lambda(\delta)$ such that the Sun never sets.

---

[2]A great circle is a section of the sphere containing a diameter of the sphere.

Solution

Consider first the Northern Hemisphere, where the Sun never sets if it is $\cos\theta \geq 0$ at all times t during the day, that is,

$$\sin\lambda\sin\delta \geq -\cos\lambda\cos\delta\cos(\omega t)$$

and, since $\cos\lambda, \cos\delta \geq 0$, we can write $\tan\lambda\tan\delta \geq -\cos(\omega t)$ at all times t. Since $-1 \leq -\cos(\omega t) \leq 1$ this inequality is satisfied at all times only if $\tan\lambda\tan\delta \geq 1$. By using the trigonometric identity

$$\tan(\lambda + \delta) = \frac{\tan\lambda + \tan\delta}{1 - \tan\lambda\tan\delta},$$

the inequality is rewritten as

$$1 - \frac{\tan\lambda + \tan\delta}{\tan(\lambda + \delta)} \geq 1.$$

Taking into account the fact that both $\tan\lambda$ and $\tan\delta$ are positive in the Northern Hemisphere, it must be $\tan(\lambda + \delta) \leq 0$ with $\lambda, \delta \geq 0$. This implies that $\lambda + \delta \geq \pi/2$, or

$$\lambda(\delta) \geq \frac{\pi}{2} - \delta :$$

at latitudes λ satisfying this inequality the Sun never sets.

Similarly, in the Southern Hemisphere it must be

$$\frac{\tan\lambda + \tan\delta}{\tan(\lambda + \delta)} \leq 0,$$

but now $-\pi/2 \leq \lambda, \delta \leq 0$ and $\tan\lambda, \tan\delta \leq 0$, so it must be

$$\tan(\lambda + \delta) \geq 0$$

and therefore

$$\lambda(\delta) \leq -\frac{\pi}{2} - \delta.$$

4 **(B)** An exceptionally well-funded geographer with a taste for mountaineering has counted all the peaks and passes on Earth, finding them to be n and m, respectively. Can you tell her how many valleys are on Earth, without counting them?

Solution

According to the Poincaré index theorem the sum of the indices of a vector field on a surface of genus γ is $2 - 2\gamma$. Consider the elevation

h as a regular function of two coordinates on the surface of an otherwise spherical Earth, and the gradient $\vec{H} = -\vec{\nabla} h$ as vector field. The index of maxima (summits) and minima (valleys) of h is +1, the index of saddle points (passes) is -1, while the genus of a sphere is $\gamma = 0$. Therefore,

$$n\,(\text{peaks}) + n\,(\text{valleys}) - n\,(\text{passes}) = 2$$

and the number of valleys on Earth is $n\,(\text{valleys}) = 2 - n + m$.

5 **(A)** Explain how the Cavendish experiment measuring the gravitational constant G yielded a value for the mass of the Earth. Compute the average density of the Earth and compare it with the average density for rocks and soil in the crust, $\rho = 2.2 \cdot 10^3\,\text{kg/m}^3$. What can one argue about the density of the Earth's core?

Solution
The Cavendish experiment provided an experimental value for the gravitational constant—the value of G is today measured as $G = 6.67 \cdot 10^{-11}\,\text{N} \cdot \text{m}^2 \cdot \text{kg}^{-2}$. The acceleration of a body on the surface of the Earth is

$$g = \frac{GM}{R^2},$$

where M and R are, respectively, the Earth's mass and radius. One then obtains the value of the mass of the Earth

$$M = \frac{gR^2}{G}.$$

Since g is easily measurable and has the average value $9.81\,\text{m} \cdot \text{s}^{-2}$ and the radius of the Earth, approximately $R = 6370$ km, has been known since the times of Erathostenes' calculation, the Cavendish experiment provides the value of M

$$M = \frac{\left(9.81\,\text{m} \cdot \text{s}^{-2}\right) \cdot \left(6.37 \cdot 10^6\,\text{m}\right)^2}{6.67 \cdot 10^{-11}\,\text{N} \cdot \text{m}^2 \cdot \text{kg}^{-2}} = 5.97 \cdot 10^{24}\,\text{kg}.$$

The average density of the Earth is then

$$\langle \rho \rangle = \frac{M}{4\pi R^3/3} = \frac{3 \cdot \left(5.97 \cdot 10^{24}\,\text{kg}\right)}{4\pi \left(6.37 \cdot 10^6\,\text{m}\right)^3} = 5.51 \cdot 10^3\,\frac{\text{kg}}{\text{m}^3}.$$

This value is considerably larger (2.5 times) than the average density of the crust; hence, the core of the planet is significantly denser than the crust. The value of $\langle \rho \rangle$ is typical of metals and one can argue that the Earth has a metallic core.

6 **(B)** Compute the moment of inertia of a homogeneous sphere with respect to an axis passing through its center and express the result as a function of the radius R and the mass M of the sphere. Apply your result to estimate the angular momentum of the spinning Earth by treating it as a sphere of constant density, mass $M = 5.978 \cdot 10^{24}$ kg, and radius $R = 6.378 \cdot 10^6$ m.

Solution
The moment of inertia is

$$\mathcal{I} = \int \int \int d^3\vec{x}\, \rho\, d^2(\vec{x}),$$

where ρ is the density of the sphere and $d(\vec{x})$ the distance of a generic point of the sphere from the rotation axis z. Since $d^2 = x^2 + y^2 = r^2 \sin^2 \theta$ in polar coordinates

$$x = r \sin \theta \cos \varphi, \quad y = r \sin \theta \sin \varphi, \quad z = r \cos \theta,$$

we have

$$\mathcal{I} = \rho \int_0^R dr \int_0^\pi d\theta \int_0^{2\pi} d\varphi\, r^2 \sin \theta\, r^2 \sin^2 \theta$$

$$= \rho \int_0^R dr\, r^4 \cdot \int_0^\pi d\theta\, \sin^3 \theta \cdot 2\pi = 2\pi \frac{R^5}{5} \rho \int_0^\pi d\theta\, \sin^3 \theta.$$

Integration by parts gives

$$\int_0^\pi d\theta\, \sin^3 \theta = -\cos\theta \sin^2\theta \big|_0^\pi - 2 \int_0^\pi d(\cos\theta) \cos^2\theta$$

$$= 2 \int_{-1}^{+1} dx\, x^2 = \frac{2x^3}{3}\Big|_{-1}^{+1} = \frac{4}{3},$$

and

$$\mathcal{I} = \frac{8\pi}{15} R^5 \rho.$$

The mass of the sphere is

$$M = \rho \cdot \frac{4\pi}{3} R^3,$$

and the moment of inertia can be written as

$$\mathcal{I} = \frac{2}{5} M R^2.$$

For the Earth spinning with the period of one day the angular momentum is

$$L = \mathcal{I}\omega = \frac{2}{5}MR^2\omega$$

$$= \frac{2}{5}\left(5.978 \cdot 10^{24}\,\text{kg}\right) \cdot \left(6.378 \cdot 10^6\,\text{m}\right)^2 \cdot \frac{2\pi}{24 \cdot 3600\,\text{s}}$$

$$= 7.074 \cdot 10^{33}\,\frac{\text{kg} \cdot \text{m}^2}{\text{s}}$$

in the approximation in which the Earth is treated as a sphere of constant density. In reality, the core density is higher than the crust and mantle density and this fact decreases the value of the moment of inertia with respect to the estimated value.

7 **(B)** Write an expression for the centrifugal force acting on a particle at a point with latitude λ on the surface of the Earth. Can the centrifugal acceleration be derived from a potential?

Solution 1
The centrifugal force is perpendicular to the rotation axis and is directed away from it. Its magnitude on a particle of mass m is $F_c = m\omega^2 r_d$, where $\omega = 2\pi/P$, P is the rotational period equal to the sidereal day[3] of 23 h 56 min 4 s $= 8.6164 \cdot 10^4$ s, and r_d is the orthogonal distance to the rotation axis. For a particle on the surface of the Earth (radius R) at latitude λ it is $r_d = R\cos\lambda$ and $F_c = m\omega^2 R\cos\lambda$. The magnitude of the centrifugal force is maximum at the equator and zero at the poles. This force can be derived from the centrifugal potential

$$V_c(x,y,z) = -\frac{1}{2}\omega^2 r_d^2 = -\frac{1}{2}\omega^2 R^2\cos^2\lambda = -\frac{1}{2}\omega^2\left(x^2+y^2\right)$$

in Cartesian coordinates (x,y,z) with the origin at the center of the Earth and with the z-axis aligned with the rotation axis. It is easy to verify that the centrifugal force is given by $\vec{F}_c = -m\vec{\nabla}V_c$.

---

[3]The sidereal day (23 h 56′4″) is the time taken for a fixed point on the Earth to regain its position with respect to the fixed stars. Because of the fact that the Earth revolves around the Sun, the sidereal day is not the same as the solar day of 24 h. The solar day is a little longer than the sidereal day because, while the Earth spins about its axis, it also moves along its orbit around the Sun and, as a result, it has to spin a little longer in order for the fixed point on the surface of the Earth to regain the position that it had with respect to the Sun.

Solution 2
A more rigorous solution is the following: use a Cartesian coordinates system with the origin at the center of the Earth and with the rotation axis as z-axis. The centrifugal acceleration is [20] $\vec{a}_c = -\vec{\omega} \times \vec{\omega} \times \vec{x}$, where the vector $\vec{\omega}$ is directed along the rotation axis and has magnitude equal to the angular velocity ω, and $\vec{x}$ is the position vector. The centrifugal force on a particle of mass m is $\vec{F}_c = m\vec{a}_c$. In order to decide whether the vector field $\vec{a}_c$ can be derived from a potential we examine its curl. If $\vec{\nabla} \times \vec{a}_c = 0$ then there exists[4] a scalar function V_c such that $\vec{a}_c = -\vec{\nabla}V_c$ (the negative sign is a mere convention). By using the vector calculus identity

$$\vec{\nabla} \times \left(\vec{a} \times \vec{b} \right) = \vec{a} \left(\vec{\nabla} \cdot \vec{b} \right) - \vec{b} \left(\vec{\nabla} \cdot \vec{a} \right) + \left(\vec{b} \cdot \vec{\nabla} \right) \vec{a} - \left(\vec{a} \cdot \vec{\nabla} \right) \vec{b}$$

with $\vec{a} = \vec{\omega}$ and $\vec{b} = \vec{\omega} \times \vec{x}$, one obtains

$$
\begin{aligned}
\vec{\nabla} \times \left(\vec{\omega} \times \vec{b} \right) &= \vec{\omega} \left(\vec{\nabla} \cdot \vec{b} \right) - \left(\vec{\omega} \cdot \vec{\nabla} \right) (\vec{\omega} \times \vec{x}) \\
&= \vec{\omega} \left[\vec{\nabla} \cdot (\vec{\omega} \times \vec{x}) \right] - \left(\vec{\omega} \cdot \vec{\nabla} \right) (\vec{\omega} \times \vec{x})
\end{aligned}
$$

because $\vec{\omega}$ is a constant vector and its derivatives vanish. We have

$$\vec{\omega} \times \vec{x} = \begin{vmatrix} \vec{e}_x & \vec{e}_y & \vec{e}_z \\ \omega_x & \omega_y & \omega_z \\ x & y & z \end{vmatrix} = \omega \left(-y\vec{e}_x + x\vec{e}_y \right),$$

by using the fact that $\vec{\omega} = (0, 0, \omega)$ and

$$
\begin{aligned}
\vec{\nabla} \times \left(\vec{\omega} \times \vec{b} \right) &= \vec{\omega} \left[\vec{\nabla} \cdot (\vec{\omega} \times \vec{x}) \right] - \left(\vec{\omega} \cdot \vec{\nabla} \right) (\vec{\omega} \times \vec{x}) \\
&= \vec{\omega} \left[\partial_x (-\omega y) + \partial_y (\omega x) \right] - \omega \frac{\partial}{\partial z} \left[-\omega y\vec{e}_x + \omega x\vec{e}_y \right] \\
&= 0.
\end{aligned}
$$

Therefore,

$$\vec{\nabla} \times \vec{a}_c = \vec{\nabla} \times (-\vec{\omega} \times \vec{\omega} \times \vec{x}) = 0$$

---

[4]Remember that the curl of a gradient is always zero.

and $\vec{a}_c$ can be represented as the gradient of a scalar function. Let us compute the components of $\vec{a}_c$. We have

$$\vec{\omega} \times (\vec{\omega} \times \vec{x}) = \begin{vmatrix} \vec{e}_x & \vec{e}_y & \vec{e}_z \\ 0 & 0 & \omega \\ -\omega y & \omega x & 0 \end{vmatrix} = -\omega^2 \left(x\vec{e}_x + y\vec{e}_y \right).$$

Then $\vec{a}_c = -\vec{\omega} \times (\vec{\omega} \times \vec{x}) = \left(\omega^2 x, \omega^2 y, 0 \right)$. By setting $\vec{a}_c = -\vec{\nabla} V_c$ we obtain

$$\frac{\partial V_c}{\partial x} = -\omega^2 x, \qquad \frac{\partial V_c}{\partial y} = -\omega^2 y, \qquad \frac{\partial V_c}{\partial z} = 0,$$

and integration yields

$$V_c(x, y) \;=\; -\frac{\omega^2 x^2}{2} + f(y),$$

$$V_c(x, y) \;=\; -\frac{\omega^2 y^2}{2} + g(x),$$

where $f(y)$ and $g(x)$ are integration functions. By equating the right-hand sides and setting to zero an arbitrary integration constant we obtain

$$V_c(x, y) = -\frac{\omega^2 \left(x^2 + y^2 \right)}{2} = -\frac{\omega^2 r_d^2}{2} = -\frac{\omega^2 R^2}{2} \cos^2 \lambda.$$

8 **(A)** Estimate the relative importance of centrifugal forces and gravity for the motion of a particle near the surface of the Earth (or for motions in the atmosphere and in the oceans) at latitude λ.

Solution
The relative importance of centrifugal forces and gravity can be measured by the ratio η of the centrifugal and the Newtonian gravitational potential. In order of magnitude, by keeping only the dominant monopole term in the gravitational potential, we obtain

$$\eta = \frac{- \left(\omega^2 R^2 \cos^2 \lambda \right)/2}{-GM/R} = \frac{\omega^2 R^3 \cos^2 \lambda}{2GM} \leq \frac{\omega^2 R^3}{2GM},$$

where ω and M are the rotational angular velocity and mass of the Earth, respectively, and G is Newton's constant. Numerically,

$$\eta \;\leq\; \frac{\left(\frac{2\pi}{24 \cdot 3600\,\text{s}} \right)^2 \cdot \left(6.37 \cdot 10^6\,\text{m} \right)^3}{2 \left(6.67 \cdot 10^{-11}\,\text{N} \cdot \text{m}^2 \cdot \text{kg}^2 \right) \left(5.978 \cdot 10^{24}\,\text{kg} \right)}$$

$$\approx \ 2 \cdot 10^{-3}.$$

The centrifugal force at the equator, where it is maximum, is almost negligible in comparison with the gravitational force. However, the centrifugal force is not the only effect of rotation: the Coriolis force, although small, is important for objects with a long flight time, and for oceanic currents and atmospheric winds, because it accumulates over the long flight time.

9 **(B)** The total geopotential (including both the gravitational and centrifugal potentials) in spherical coordinates near the surface of the Earth is

$$U(\vec{x}) = -\frac{GM}{r} + \frac{G}{2r^3}(C - A)\left(3\sin^2\lambda - 1\right) - \frac{\omega^2 r^2}{2}\cos^2\lambda,$$

where M is the Earth mass, A and C are its moments of inertia with respect to the x- and z- axes, λ is the geographic latitude, and ω is the angular velocity. The *flattening* of the Earth is

$$\epsilon \equiv \frac{R_e - R_p}{R_e},$$

where R_e and R_p are the equatorial $(\lambda = 0)$ and polar $(\lambda = \pi/2)$ radii, respectively. Find an approximate expression for the flattening ϵ as a function of M, A, C, ω, and R_e by using the fact that $\epsilon \ll 1$.

Solution
The geoid (equipotential surface of U) passing through the equator and the poles defines the surface of the Earth and satisfies

$$U_0 = -\frac{GM}{R_e} - \frac{G}{2R_e^3}(C - A) - \frac{\omega^2 R_e^2}{2}$$

at the equator and

$$U_0 = -\frac{GM}{R_p} + \frac{G}{R_p^3}(C - A)$$

at the poles, where U_0 is constant over the surface of the Earth. By equating the right-hand sides and substituting the value of $R_p = R_e(1 - \epsilon)$, we obtain

$$-\frac{1}{R_e} + \frac{(A - C)}{2MR_e^3} - \frac{\omega^2 R_e^2}{2GM} = -\frac{(1 + \epsilon)}{R_e} + \frac{(C - A)}{R_e^3 M}(1 + 3\epsilon)$$

Figure 2.1.   Tides on the Earth due to the Moon.

to first order in ϵ and

$$\epsilon \left[1 - \frac{3\,(C-A)}{MR_e^2} \right] = \frac{3\,(C-A)}{2R_e^2 M} + \frac{\omega^2 R_e^3}{2GM}$$

and since $C \approx A \approx MR_e^2$ with

$$\frac{C-A}{MR_e^2} = O(\epsilon) \ll 1,$$

we obtain the expression of the Earth's flattening

$$\epsilon = \frac{3\,(C-A)}{2MR_e^2} + \frac{\omega^2 R_e^3}{2GM} + \cdots$$

10 **(A, B)** Terrestrial tides are due to the action of the Moon and the Sun. Consider two points of the Earth's surface located, respectively (Fig. 2.1),
1) on the line joining the Earth and the Moon's centers, and closest to the Moon, and
2) at the antipodes of the first point. Compute the difference $\Delta g = g_1 - g_2$ in the magnitude of the gravitational acceleration due to the Moon between these two points. Neglect the effect of the Sun.

Solution 1 (level A)
At point 1, closest to the Moon, the gravitational acceleration due to

the Moon has magnitude

$$g_1 = \frac{GM_M}{(d-R)^2},$$

where R is the Earth's radius, M_M is the Moon mass, and d is the Earth–Moon distance measured from the center of the Earth to the center of the Moon. At the antipodal point 2 the gravitational acceleration due to the Moon has magnitude

$$g_2 = \frac{GM_M}{(d+R)^2}.$$

The acceleration difference is

$$\Delta g \equiv g_1 - g_2 = GM_M \left(\frac{1}{(d-R)^2} - \frac{1}{(d+R)^2} \right)$$

$$= \frac{4GM_M R}{d^3(1-R/d)^2(1+R/d)^2} = \frac{4GM_M R}{d^3 [1-(R/d)^2]^2} \simeq \frac{4GM_M R}{d^3},$$

where the expression $1 - (R/d)^2$ in the denominator has been approximated with unity because $R/d \ll 1$.

Solution 2 (level B)
The gravitational acceleration due the Moon at a point at distance r from its center is $g = -GM_M/r^2$. Tides are due to the differential acceleration at different points, i.e., to the *acceleration gradient*

$$\frac{dg}{dr} = \frac{d}{dr} \left(\frac{-GM_M}{r^2} \right) = \frac{2GM_M}{r^3}.$$

At point 2 on the surface of the Earth, the farthest away from the Moon, we have

$$g(d+R) = g(d-R) + \left. \frac{dg}{dr} \right|_{d-R} (2R) + \ldots$$

and

$$\Delta g \equiv g(d+R) - g(d-R) \simeq \frac{4GM_M R}{(d-R)^3} \simeq \frac{4GM_M R}{d^3}.$$

11 **(A)** The differential accelerations due to the Moon on different points of the ocean are responsible for tides on the Earth. The Sun also contributes to terrestrial tides. Assume for simplicity that the Moon, the Sun, and the Earth are aligned, and consider two points 1 and 2 on this line, with point 1 being closest to the Moon. Denote by $(\Delta g)_M - g_1 - g_2$ the differential acceleration due to the Moon between these two points, and by $(\Delta g)_S$ the same quantity due to the

Sun. Compute the ratio $(\Delta g)_M/(\Delta g)_S$. The mass of the Moon is
$7.35 \cdot 10^{22}$ kg, the mass of the Sun is $1.99 \cdot 10^{30}$ kg, the average
Earth–Sun distance is $d_{ES} = 1.50 \cdot 10^{11}$ m, and the average Earth–
Moon distance is $d_{EM} = 3.84 \cdot 10^8$ m.

Solution
From the previous problem, we have

$$(\Delta g)_M = \frac{4GM_MR}{d_{EM}^3}, \qquad (\Delta g)_S = \frac{4GM_\odot R}{d_{ES}^3},$$

where M_M and $M_\odot$ are the Moon and the Sun masses, respectively.
The ratio $(\Delta g)_M/(\Delta g)_S$ is

$$\frac{(\Delta g)_M}{(\Delta g)_S} = \frac{M_M}{M_\odot}\left(\frac{d_{ES}}{d_{EM}}\right)^3 = \frac{7.35 \cdot 10^{22}\,\text{kg}}{1.99 \cdot 10^{30}\,\text{kg}}\left(\frac{1.50 \cdot 10^{11}\,\text{m}}{3.84 \cdot 10^8\,\text{m}}\right)^3 = 2.20.$$

The effect of the gravitational acceleration of the Moon on terrestrial
tides is approximately twice that due to the Sun.

12 **(B)** Compute the area that the Earth presents perpendicular to the
rays coming from the Sun, assuming that these rays are all parallel
to each other. What fraction of the surface area of the planet is this?
Interpret your result geometrically.

Solution
Take the direction of propagation of the solar rays as the z-axis and
use polar coordinates (θ, φ) to describe the position on the Earth
surface, assumed to be a sphere (Fig. 2.2). An element of area dS
with unit normal $\vec{n}$ on the surface of the Earth presents to the solar
rays the normal area

$$dS_\perp = dS\,\vec{e}_z \cdot \vec{n} = dS\,\cos\theta = \left(R^2 \sin\theta d\theta d\varphi\right)\cos\theta,$$

where $\vec{e}_z$ is the unit vector in the direction of the z-axis and R is
the radius of the Earth. By integrating over the hemisphere that is
illuminated by the Sun, one obtains the total normal area

$$S_\perp = \int_0^{\pi/2} d\theta \int_0^{2\pi} d\varphi\, R^2 \sin\theta \cos\theta = 2\pi R^2 \int_0^{\pi/2} d\theta\, \frac{\sin(2\theta)}{2}$$

$$= \pi R^2 \left[-\frac{\cos(2\theta)}{2}\right]_0^{\pi/2} = \pi R^2.$$

Thus, $S_\perp$ is one quarter of the total area of the planet $4\pi R^2$; the
geometrical meaning of this result is that the normal area presented

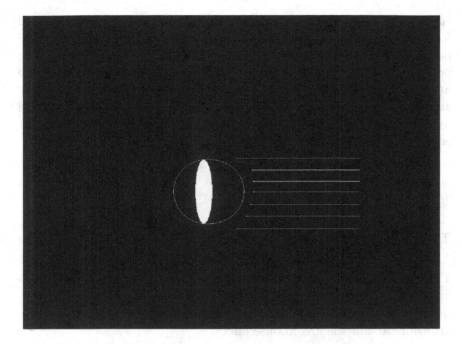

Figure 2.2. Normal area presented by the Earth to the solar rays.

by the Earth to the rays coming from the Sun coincides with the area
of the equatorial cross section, a circle of radius R.

13 **(A)** Derive a simple equation describing the energy balance for the
Earth in space. Neglect the presence of the atmosphere and the heating due to the decay of radioactive elements inside the Earth.[5] The
average albedo of the Earth is $a = 0.34$. Compare your result with
the measured values of the average surface temperature $T_s = 288$ K
and the average atmospheric temperature $T_a = 255$ K.

Solution
The power $\pi R^2 S$ reaching the Earth in the form of electromagnetic
radiation from the Sun is the solar constant S (in W/m^2) times the
effective cross-sectional area presented to the Sun πR^2, where R is the
Earth radius (cf. the previous problem). A fraction a of this power

---

[5]This is only important on astronomical time scales, e.g., if one wants to compute the cooling
time of a newly formed planet.

(*albedo*) is reflected back into space, hence the total power received from the Sun is $(1 - a)\pi R^2 S$. At equilibrium, this power equals the power radiated into space by the Earth as blackbody radiation. The power radiated by a blackbody per unit of surface area is given by the Stefan–Boltzmann law σT^4, where σ is the Stefan–Boltzmann constant and T is the Kelvin temperature, hence the total power radiated by the Earth is $4\pi R^2 \sigma T_s^4$, where T_s is the surface temperature. At equilibrium it must be

$$(1 - a)\pi R^2 S = 4\pi R^2 \sigma T_s^4,$$

and the surface temperature of the Earth is

$$T_s = \left[\frac{(1-a)\,S}{4\sigma}\right]^{1/4} = \left[\frac{(1 - 0.34)\,(1.37 \cdot 10^3\,\text{W/m}^2)}{4\,(5.761 \cdot 10^{-8}\,\text{W} \cdot \text{m}^{-2} \cdot \text{K}^{-4})}\right]^{1/4} = 250\,\text{K}.$$

This number is close to the measured average value 255 K of the atmospheric temperature, but different from the average surface temperature of 288 K. This discrepancy is due to the presence of the atmosphere and the greenhouse effect.

14 **(B)** Assume that the Earth receiving energy from the Sun and emitting into space is not a system in equilibrium, and that its surface temperature is allowed to change with time. Neglecting the presence of the atmosphere, derive an equation describing this energy balance, and study it qualitatively. For simplicity, assume that the albedo does not vary with T.

Solution
The power reaching the Earth from the Sun is the product of the solar constant S and of the cross-sectional area πR^2 presented to the solar rays. A fraction a (*albedo*) of this power is reflected back into space, and the total power received by the Earth is $(1 - a)\pi R^2 S$. According to the Stefan–Boltzmann law, the power emitted as a blackbody is $4\pi R^2 \sigma T^4$, where T is the surface temperature of the Earth. The heat absorbed by the Earth during the temperature change dT is $C\,dT$, where C is the heat capacity of the planet. The heat balance for the planet is therefore

$$C\frac{dT}{dt} = (1 - a)\pi R^2 S - 4\pi R^2 \sigma T^4;$$

this can be written as the nonlinear ordinary differential equation

$$\frac{dT}{dt} = \alpha\left(T_*^4 - T^4\right),$$

where

$$\alpha \equiv \frac{4\pi R^2 \sigma}{C}, \qquad T_* \equiv \left[\frac{(1-a)\,S}{4\sigma} \right]^{1/4}.$$

By assuming that the albedo does not change with time,[6] one finds by inspection the steady-state equilibrium solution $T = T_*$. This a stable solution: in fact, if $T > T_*$ initially, the temperature will always be larger than T_*, otherwise the curve representing the solution $T(t)$ will cross the line representing the exact solution $T = T_*$, which is forbidden by the uniqueness theorems for the solutions of ODEs. Then, $dT/dt = \alpha\left(T_*^4 - T^4\right) < 0$ and the solution $T(t)$ is forced to decrease monotonically and approaches asymptotically its lower bound T_*. Similarly, if $T < T_*$ initially, then $dT/dt > 0$ and the solution $T(t)$ approaches its upper bound T_* asymptotically without crossing it—we conclude that the solution $T = T_*$ is stable and is an attractor in the phase space.

The nonlinear differential equation can be integrated implicitly: by using the variable $\theta(T) \equiv T(t)/T_*$, it can be written as

$$\frac{1}{1-\theta^4} \frac{d\theta}{dt} = \alpha T_*^3,$$

from which it follows that

$$\int \frac{d\theta}{1-\theta^4} = \alpha T_*^3\, t,$$

and, using a table of integrals [21], the implicit solution

$$\ln\left| \frac{\theta+1}{\theta-1} \right| + 2\tan^{-1}\theta = 4\alpha T_*^3\, t + \text{const.}$$

is found.

---

[6] This simplifying assumption is not very realistic—the albedo will change with the temperature T, which is a function of time.

Chapter 3

OCEAN AND ATMOSPHERIC PHYSICS

If God had consulted me before embarking on the creation, I would have suggested something simpler.

—Alfonso de Castile

Approximately seventy percent of the surface of the planet is covered by oceans, and not much is really known about these, especially below the air–water interface, because of the difficulty in obtaining systematic and reliable data. The fact that the oceans are studied in physics, geology, chemistry, biochemistry, and biology testifies to their complexity. The atmosphere surrounds the Earth like a blanket and allows for the presence of aerobic life and for a host of related phenomena. The oceans and the atmosphere are not separate entities but they interact in many ways to determine the global circulation and climate, as well as many local phenomena. For example, winds drive oceanic currents and generate waves.

From the physical point of view, the dynamics of both the oceans and the atmosphere are regulated by fluid dynamics and thermodynamics, which are necessarily complicated by the practical need of referring theory and observations to a noninertial reference frame connected with the rotating Earth. While ocean water can usually be treated as an incompressible fluid, atmospheric air is very compressible and is mostly concentrated in the lowest 10 kilometers of the atmosphere. The presence of water vapor in this lower layer is responsible for clouds, precipitation, transport of energy stored as latent heat, and other phenomena.

The study of the atmosphere and the oceans, based on the nonlinear laws of fluid dynamics, is complicated. In this chapter we touch upon the most elementary aspects, which are usually covered in introductory courses in environmental science. Reference [55] is suggested for descriptive qualitative oceanography and Ref. [56] for quantitative dynamic oceanography, while Refs. [13, 19, 48] are more advanced. As introductory readings on atmospheric physics and meteorology, we recommend Refs. [34, 67, 71], and more advanced texts are [31, 7].

3.1   The blue planet

This section has the purpose of familiarizing the student with basic physical aspects of the oceans.

1 **(A)** The oceans cover 70% of the Earth's surface and have an average depth of 3.8 km. What fraction of the mass of the planet is contained in the oceans? The mass of the Earth is $M_E = 5.98 \cdot 10^{24}$ kg, its radius is 6370 km, and the average density of saltwater is $1.03 \cdot 10^3$ kg/m^3.

Solution
The volume occupied by the oceans is $V = 0.7 \cdot 4\pi R^2 d$, where R is the radius of the Earth and d is the average depth of the oceans. The mass of the oceans is ρV, where ρ is the density of water; the mass of the oceans in units of the Earth's mass is given by the fraction

$$\frac{\rho \cdot (0.7 \cdot 4\pi R^2)d}{M_E}$$

$$= \frac{(1.03 \cdot 10^3 \,\text{kg/m}^3) \cdot 0.7 \cdot 4\pi (6.37 \cdot 10^6 \,\text{m})^2 \cdot (3.8 \cdot 10^3 \,\text{m})}{5.98 \cdot 10^{24} \,\text{kg}}$$

$$= 2.34 \cdot 10^{-4}.$$

2 **(A)** On average, 2.4% by mass of sea water consists of dissolved salts. If the Earth's oceans could be completely evaporated, the remaining salts would cover the entire surface of the Earth with a layer of depth h. Given that the oceans cover 70% of the Earth's surface and their average depth is $d = 3.8$ km, compute the value of h. The density of sea water is $\rho_w = 1.025 \cdot 10^3$ kg/m^3 and the average density of the salts in it is $\rho_s = 25$ kg/m^3. What would be the height h' of the salts if they were spread over the continental surface instead?

Solution
The total mass of the oceans is

$$M = \rho_w V = \rho_w \cdot 0.7 \cdot 4\pi R_E^2 d,$$

where R_E is the Earth's radius and V is the volume occupied by the oceans. The total mass of salts dissolved in sea water is then $M_s = 2.4 \cdot 10^{-2} M$. If these salts were spread over the entire surface of the Earth (with area $A = 4\pi R_E^2$), they would occupy the volume $V' = Ah$. The value of h is therefore

$$h = \frac{V'}{A} = \frac{M_s}{\rho_s A} = 2.4 \cdot 10^{-2} \frac{\rho_w}{\rho_s} \frac{0.7 \cdot \left(4\pi R_E^2 d\right)}{4\pi R_E^2}$$

$$= 2.4 \cdot 10^{-2} \frac{0.7 \cdot \left(1.025 \cdot 10^3 \, \text{kg/m}^3\right) \cdot \left(3.8 \cdot 10^3 \, \text{m}\right)}{25 \, \text{kg/m}^3} = 2.6 \, \text{km}.$$

If instead the salts were spread over the continental surface, they would occupy the area $A' = 0.3 \cdot 4\pi R_E^2 = 0.3A$ and they would reach the height $h' = h/0.3 = 8.7 \, \text{km}$, approximately the height of Mt. Everest.

3 **(B)** The Marianas trench has a depth of 11.0 km below sea level. Compute the pressure at the bottom of the trench
a) by treating sea water as an incompressible fluid,
b) by taking into account the compressibility of water.
The density of water at the surface is $\rho_0 = 1.0 \cdot 10^3 \, \text{kg/m}^3$ and its bulk modulus is $B = 2.1 \cdot 10^9 \, \text{Pa}$.

Solution
a) If the compressibility of water is ignored, the pressure at the bottom of the trench is simply

$$P = P_0 + \rho g z,$$

where P_0 is the atmospheric pressure at the surface, and z is the depth. Numerically,

$$P = \left(1.01 \cdot 10^5 \, \text{Pa}\right) + \left(1.0 \cdot 10^3 \, \frac{\text{kg}}{\text{m}^3}\right) \cdot \left(9.81 \, \frac{\text{m}}{\text{s}^2}\right) \cdot \left(1.10 \cdot 10^4 \, \text{m}\right)$$

$$= 1.1 \cdot 10^8 \, \text{Pa}.$$

b) The hydrostatic pressure obeys the equation

$$\frac{dP}{dz} = \rho g, \tag{3.1}$$

while the compression (relative change of volume) of sea water due to the weight of the column above is described by the equation[1]

$$\frac{dV}{V} = -\frac{1}{B} dP.$$

Since the density of water is $\rho = M/V$, where M is the mass of water, we have

$$dV = d\left(\frac{M}{\rho}\right) = -\frac{M}{\rho^2} d\rho$$

and

$$-\frac{M}{\rho^2} d\rho \frac{\rho}{M} = -\frac{1}{B} dP,$$

or

$$B \frac{d\rho}{\rho} = dP. \qquad (3.2)$$

Comparison of Eqs. (3.1) and (3.2) yields

$$\frac{d\rho}{\rho^2} = -d\left(\frac{1}{\rho}\right) = \frac{g}{B} dz,$$

which is immediately integrated, giving

$$\rho(z) = \frac{1}{A - gz/B}.$$

Here it is assumed that g is constant and A is an integration constant determined by the initial condition at the surface

$$\rho(z = 0) = \rho_0 = 1.0 \cdot 10^3 \, \frac{\text{kg}}{\text{m}^3},$$

which yields

$$\rho(z) = \frac{\rho_0}{1 - z/z_*},$$

where

$$z_* \equiv \frac{B}{g\rho_0} = \frac{(2.1 \cdot 10^9 \, \text{Pa})}{(9.81 \, \text{m} \cdot \text{s}^{-2}) \cdot (1.0 \cdot 10^3 \, \text{kg/m}^3)} = 2.14 \cdot 10^5 \, \text{m}$$

is a characteristic length. Equation (3.2) now yields

$$dP = B d(\ln \rho)$$

---

[1]The (isothermal) bulk modulus B is the inverse of the isothermal compressibility $\kappa \equiv -\frac{1}{V}\left(\frac{\partial V}{\partial P}\right)_T$.

and integrating between the surface of the sea and the depth z we obtain

$$P(z) = P_0 - B \ln\left(1 - z/z_*\right) = \left(1.01 \cdot 10^5\,\text{Pa}\right) - \left(2.1 \cdot 10^9\,\text{Pa}\right)$$
$$\cdot \ln\left(1 - \frac{1.10 \cdot 10^4\,\text{m}}{2.14 \cdot 10^5\,\text{m}}\right) = 1.1 \cdot 10^8\,\text{Pa}.$$

There is no difference between the two calculations of the pressure to two significant figures.

4 **(C)** The difference between water levels at low and high tide (*tidal range*) in the open ocean is approximately 1 m. Why can the tidal range assume values much larger than 1 m along the coast?

Solution
Along coastlines water can be funnelled, pile up, and surge in channels, inlets, fjords, and other natural constrictions. An extreme case is the Bay of Fundy on the Atlantic coast of Canada, in which the tidal range can reach 15 m.

3.2   Oceanic circulation

Oceanic currents include geostrophic currents, wind-driven currents, and tidal currents. Our description of these phenomena in a reference frame connected to the rotating Earth must include noninertial forces, and the scarcity of reliable data limits our knowledge of deep oceanic circulation.[2]

1 **(B)** Consider two Cartesian coordinate systems with common origin, the first being an inertial system and the second rotating with respect to the first one with angular velocity $\vec{\Omega}$ (this vector is pointing along the rotation axis and has magnitude equal to the angular velocity Ω). It can be shown [20] that the relation between the time derivatives of a vector $\vec{A}$ in the two frames is

$$\frac{d\vec{A}}{dt} = \left(\frac{d\vec{A}}{dt}\right)_{\text{rot}} + \vec{\Omega} \times \vec{A},$$

where the subscript "rot" refers to the rotating frame and the derivative with no subscript is taken in the inertial frame. Find $d^2\vec{A}/dt^2$ in

---

[2]A popular science book written by H. Stommel [66], one of the pioneers in this field, is recommended as a side reading in addition to the more technical references [56, 55].

the rotating frame and interpret your result physically. Apply your result to the case in which $\vec{A} = \vec{x}$ is the position vector of a point particle moving on the surface of the Earth, which is rotating with respect to the inertial frame of the fixed stars.

Solution
We just need to apply the differential operator

$$\frac{d(...)}{dt} = \left(\frac{d\,...}{dt}\right)_{\text{rot}} + \vec{\Omega} \times \, ...$$

to the vector $d\vec{A}/dt$. The result is

$$\frac{d^2\vec{A}}{dt^2} = \frac{d}{dt}\left(\frac{d\vec{A}}{dt}\right) = \left[\left(\frac{d}{dt}\right)_{\text{rot}} + \vec{\Omega}\times\right]\left[\left(\frac{d\vec{A}}{dt}\right)_{\text{rot}} + \vec{\Omega} \times \vec{A}\right]$$

$$= \left(\frac{d^2\vec{A}}{dt^2}\right)_{\text{rot}} + \left(\frac{d\vec{\Omega}}{dt}\right)_{\text{rot}} \times \vec{A} + 2\vec{\Omega} \times \left(\frac{d\vec{A}}{dt}\right)_{\text{rot}}$$

$$+\vec{\Omega} \times \vec{\Omega} \times \vec{A}.$$

The second time derivative of $\vec{A}$ in the rotating frame is

$$\left(\frac{d^2\vec{A}}{dt^2}\right)_{\text{rot}} = \frac{d^2\vec{A}}{dt^2} - 2\vec{\Omega} \times \left(\frac{d\vec{A}}{dt}\right)_{\text{rot}} - \vec{\Omega} \times \vec{\Omega} \times \vec{A} + \vec{A} \times \frac{d\vec{\Omega}}{dt},$$

where in the last term we used the fact that

$$\left(\frac{d\vec{\Omega}}{dt}\right)_{\text{rot}} = \frac{d\vec{\Omega}}{dt}$$

because $\Omega \times \Omega = 0$. The physical interpretation of this result is as follows. The terms containing the angular velocity $\vec{\Omega}$ describe noninertial effects due to the fact that the derivative on the left-hand side is taken in a noninertial (rotating) frame. The last term on the right-hand side only occurs if the angular velocity $\vec{\Omega}$ changes direction or magnitude with time.

In the case in which $\vec{A} = \vec{x}$ is the position of an object and the rotating frame is connected to the Earth, which is the situation occurring in oceanography or atmospheric physics, we have the relation between velocities and accelerations in the rotating frame fixed to the Earth and the inertial frame of the fixed stars

$$\vec{v}_{\text{rot}} = \vec{v} - \vec{\Omega} \times \vec{x},$$

$$\vec{a}_{\text{rot}} = \vec{a} - 2\vec{\Omega} \times \vec{v}_{\text{rot}} - \vec{\Omega} \times \vec{\Omega} \times \vec{x},$$

where the term proportional to $d\vec{\Omega}/dt$ is dropped due to the fact that the angular velocity of the Earth is nearly constant.[3]

The first term on the right-hand side is the acceleration in the non-rotating frame caused by real forces; the second term is the Coriolis acceleration proportional to the speed of the object and the sine of the angle between $\vec{\Omega}$ and $\vec{v}$ (remember that $\left|\vec{\Omega} \times \vec{v}\right| = \left|\vec{\Omega}\right| |\vec{v}| \sin\theta$), which is equal to the latitude λ. The last term is the centrifugal acceleration directed radially away from the center of the Earth and with magnitude $\left|\vec{\Omega} \times \vec{\Omega} \times \vec{x}\right| = \Omega^2 R \sin(\pi/2 - \lambda) = \Omega^2 R \cos\lambda$, where R is the Earth radius.

Numerically, the Coriolis and the centrifugal accelerations are small because Ω is small. The centrifugal acceleration is counteracted by the much larger acceleration of gravity $\vec{g}$. The Coriolis acceleration is not counteracted by gravity: although it is small its effect cumulate over long (say, larger than one hour) periods of time and it is not negligible for objects with a long flight time. For oceanic currents and atmospheric winds, which have a virtually infinite "flight time," the Coriolis force is significant, giving rise to geostrophic currents and winds. For example, at the latitude $\lambda = 38°$ and for a speed $v \approx 1\,\mathrm{m/s}$ for the Gulf Stream we get

$$\left|-2\vec{\Omega} \times \vec{v}_{\mathrm{rot}}\right| = 2\Omega v \sin\lambda = 2 \left(\frac{2\pi}{24 \cdot 3600\,\mathrm{s}}\right)\left(1\,\frac{\mathrm{m}}{\mathrm{s}}\right)\sin 38° \approx 10^{-4}\,\frac{\mathrm{m}}{\mathrm{s}^2}.$$

After one hour the lateral displacement is of order

$$\Delta x \simeq \frac{1}{2}\left|-2\vec{\Omega} \times \vec{v}_{\mathrm{rot}}\right| t^2 \approx \frac{1}{2}\left(10^{-4}\,\frac{\mathrm{m}}{\mathrm{s}^2}\right)(3600\,\mathrm{s})^2 \approx 650\,\mathrm{m}.$$

2 **(A)** The Kuroshio current has speed of order $v = 1\,\mathrm{m/s}$ to $1.2\,\mathrm{m/s}$ between latitudes 31 and 33 degrees north. The dynamic viscosity coefficient and the density of sea water in the relevant range of temperature and salinity ($T \simeq 25°\mathrm{C}$ and $S = 35$) are $\eta = 9 \cdot 10^{-4}\,\mathrm{kg \cdot m^{-1} \cdot s^{-1}}$ and $\rho = 1022\,\mathrm{kg \cdot m^{-3}}$. Do you expect turbulence in this current?

Solution
Whether the flow is turbulent or not can be decided by looking at the value of the Reynolds number $Re = vL/\nu$, where v is the typical speed, L is the length scale over which the flow varies, and $\nu \equiv \eta/\rho$

---

[3]This is not exactly true on long (astronomical) time scales due, for example, to the precession of the rotation axis of the Earth, or the tiny dissipation of rotational energy due to friction caused by tides.

is the kinematic viscosity coefficient. The scale of variation L corresponds to three degrees of latitude λ (corresponding to $\pi/60$ radians) and is related to the Earth radius R by

$$L = \lambda R = \frac{\pi}{60} \left(6.37 \cdot 10^6 \, \text{m}\right) \simeq 3 \cdot 10^5 \, \text{m} = 300 \, \text{km}.$$

The kinematic viscosity coefficient is

$$\nu \equiv \frac{\eta}{\rho} = \frac{9 \cdot 10^{-4} \, \text{kg} \cdot \text{m}^{-1} \cdot \text{s}^{-1}}{1022 \, \text{kg} \cdot \text{m}^{-3}} = 8.8 \cdot 10^{-7} \, \frac{\text{m}^2}{\text{s}}.$$

The Reynolds number is therefore

$$Re = \frac{\nu L}{\nu} = \frac{(1 \, \text{m/s}) \left(3 \cdot 10^5 \, \text{m}\right)}{8.8 \cdot 10^{-7} \, \text{m}^2/\text{s}} \simeq 3 \cdot 10^{11}.$$

The Reynolds number is much larger than 3000 and the flow is certainly turbulent.

3 **(B)** In oceanography the specific volume $\alpha = 1/\rho$ is often used instead of the density ρ. Express conservation of mass of sea water in terms of α instead of ρ when sea water flows with velocity $\vec{v}$. What is the form of this equation when sea water is treated as incompressible, which is a satisfactory approximation for many purposes in dynamic oceanography?

Solution
Conservation of mass in the absence of sources or sinks is expressed by the continuity equation

$$\frac{\partial \rho}{\partial t} + \vec{\nabla} \cdot (\rho \vec{v}) = 0.$$

By using the definition of specific volume $\alpha \equiv \rho^{-1}$, one obtains

$$\frac{\partial}{\partial t}\left(\frac{1}{\alpha}\right) + \vec{v} \cdot \vec{\nabla}\left(\frac{1}{\alpha}\right) + \frac{1}{\alpha}\vec{\nabla} \cdot \vec{v} = 0,$$

or

$$-\frac{1}{\alpha^2}\frac{\partial \alpha}{\partial t} - \frac{1}{\alpha^2}\vec{v} \cdot \vec{\nabla}\alpha + \frac{1}{\alpha}\vec{\nabla} \cdot \vec{v} = 0,$$

and finally

$$\frac{\partial \alpha}{\partial t} + \vec{v} \cdot \vec{\nabla}\alpha - \alpha\vec{\nabla} \cdot \vec{v} = 0.$$

By treating sea water as an incompressible fluid ρ, and therefore also α, are assumed to be constants and conservation of mass takes the form $\vec{\nabla} \cdot \vec{v} = 0$, or the velocity field is solenoidal.

4 **(B)** In open ocean the horizontal velocity field at the surface ($z = 0$)
is

$$(v_x, v_y) = (1 + 0.01\,(x + y)\,, 0.1y) \cdot 10^{-2}\,\frac{\text{m}}{\text{s}}.$$

Estimate the vertical velocity at a depth of 60 m (this is usually much
smaller than the horizontal velocity and harder to measure).

Solution
By treating sea water as an incompressible fluid the continuity equation expressing mass conservation becomes $\vec{\nabla} \cdot \vec{v} = 0$ and therefore,
at the surface $z = 0$

$$\left.\frac{\partial v_z}{\partial z}\right|_{z=0} = -\left(\frac{\partial v_x}{\partial x} + \frac{\partial v_y}{\partial y}\right)\bigg|_{z=0} = -\left(1 \cdot 10^{-4} + 1 \cdot 10^{-3}\right)\frac{\text{m}}{\text{s}}$$

$$= -1.1 \cdot 10^{-3}\,\frac{\text{m}}{\text{s}}.$$

In order of magnitude, the vertical velocity at depth $z = -60$ m is

$$v(z) \simeq \frac{\partial v_z}{\partial z}\Delta z = \left(-1.1 \cdot 10^{-3}\,\frac{\text{m}}{\text{s}}\right)(-60\,\text{m}) = 6.6 \cdot 10^{-2}\,\frac{\text{m}}{\text{s}}.$$

5 **(B)** A strong wind sets ocean water in motion and then dies off. The
surface of the sea is now horizontal, the pressure $P(z)$ does not depend
on the horizontal coordinates x and y, and all the forces (including
friction) other than gravity and the Coriolis force can be neglected.
Assume also that sea water is incompressible and that the circulation
is horizontal, i.e., the velocity of the water is $\vec{v} = (v_x, v_y, 0)$. Write the
equations for this situation (*inertial motion*) and solve them. What
does your solution represent? How does the pressure vary with the
depth?

Solution
The full equations of motion for oceanic circulation are

$$\frac{dv_x}{dt} = -\alpha\frac{\partial P}{\partial x} + 2\Omega v_y \sin\lambda - 2\Omega v_z \cos\lambda + f_x,$$

$$\frac{dv_y}{dt} = -\alpha\frac{\partial P}{\partial y} - 2\Omega v_x \sin\lambda + f_y,$$

$$\frac{dv_z}{dt} = -\alpha\frac{\partial P}{\partial z} + 2\Omega v_x \cos\lambda - g + f_z,$$

where α is the specific volume, λ is the latitude, Ω is the rotational
angular velocity of the Earth, and $\vec{f}$ is the force density per unit

volume including frictional, tidal, and other forces. In the special case of inertial motion under consideration these equations reduce to

$$\frac{dv_x}{dt} = 2\Omega v_y \sin \lambda,$$

$$\frac{dv_y}{dt} = -2\Omega v_x \sin \lambda,$$

$$\frac{\partial P}{\partial z} = \alpha^{-1} (g - 2\Omega v_x \cos \lambda).$$

By differentiating the first equation with respect to time, we find the second-order ODE for v_x

$$\frac{d^2 v_x}{dt^2} - 2\Omega \sin \lambda \frac{dv_y}{dt} = 0,$$

and by substituting dv_y/dt from the second equation, the decoupled equation for v_x follows,

$$\frac{d^2 v_x}{dt^2} + (2\Omega \sin \lambda)^2 v_x = 0.$$

This is recognized as the harmonic oscillator equation for v_x, with angular frequency $2\Omega \sin \lambda$, and its general solution is

$$v_x(t) = C_1 \cos \left[(2\Omega \sin \lambda)\, t \right] + C_2 \sin \left[(2\Omega \sin \lambda)\, t \right],$$

where $C_{1,2}$ are integration constants. By substituting this solution into the ODE for v_y, one now obtains

$$\frac{dv_y}{dt} = -2\Omega \sin \lambda \left\{ C_1 \cos \left[(2\Omega \sin \lambda)\, t \right] + C_2 \sin \left[(2\Omega \sin \lambda)\, t \right] \right\},$$

which is immediately integrated to

$$v_y(t) = C_2 \cos \left[(2\Omega \sin \lambda)\, t \right] - C_1 \sin \left[(2\Omega \sin \lambda)\, t \right].$$

Suppose that we choose the origin of time so that $C_2 = 0$: then

$$v_x(t) = \frac{dx}{dt} = C_1 \cos \left[(2\Omega \sin \lambda)\, t \right],$$

$$v_y(t) = \frac{dy}{dt} = -C_1 \sin \left[(2\Omega \sin \lambda)\, t \right],$$

where $x(t)$ and $y(t)$ are the positions of a small parcel of sea water. These equations are integrated, obtaining

$$x(t) = \frac{C_1}{2\Omega \sin \lambda} \sin [(2\Omega \sin \lambda) t],$$

$$y(t) = \frac{C_1}{2\Omega \sin \lambda} \cos [(2\Omega \sin \lambda) t],$$

which constitute the parametric representation of a horizontal circle of radius

$$R = \left| \frac{C_1}{2\Omega \sin \lambda} \right|$$

away from the equator. At the equator ($\lambda = 0$) the ODEs for v_x and v_y yield constant v_x and v_y because the Coriolis force is absent there. The circle is traveled clockwise in the Northern Hemisphere with angular velocity $2\Omega \sin \lambda$ maximum at the poles and zero at the equator. The corresponding period, called the *inertial period*, is

$$P_{\text{circ}} = \frac{2\pi}{2\Omega \sin \lambda} = \frac{P_{\text{rot}}}{2 \sin \lambda},$$

where P_{rot} is the rotational period of the Earth (the sidereal day of 23 h 56' 4"). The Coriolis force keeps water parcels on their circular paths.

The vertical velocity is zero and the vertical equation of motion yields

$$\frac{dP}{dz} = \frac{1}{\alpha} \left(g - 2\Omega v_x \cos \lambda \right),$$

which is integrated to

$$P(z) = \frac{1}{\alpha} \left(g - 2\Omega v_x \cos \lambda \right) z$$

by imposing the initial condition that the pressure vanishes at the surface[4] $z = 0$. Note that the z-axis is pointing upward and that $z < 0$ under water.

6 **(B)** Consider a two-dimensional irrotational flow around an island, which can be modeled by a vertical cylinder of radius R and the

---

[4]P is a gauge pressure, i.e., the difference between the absolute pressure and the atmospheric pressure. The absolute pressure at the surface is equal to the atmospheric pressure.

z-axis as symmetry axis. Verify that

$$\Phi\left(r,\varphi\right) \;=\; v\left(r+\frac{R^2}{r}\right)\cos\varphi,$$

$$\Psi\left(r,\varphi\right) \;=\; v\left(r-\frac{R^2}{r}\right)\sin\varphi,$$

are the velocity potential and the stream function, respectively, corresponding to the boundary conditions

$$\frac{\partial\Phi}{\partial r}\left(R,\varphi\right)=0,$$

$$\Psi\left(R,\varphi\right)=0,$$

where the first boundary condition corresponds to stream lines parallel to the surface of the cylindrical island. Here v is a constant.

Solution
Let us verify that both Φ and Ψ satisfy the Laplace equation $\nabla^2 f = 0$. The Laplacian operator in two dimensions and in cylindrical coordinates (r,φ) is

$$\nabla^2 = \frac{1}{r}\frac{\partial}{\partial r}\left(r\frac{\partial}{\partial r}\right)+\frac{1}{r^2}\frac{\partial}{\partial\varphi^2}$$

and therefore

$$\begin{aligned}\nabla^2\Phi &= \frac{1}{r}\frac{\partial}{\partial r}\left(r\frac{\partial\Phi}{\partial r}\right)+\frac{1}{r^2}\frac{\partial\Phi}{\partial\varphi^2}\\[2mm] &= \frac{v}{r}\left(1+\frac{R^2}{r^2}\right)\cos\varphi-\frac{v}{r}\left(1+\frac{R^2}{r^2}\right)\cos\varphi=0,\\[2mm]\nabla^2\Psi &= \frac{1}{r}\frac{\partial}{\partial r}\left(r\frac{\partial\Psi}{\partial r}\right)+\frac{1}{r^2}\frac{\partial\Psi}{\partial\varphi^2}\\[2mm] &= \frac{v}{r}\left(1-\frac{R^2}{r^2}\right)\sin\varphi-\frac{v}{r}\left(1-\frac{R^2}{r^2}\right)\sin\varphi=0.\end{aligned}$$

In addition,

$$\begin{aligned}\vec{\nabla}\Phi &= \frac{\partial\Phi}{\partial r}\vec{e}_r+\frac{1}{r}\frac{\partial\Phi}{\partial\varphi}\vec{e}_\varphi\\[2mm] &= v\left(1-\frac{R^2}{r^2}\right)\cos\varphi\,\vec{e}_r-v\left(1+\frac{R^2}{r^2}\right)\sin\varphi\,\vec{e}_\varphi,\end{aligned}$$

$$\vec{\nabla}\Psi \;=\; \frac{\partial\Psi}{\partial r}\,\vec{e}_r + \frac{1}{r}\frac{\partial\Psi}{\partial\varphi}\,\vec{e}_\varphi$$

$$= \; v\left(1+\frac{R^2}{r^2}\right)\sin\varphi\,\vec{e}_r + v\left(1-\frac{R^2}{r^2}\right)\cos\varphi\,\vec{e}_\varphi,$$

and using the fact that $\vec{e}_r\cdot\vec{e}_r = \vec{e}_\varphi\cdot\vec{e}_\varphi = 1, \vec{e}_r\cdot\vec{e}_\varphi = 0$ one finds that

$$\vec{\nabla}\Phi\cdot\vec{\nabla}\Psi \;=\; v^2\left(1-\frac{R^4}{r^4}\right)\sin\varphi\cos\varphi\,(\vec{e}_r\cdot\vec{e}_r)$$

$$-v^2\left(1-\frac{R^4}{r^4}\right)\sin\varphi\cos\varphi\,(\vec{e}_\varphi\cdot\vec{e}_\varphi) = 0,$$

i.e., $\vec{\nabla}\Phi$ and $\vec{\nabla}\Psi$ are orthogonal.

Finally, let us verify that the boundary conditions are satisfied:

$$\Psi\,(R,\varphi) = v\left(R-\frac{R^2}{R}\right)\sin\varphi \qquad = 0,$$

$$\left.\tfrac{\partial\Phi}{\partial r}\,(R,\varphi) = v\left(1-\frac{R^2}{r^2}\right)\cos\varphi\right|_{r=R} \; = 0.$$

3.3 Ocean waves

Ocean waves can be classified as follows:

- *Surface waves* such as ripples, wind waves, and swells due to the wind action,

- *Internal waves* that travel below the interface between sea water and the atmosphere,

- *Gravity waves*, which can be both surface or internal waves. If their period is sufficiently long, the Coriolis force becomes important and they are called *gyroscopic-gravity* waves.

- *Planetary* or *Rossby waves* with wavelength comparable to the radius of the planet. These are really oscillating geostrophic currents propagating westward due to the periodic shrinking and stretching of the surrounding masses of water in the north–south direction.

- *Tsunamis*, long waves caused by sudden vertical motions of the ocean floor and, from the human point of view, responsible for many disasters.

Tides are also sometimes classified as ocean waves. They are periodic phenomena due to the time-varying gradients of the gravitational acceleration due to the Moon and Sun.

For this section we recommend [44, 54, 42, 56] as references.

1 **(A)** Can the following waves be considered as linear waves?
   a) a ripple with wavelength $\lambda = 3$ cm and height $h = 0.5$ cm
   b) a ripple with wavelength $\lambda = 3$ cm and height $h = 1$ mm
   c) a swell with wavelength $\lambda = 300$ m and height $h = 1$ m.

Solution
In order for a wave to be linear the height h of the wave must be much smaller than its wavelength λ. A practical criterion for linearity is $h/\lambda < 1/20 = 0.05$. In case a) the ratio of height to wavelength is

$$\frac{h}{\lambda} = \frac{0.5 \, \text{cm}}{3 \, \text{cm}} = \frac{1}{6} \simeq 0.1667 > 0.05$$

and this wave is nonlinear. In case b)

$$\frac{h}{\lambda} = \frac{0.1 \, \text{cm}}{3 \, \text{cm}} \simeq 0.033 < 0.05,$$

and this wave is linear. In case c)

$$\frac{h}{\lambda} = \frac{1 \, \text{m}}{300 \, \text{m}} \simeq 0.0033 \ll 0.05,$$

and the wave is definitely linear.

2 **(B)** In the literature, two expressions are found for the phase velocity of a wave in a dispersive medium: $v_p = \lambda \nu$ and $v_p = \omega/k$, where $\omega = 2\pi\nu$ is the angular frequency and $k = 2\pi/\lambda$ is the wave vector. Show that the two expressions coincide and that the group velocity $v_g = d\Omega/dk$ can be written as $-\lambda^2 d\nu/d\lambda$.

Solution
We have

$$\frac{\omega}{k} = \frac{2\pi\nu}{2\pi/\lambda} = \lambda\nu.$$

The group velocity is

$$v_g = \frac{d\omega}{dk} = 2\pi \frac{d\nu}{d\lambda} \frac{d\lambda}{dk} = 2\pi \frac{d\nu}{d\lambda} \left(\frac{-2\pi}{k^2} \right) = -\left(\frac{2\pi}{k} \right)^2 \frac{d\nu}{d\lambda} = -\lambda^2 \frac{d\nu}{d\lambda}.$$

3 **(B)** Show that the ratio between group and phase velocity for a wave propagating in a dispersive medium can be written as

$$\frac{v_g}{v_p} = -\frac{d(\ln \nu)}{d(\ln \lambda)}. \tag{3.3}$$

Solution
By remembering that the angular frequency and the wave vector are defined, respectively, by $\omega = 2\pi\nu$ and $k = 2\pi/\lambda$, we have

$$\frac{v_g}{v_p} = \frac{d\omega/dk}{\omega/k} = \frac{k}{\omega}\frac{d\omega}{dk} = \frac{1}{\lambda\nu}2\pi\frac{d\nu}{d\lambda}\frac{d\lambda}{dk} = \frac{1}{\lambda\nu}\frac{d\nu}{d\lambda}\frac{d}{dk}\left(\frac{4\pi^2}{k}\right)$$

$$= -\frac{\lambda}{\nu}\frac{d\nu}{d\lambda} = -\frac{d(\ln \nu)}{d(\ln \lambda)}.$$

4 **(A)** The water's surface is characterized by its surface tension γ and density ρ. Using dimensional considerations, derive an approximate dispersion relation $\omega = \omega(k)$ for short wavelength waves (ripples) generated by wind drag and propagating on the surface of deep water.

Solution
The dimensions of surface tension (a force per unit length) and mass density are

$$[\gamma] = \left[MT^{-2}\right],$$
$$[\rho] = \left[ML^{-3}\right],$$

and the dimensions of angular frequency and wave vector are $[\omega] = \left[T^{-1}\right]$, $[k] = \left[L^{-1}\right]$. By requiring that

$$\omega = A\gamma^\alpha \rho^\beta k^\gamma,$$

where A is a dimensionless coefficient, we obtain

$$\left[T^{-1}\right] = \left[M^\alpha T^{-2\alpha}\right]\left[M^\beta L^{-3\beta}\right]\left[L^{-\gamma}\right] = \left[M^{\alpha+\beta}T^{-2\alpha}L^{-3\beta-\gamma}\right].$$

The linear system

$$\alpha + \beta = 0, \tag{3.4}$$

$$-3\beta - \gamma = 0, \tag{3.5}$$

$$-2\alpha = -1, \tag{3.6}$$

must be satisfied, and the solutions are $\alpha = -\beta = 1/2$, $\gamma = 3/2$. The approximate dispersion relation is

$$\omega(k) = A \sqrt{\frac{\gamma}{\rho}} \, k^{3/2}.$$

The exact dispersion relation is given by $A = 1$ and is used in the following problem.

5 **(B)** Short waves (ripples) at the surface of deep water propagate according to the relation between frequency ν and wavelength λ

$$\nu = \left(\frac{2\pi \gamma}{\rho \lambda^3}\right)^{1/2}, \tag{3.7}$$

where γ and ρ are, respectively, the surface tension and the density of water. Derive the dispersion relation $\omega = \omega(k)$ between the angular frequency and the wave vector, the phase and group velocities, and their ratio.

Solution
The angular frequency is $\omega = 2\pi\nu$ and the wave vector is $k = 2\pi/\lambda$; we have

$$\omega = \sqrt{\frac{\gamma}{\rho}} \left(\frac{2\pi}{\lambda}\right)^{3/2}.$$

The dispersion relation is

$$\omega(k) = \sqrt{\frac{\gamma}{\rho}} \, k^{3/2},$$

the group velocity is

$$v_g = \frac{d\omega}{dk} = \frac{3}{2}\sqrt{\frac{\gamma}{\rho}} \, k^{1/2}, \tag{3.8}$$

and the phase velocity is

$$v_p = \frac{\omega}{k} = \lambda\nu = \sqrt{\frac{\gamma}{\rho}} \, k^{1/2}. \tag{3.9}$$

The ratio between group and phase velocity is $v_g/v_p = 3/2$ (the situation in which $v_g > v_p$ is called *anomalous dispersion*), and shorter waves propagate faster.

6 **(A)** Waves propagating in deep water are dominated by gravity. The quantities relevant to the propagation of gravity waves should be

the acceleration of gravity g and the density ρ of the fluid. Using dimensional considerations, derive an approximate dispersion relation $\omega = \omega(k)$ for deep-water waves.

Solution
The dimensions of g and ρ are $[g] = [LT^{-2}]$ and $[\rho] = [ML^{-3}]$, while the dimensions of angular frequency and wave vector are $[\omega] = [T^{-1}]$, $[k] = [L^{-1}]$. By requiring that

$$\omega = A g^\alpha \rho^\beta k^\gamma,$$

where A is a dimensionless coefficient, we obtain

$$[T^{-1}] = [L^\alpha T^{-2\alpha}] [M^\beta L^{-3\beta}] [L^{-\gamma}] = [L^{\alpha - 3\beta - \gamma} T^{-2\alpha} M^\beta].$$

The linear system

$$\alpha - 3\beta - \gamma = 0, \tag{3.10}$$

$$\beta = 0, \tag{3.11}$$

$$-2\alpha = -1, \tag{3.12}$$

must be satisfied, and its solutions are $\alpha = \gamma = 1/2$, $\beta = 0$. The approximate dispersion relation is

$$\omega(k) = A \sqrt{g}\, k^{1/2};$$

with no dependence on the density ρ. The exact dispersion relation corresponds to $A = 1$, which is used in the following problem.

7 **(B)** Consider the dispersion relation for linear monochromatic water waves

$$\omega(k) = \left[\left(g + \frac{\gamma k^2}{\rho} \right) k \, \mathrm{tgh}(kh) \right]^{1/2},$$

where g is the acceleration of gravity, γ is the surface tension, h is the water's depth, and $k = 2\pi/\lambda$ is the wave vector. Derive approximate dispersion relations
a) for surface waves in deep-water
b) for gravity waves in deep-water.
Discuss the meaning of "deep" and "shallow" water.

Solution

"Deep" or "shallow" water are concepts relative to the wavelength of the waves: deep-water means that $h \gg \lambda$, or $kh \gg 1$; shallow water means that $h \ll \lambda$, or $kh \ll 1$.

In the deep-water approximation $kh \gg 1$, we have

$$\text{tgh}(kh) = \frac{e^{kh} - e^{-kh}}{e^{kh} + e^{-kh}} \simeq 1;$$

the dependence on the depth h disappears from the dispersion relation and

$$\omega^2 \simeq \left(g + \frac{\gamma k^2}{\rho}\right) k. \tag{3.13}$$

If the first term in Eq. (3.13) dominates, we have *gravity waves* while if the second term dominates we have *surface waves* instead. The two terms have equal magnitudes when $g = \gamma k^2/\rho$, or when

$$\lambda = \lambda^* \equiv 2\pi \sqrt{\frac{\gamma}{g\rho}}.$$

a) For *surface waves* in deep-water, the term $\gamma k^2/\rho$ containing the surface tension dominates over g and we have the dispersion relation

$$\omega(k) = \sqrt{\frac{\gamma}{\rho}}\, k^{3/2},$$

and short waves travel faster than long waves.
b) For *gravity waves* in the deep-water approximation $g > \gamma k^2/\rho$, and we have instead the dispersion relation

$$\omega(k) = \sqrt{g}\, k^{1/2},$$

and long waves travel faster than short waves.

8 **(B)** The dispersion relation for linear water waves with wavelength $\lambda > 5$ cm (which allows surface tension effects to be neglected) is

$$\omega(k) = [gk \tanh (kh)]^{1/2},$$

where g is the acceleration of gravity and h is the water depth, respectively. What are the phase and group speeds?

Solution
The phase speed is

$$v_p \equiv \frac{\omega}{k} = \left[\frac{g}{k} \tanh (kh)\right]^{1/2},$$

while the group speed is

$$
v_g \equiv \frac{d\omega}{dk} = \sqrt{g}\,\frac{\tanh{(kh)} + \frac{kh}{\cosh^2{(kh)}}}{2\sqrt{k}\tanh{(kh)}}
$$

$$
= \frac{1}{2}\sqrt{\frac{g}{k}\tanh{(kh)}}\left[1 + \frac{2kh}{\sinh{(2kh)}}\right]
$$

$$
= \frac{v_p}{2}\left[1 + \frac{2kh}{\sinh{(2kh)}}\right].
$$

9 **(B)** A storm far out at sea generates a swell with typical wavelengths λ of order 100 m. What is the phase velocity? What is the group velocity? Will short or long waves announce the storm to a boat out on the sea? How long does it take for 100 m waves to travel the 100 km distance between the storm and the boat? The density of water is $\rho = 1.0 \cdot 10^3\,\text{kg/m}^3$ and the surface tension is $\gamma \simeq 7.28 \cdot 10^{-2}\,\text{N/m}$ at 20°C. Use the fact that the general dispersion relation for water waves is

$$
\omega = \left[\left(g + \frac{\gamma k^2}{\rho}\right)k\,\text{tgh}(kh)\right]^{1/2},
$$

where g is the acceleration of gravity and h is the water's depth.

Solution
Far out at sea, the depth is much larger than the typical wavelength ($h \gg 100$ m) and one can use the approximation $\text{tgh}(kh) \simeq 1$ for $kh \gg 1$, obtaining

$$
\omega(k) \simeq \left[\left(g + \frac{\gamma k^2}{\rho}\right)k\right]^{1/2}.
$$

The g-term and the γ-term in the dispersion relation have equal magnitudes when

$$
\lambda = \lambda^* \equiv 2\pi\sqrt{\frac{\gamma}{g\rho}};
$$

from the given values of ρ and γ one derives $\lambda^* = 1.7$ cm. The gravity term dominates the dispersion relation for $\lambda \simeq 100\,\text{m} \gg \lambda^*$, hence the dispersion relation for the long waves generated in the storm is

$$
\omega(k) = \sqrt{g}\,k^{1/2}.
$$

The phase velocity is $v_p = \omega/k = \sqrt{g}\,k^{-1/2}$ and the group velocity is

$$
v_g = \frac{d\omega}{dk} = \frac{\sqrt{g}}{2}\,k^{-1/2} = \frac{1}{2}\,v_p.
$$

We have $v_p = 2v_g$ ($v_p > v_g$ is called *normal dispersion*); since $v_p = 2v_g \propto \sqrt{\lambda}$, long waves travel faster than short ones and will bring the news of the storm to the boat—the energy of the waves travels with the group velocity v_g. For waves of wavelength $\lambda = 100$ m the phase velocity is

$$v_p = \left[\frac{(9.81 \, \text{m} \cdot \text{s}^{-2}) \cdot (100 \, \text{m})}{2\pi}\right]^{1/2} \simeq 12.5 \, \frac{\text{m}}{\text{s}} \simeq 45 \, \frac{\text{km}}{\text{h}}.$$

The waves cover the distance $L = 100$ km between the storm and the boat in the time

$$t = \frac{L}{v_p} = \frac{10^5 \, \text{m}}{12.5 \, \text{m} \cdot \text{s}^{-1}} = 8 \cdot 10^3 \, \text{s} = 2.2 \, \text{hours}.$$

10 **(B)** When two layers of water with different densities ρ and ρ' are in contact, deep-water waves propagate along the interface with group velocity

$$v_g = \frac{1}{2}\left[\frac{g}{k}\frac{\rho - \rho'}{\rho + \rho'}\right]^{1/2},$$

where k is the wave vector and g is the acceleration of gravity. Find the dispersion relation and the phase velocity for these waves.

Solution
The group velocity is $v_g = d\omega/dk$, where ω is the angular frequency, and the dispersion relation $\omega = \omega(k)$ is obtained integrating v_g with respect to k,

$$\omega(k) = \int dk \, v_g(k) = \sqrt{g\frac{\rho - \rho'}{\rho + \rho'}}\int \frac{dk}{2\sqrt{k}} = \sqrt{gk\frac{\rho - \rho'}{\rho + \rho'}}.$$

The phase velocity is

$$v_p(k) \equiv \frac{\omega}{k} = \sqrt{\frac{g}{k}\frac{\rho - \rho'}{\rho + \rho'}}.$$

3.4   General features of the atmosphere

The lowest 10 kilometers of the atmosphere—the troposphere—contain most of the mass of air around the planet and are the arena for weather-related phenomena. The power from the Sun drives weather and climate, and the water vapor present in the troposphere is responsible for phenomena such as clouds and precipitation, and concurs in determining the stability or instability of air masses. This section reviews basic properties of the atmosphere.

1 **(A)** Given equal conditions of temperature, volume, and pressure, is moist air heavier or lighter than dry air? Is this a reason why water vapor is only found at the bottom of the atmosphere?

Solution
Dry air is composed of approximately 75% of N_2 and 25% of O_2, with average molecular mass

$$m_{\text{dry}} = 0.75\, m_{N_2} + 0.25\, m_{O_2} = 2\,(0.75\, m_N + 0.25\, m_O)$$

$$= 2\,(0.75 \cdot 7 + 0.25 \cdot 8)\ \text{a.m.u.} = 14.5\,\text{a.m.u.}$$

The mass of water molecules (H_2O) is

$$m_{H_2O} = 2\,m_H + m_O = (2 \cdot 1 + 8)\ \text{a.m.u.} = 10\,\text{a.m.u.} < m_{\text{dry}}.$$

By replacing part of the dry air in a given volume with water vapor (i.e., by considering wet air instead of dry air) one obtains a lighter mixture. In spite of this, water vapor is mostly concentrated within the first two kilometers of the atmosphere because water vapor enters the atmosphere through the air–surface or air–water boundary layer.

2 **(B)** Approximately, what is the thickness of the atmosphere? Compare it with the radius of the Earth. Given that the average atmospheric pressure at sea level is $P_0 = 1.01 \cdot 10^5$ Pa, compute the average density of the atmosphere and its total mass.

How does pressure depend on the elevation z? How does the particle density change with z? State the barometric formula. Where does most of the atmosphere's mass reside? Compute the fraction η of the total mass of the atmosphere found in a layer between sea level ($z = 0$) and z.

Solution
The atmosphere is approximately $h = 100$ km thick. The ratio between the size of the atmosphere and the Earth's radius is

$$\frac{h}{R} = \frac{100\,\text{km}}{6370\,\text{km}} = 1.57 \cdot 10^{-2}.$$

The atmospheric pressure at sea level is due to a column of air with height $h = 100$ km and is given by the formula for hydrostatic pressure

$$P_0 = \bar{\rho}\, g\, h,$$

where

$$\bar{\rho} = \frac{P_0}{gh} = \frac{1.01 \cdot 10^5 \, \text{Pa}}{(9.81 \, \text{m} \cdot \text{s}^{-2}) \cdot (1.00 \cdot 10^5 \, \text{m})} = 0.10 \, \frac{\text{kg}}{\text{m}^3}$$

is the atmospheric average density. The total mass of the atmosphere is

$$
\begin{aligned}
M &= \bar{\rho} V = \bar{\rho} \left(4\pi R_E^2 \right) h \\[2mm]
&= \left(0.103 \, \text{kg} \cdot \text{m}^{-3} \right) \cdot 4\pi \cdot \left(6.37 \cdot 10^6 \, \text{m} \right)^2 \cdot \left(1.00 \cdot 10^5 \, \text{m} \right) \\[2mm]
&= 5.25 \cdot 10^{18} \, \text{kg}.
\end{aligned}
$$

Over spatial scales sufficiently small that the temperature can be approximated by an average temperature $\bar{T}$, the pressure decreases exponentially with the elevation z,

$$P(z) = P_0 \, e^{-z/H},$$

where

$$H = \frac{kT}{mg}$$

is a length scale characteristic of the atmosphere and m is the mass of the average atmospheric gas molecule.

By treating the mixture of atmospheric gases as an ideal gas, the number density of particles also decreases exponentially with elevation,

$$n_d(z) = n_d^{(0)} \, e^{-z/H}$$

(*barometric formula*), where $n_d^{(0)} = P_0/(kT) = n_d \, (z = 0)$. In fact, the ideal gas law

$$PV = NkT,$$

where N is the number of gas particles in the volume V, yields $P = n_d kT$, where $n_d \equiv N/V$ is the particle number density and

$$n_d(z) = \frac{P(z)}{kT} = n_d^{(0)} \, e^{-z/H}.$$

The atmospheric density is then given by

$$\rho(z) = \rho_0 \, e^{-z/H},$$

where $\rho_0 = m \, n_d^{(0)} = m P_0/(kT) = \rho(z = 0)$.

The mass contained in a layer of the atmosphere between sea level
($z = 0$) and the elevation z is

$$
M(z) = \int_0^{2\pi} d\varphi \int_0^{\pi} d\theta \int_R^{R+z} dr\, r^2 \sin\theta\, \rho
$$

$$
= 4\pi \int_R^{R+z} dr\, r^2 \rho_0\, e^{-(r-R)/H}
$$

$$
= 4\pi \rho_0\, e^{R/H} \int_R^{R+z} dr\, r^2 e^{-r/H}
$$

$$
= 4\pi \rho_0 H^3\, e^{R/H} \int_{R/H}^{(R+z)/H} dx\, x^2\, e^{-x},
$$

where $x \equiv r/H$ and the elevation is $z = r - R$. Integrating by parts,
we obtain the integral

$$
\int dx\, x^2\, e^{-x} = -e^{-x}\left(x^2 + 2x + 2\right),
$$

and

$$
M(z) = 4\pi \rho_0 \left[\left(1 - e^{-z/H}\right)\left(R^2 H + 2RH^2 + 2H^3\right)\right.
$$
$$
\left. - zH\, e^{-z/H}\left(z + 2H + 2R\right)\right].
$$

The fraction of mass contained in the shell $R \le r \le R + z$ is then

$$
\eta \equiv \frac{M(z)}{M} = 4\pi \frac{\rho_0}{M}\left[\left(1 - e^{-z/H}\right)\left(R^2 H + 2RH^2 + 2H^3\right)\right.
$$
$$
\left. - zH\, e^{-z/H}\left(z + 2H + 2R\right)\right].
$$

In order to compute the elevation z such that a given fraction η of
the atmosphere resides in the shell $(R, R+z)$, we must invert the
previous equation numerically.

3 **(C)** What is, approximately, the composition of the atmosphere?
What would happen to human and animal life if the percentages of
nitrogen and oxygen were suddenly reversed?

Solution
The atmosphere consists of approximately 78% of N_2 by volume, 21%
of O_2, and 1% of argon. In addition to this gas mixture (*dry air*),

water vapor is present in the bottom part of the atmosphere (mainly below 2 km) together with CO_2 and particles of dust and pollution. CO_2 is the main greenhouse gas and is an essential component of plant and animal life on the planet. Water vapor is the source of clouds and precipitation and is a very effective absorber of infrared radiation radiated by the Earth, and of some radiation from the Sun.

From the point of view of human and animal life, the oxygen contained in the atmosphere is conveniently diluted by nitrogen. If the proportions of N_2 and O_2 were suddenly reversed, the breathing apparatus of virtually all forms of aerobic life would be burned by the oxygen excess.

4 (A) Compute the escape speed v_e at the top of the atmosphere, $z \simeq 110$ km; at what temperature is the thermal speed of hydrogen molecules comparable with v_e? Compare the result with the escape speed from the surface of the Moon. The Earth has mass $M_E = 5.98 \cdot 10^{24}$ kg and radius $R_E = 6370$ km, and the Moon has mass $M_M = 7.35 \cdot 10^{22}$ kg and radius $R_M = 1740$ km.

Solution
The escape speed at a distance r from the center of a planet is

$$v_e = \left(\frac{2GM}{r}\right)^{1/2} = \left(\frac{2GM}{R+h}\right)^{1/2},$$

where R is the radius of the planet and h the elevation above its surface in the radial direction. The escape speed at $z = 110$ km is

$$v_e = \left[\frac{2 \cdot (6.67 \cdot 10^{-11}\,\text{N} \cdot \text{m}^2 \cdot \text{kg}^{-2}) \cdot (5.98 \cdot 10^{24}\,\text{kg})}{(6370 + 110) \cdot (10^3\,\text{m})}\right]^{1/2}$$

$$= 11.1\,\text{km} \cdot \text{s}^{-1}.$$

The temperature at which the thermal speed of hydrogen molecules equals the escape speed is given by

$$\frac{1}{2} m v_e^2 = \frac{3}{2} kT,$$

which yields

$$T = \frac{m\,v_e^2}{3k},$$

where $m = 2\,\text{a.m.u.} = 3.32 \cdot 10^{-27}$ kg is the mass of the hydrogen molecule H_2. Numerically,

$$T = \frac{\left(3.32 \cdot 10^{-27}\,\text{kg}\right) \cdot \left(1.11 \cdot 10^4\,\text{m} \cdot \text{s}^{-1}\right)^2}{3 \cdot \left(1.38 \cdot 10^{-23}\,\text{J} \cdot \text{K}^{-1}\right)} = 9.88 \cdot 10^3\,\text{K}.$$

Since this temperature is much higher than the average atmospheric temperature the Earth retains its atmosphere mainly composed of nitrogen (N_2) and oxygen (O_2) molecules, which are heavier and slower than H_2.

The escape speed from the surface of the Moon is only

$$\begin{aligned}
v_e &= \left(\frac{2GM_M}{R_M}\right)^{1/2} \\
&= \left[\frac{2 \cdot \left(6.67 \cdot 10^{-11}\,\text{N} \cdot \text{m}^2 \cdot \text{kg}^{-2}\right) \cdot \left(7.35 \cdot 10^{22}\,\text{kg}\right)}{1.74 \cdot 10^6\,\text{m}}\right]^{1/2} \\
&= 2.37\,\text{km} \cdot \text{s}^{-1},
\end{aligned}$$

due to the weaker gravity of the Moon. The temperature needed for hydrogen molecules to escape from the surface of the Moon is

$$T = \frac{m\,v_e^2}{3k} = \frac{\left(3.32 \cdot 10^{-27}\,\text{kg}\right) \cdot \left(2.37 \cdot 10^3\,\text{m} \cdot \text{s}^{-1}\right)^2}{3 \cdot \left(1.38 \cdot 10^{-23}\,\text{J} \cdot \text{K}^{-1}\right)} = 450\,\text{K};$$

most hydrogen molecules—not only those in the tails of the Maxwell–Boltzmann distribution—can escape. These figures explain why the Moon lost its atmosphere while the Earth did not.

5 **(A)** The atmospheric pressure at sea level is $P_0 = 1$ atm$= 1.01 \cdot 10^5$ Pa. What is the *average* density of the atmosphere, given that it extends to an elevation of 100 km? Compare it with the density at sea level, $\rho_0 = 1.29\,\text{kg} \cdot \text{m}^{-3}$. What is the height of the equivalent water column, i.e., the column of water that causes the same pressure?

Solution
The pressure at depth h in a fluid of density ρ is $P = \rho g h$, hence the average atmospheric density is

$$\bar{\rho} = \frac{P}{gh} = \frac{1.01 \cdot 10^5\,\text{Pa}}{\left(9.81\,\text{m} \cdot \text{s}^{-2}\right) \cdot \left(1.0 \cdot 10^5 \text{m}\right)} = 0.1\,\text{kg} \cdot \text{m}^{-3}.$$

The ratio between the average atmospheric density and the density at sea level is

$$\frac{\bar{\rho}}{\rho_0} = \frac{0.1 \, \text{kg} \cdot \text{m}^{-3}}{1.29 \, \text{kg} \cdot \text{m}^{-3}} = 7.8 \cdot 10^{-2}.$$

A water column with the same pressure must satisfy $P = \rho_{\text{water}} \, g \, z = P_0$, that gives

$$z = \frac{P_0}{\rho_{\text{water}} \, g} = \frac{(1.01 \cdot 10^5 \, \text{Pa})}{(1.0 \cdot 10^3 \, \text{kg/m}^3) \cdot (9.81 \, \text{m/s}^2)} = 10.3 \, \text{m}.$$

It is a rule of thumb among scuba divers that pressure increases by approximately one atmosphere every 10 meters of depth.

3.5   Temperature and pressure

Temperature and pressure are two basic physical observables that determine the dynamics of the atmosphere and are used to describe it. To these, we must add the water vapor content of atmospheric air and the velocity field of air parcels.

1 **(A)** Using dimensional considerations, derive the vertical temperature gradient of dry atmospheric air as a function of the acceleration of gravity g and of the specific heat at constant pressure c_P.

Solution

The dimensions of the vertical temperature gradient are $[\partial T/\partial z] = [K][L^{-1}]$ where K is the (Kelvin) temperature. The dimensions of g are $[g] = [LT^{-2}]$ and the dimensions of c_P are obtained from the formula $\delta Q = c_P \, m \, \delta T$, where δQ is the heat supplied to a mass m to change its temperature by the amount δT. Hence, $[c_P] = [L^2 T^{-2} K^{-1}]$. By imposing that

$$\frac{\partial T}{\partial z} = A \, g^\alpha c_P^\beta,$$

where A is a dimensionless coefficient, we have

$$\left[\frac{\partial T}{\partial z}\right] = [g^\alpha] \left[c_P^\beta\right],$$

or

$$[K][L^{-1}] = [L^\alpha T^{-2\alpha}] \left[L^{2\beta} T^{-2\beta} K^{-\beta}\right],$$

and we obtain the equations

$$\begin{aligned}
-\beta &= 1, \\
\alpha + 2\beta &= -1, \\
-2\alpha - 2\beta &= 0,
\end{aligned}$$

with the solutions $\alpha = 1$ and $\beta = -1$; therefore, $\partial T/\partial z = Ag/c_P$. The exact expression is

$$\frac{\partial T}{\partial z} = -\frac{g}{c_P} \equiv -\Gamma_d,$$

where $\Gamma_d \equiv g/c_P$ is called the *dry adiabatic lapse rate*.

2 **(A)** You are planning to climb to the top of Mt. Robson (m. 3954) in the Canadian Rockies from a camp at 1000 m of elevation where the temperature is 5.0°C. What temperature should you expect at the top? Consider dry air and assume that the variation of the heat coefficient at constant pressure c_P with elevation is negligible. The dry adiabatic lapse rate is $\Gamma_d = 0.01$°C·m^{-1}. What is the physical meaning of a negative Γ_d?

Solution
The vertical temperature gradient is given by $\partial T/\partial z = -\Gamma_d$, where $\Gamma_d = g/c_P = \text{const.} = 0.01$°C·m^{-1} is the adiabatic lapse rate for dry air. Therefore, the temperature is a linear decreasing function of elevation, $T(z) = -\Gamma_d z + T_0$, where $T_0 = T(z=0)$. At camp (elevation z_1), $T(z_1) = -\Gamma_d z_1 + T_0$, while at the top $T(z_2) = -\Gamma_d z_2 + T_0$; hence

$$T(z_2) - T(z_1) = -\Gamma_d(z_2 - z_1)$$

and

$$T(z_2) = -(0.01°\text{C} \cdot \text{m}^{-1})(3954\,\text{m} - 1000\,\text{m}) + 5.0°\text{C} = -24.5°\text{C}$$

at the top of the mountain. In practice the temperature gradient often changes to assume a milder slope and the temperature at the top will be somehow higher than -25°C. This happens because air is never dry and as moisture rises it condenses, releasing latent heat that raises the temperature. Remember that the latent heat of evaporation of water, $2.26 \cdot 10^6$ J/kg is unusually high.

If $\Gamma_d < 0$, then $\partial T/\partial z > 0$ and the temperature increases with the elevation. This phenomenon is known as *thermal inversion*.

3 **(A)** A mass of air rises along a mountain slope. Assuming that air is an ideal gas and that the expansion is sufficiently fast to be considered adiabatic, compute the temperature change and the final volume of the air mass by knowing the initial and final pressures P_i and P_f, the initial volume V_i, and the adiabatic index γ.

Solution

Let T_i, T_f, V_i, and V_f be the initial and final temperatures and volumes; the equation of an adiabatic transformation of an ideal gas is

$$PV^\gamma = \text{constant},$$

and therefore

$$P_i V_i^\gamma = P_f V_f^\gamma,$$

from which one deduces that

$$V_f = V_i \left(\frac{P_i}{P_f}\right)^{1/\gamma}.$$

The ideal gas equation of state yields

$$P_i V_i = nRT_i,$$

$$P_f V_f = nRT_f.$$

Division of these two equations term to term and the use of the expression for V_i then yield

$$T_f = T_i \frac{P_f}{P_i} \frac{V_f}{V_i} = T_i \frac{P_f}{P_i} \left(\frac{P_i}{P_f}\right)^{1/\gamma}$$

$$= T_i \left(\frac{P_f}{P_i}\right)^{\frac{\gamma-1}{\gamma}}.$$

3.6　　Atmospheric circulation

A complete treatment of atmospheric circulation is beyond the scope of this exercise book—we include a few problems that can be approached by students beginning their studies in environmental science.

1 **(A)** A tornado can be approximated by a cylinder with an 80m radius rotating with angular velocity $\omega = 2.5\,\text{rad} \cdot \text{s}^{-1}$. Estimate the wind speed at the outer edge of the tornado.

Solution

Assuming rigid rotation for simplicity, the linear velocity at the edge of the tornado has magnitude

$$v = \omega r = \left(2.5\,\frac{\text{rad}}{\text{s}}\right) \cdot (80\,\text{m}) = 200\,\frac{\text{m}}{\text{s}} = 720\,\frac{\text{km}}{\text{h}}.$$

2 (C) You are looking at a picture taken from space of a hurricane rotating clockwise. In what hemisphere is the hurricane?

Solution
In the Southern Hemisphere. Because of the Coriolis force, objects in motion including winds deviate to their left in the Southern Hemisphere. Hurricanes are created by air masses moving toward a low pressure center and deviated by the Coriolis force. Therefore, hurricanes rotate clockwise in the Southern Hemisphere. In the Northern Hemisphere objects are deviated to their right and hurricanes rotate counterclockwise instead.

3 (A) When there is large scale rotation in a thunderstorm a tornado can form. A *mesocyclone* (rotation on the scale of the whole thunderstorm updraft) has a spatial scale $R = 6.00$ km and is slowly rotating with angular velocity $w_m = 5.55 \cdot 10^{-4}$ rad $\cdot$ s^{-1}. It degenerates into a tornado of radius $r = 200$ m. Estimate the angular velocity of the tornado and the wind speed at its boundary.

Hint: Assume that the mass of air is conserved when the tornado forms and that the air mass is isolated.

Solution
The angular momentum is $L = \mathcal{I}w$, where $\mathcal{I} = \alpha M l^2$ is the moment of inertia of the system, M is the mass of the air, l the spatial scale of rotation, and α a dimensionless coefficient assumed to be constant during the formation of the tornado. In other words, it is assumed that the density profile inside the tornado is the same as in the initial mesocyclone. Since there are no torques acting on the air mass, its angular momentum is conserved, yielding

$$\alpha M R^2 w_m = \alpha M r^2 w_T, \qquad (3.14)$$

where w_T is the angular velocity of the tornado. Therefore, we obtain

$$w_T = w_m \left(\frac{R}{r}\right)^2 = \left(5.55 \cdot 10^{-4} \frac{\text{rad}}{\text{s}}\right)\left(\frac{6.00 \cdot 10^3 \, \text{m}}{2.00 \cdot 10^2 \, \text{m}}\right)^2 = 0.50 \frac{\text{rad}}{\text{s}}.$$

The linear speed at the edge of the tornado is

$$v = w_T \, r = \left(0.50 \frac{\text{rad}}{\text{s}}\right) \cdot (200 \, \text{m}) = 100 \frac{\text{m}}{\text{s}} = 360 \frac{\text{km}}{\text{h}}.$$

4 (B) The Himalayan mountains cause small perturbations in a jet stream that crosses them. These perturbations have wavelike char-

acter (*Rossby* or *planetary* waves) and propagate in the direction opposite to the jet stream with wavelengths comparable to the Earth's radius. The group velocity of Rossby waves is given by

$$v_g = -\frac{2\Omega \cos\theta}{R k^2},$$

where Ω and R are the Earth's angular velocity and radius, θ is the latitude, and the negative sign describes the fact that Rossby waves propagate in the direction opposite to the jet stream flow. Derive the dispersion relation $\omega = \omega(k)$ for planetary waves. Do longer or shorter Rossby waves propagate faster?

Solution
The group velocity of Rossby waves is

$$v_g = \frac{d\omega}{dk} = -\frac{\alpha}{k^2},$$

where $\alpha = (2\Omega \cos\theta)/R$. Integration with respect to k gives the dispersion relation

$$\omega(k) = \frac{\alpha}{k} = \frac{2\Omega \cos\theta}{R k}.$$

Since $v_g \propto \lambda^2$, longer waves propagate faster.

3.7   Precipitation

The purpose of this section is to make the student realize how much physical intuition and a knowledge of basic physics can help in understanding the physical properties of the natural world.

1 **(C)** Why do cross sections of hailstones often show approximately concentric layers of ice in them?

Solution
Because when hailstones form they do not immediately fall to the ground but are carried upward by winds and undergo successive cycles of melting and freezing during their vertical motions.

2 **(A)** Fog occurs because of natural condensation—minuscule droplets of water form because of the condensation of atmospheric water vapor caused by a temperature drop, a process assisted by the presence of condensation nuclei. Water droplets are suspended in air. Compute the terminal speed of a spherical droplet of radius $r = 1.0 \cdot 10^{-3}$ cm, given that the air (dynamic) viscosity coefficient is

$\eta = 1.0 \cdot 10^{-5}\,\mathrm{kg \cdot m^{-1} \cdot s^{-1}}$, its density is $\rho' = 1.29\,\mathrm{kg/m^3}$, and the density of water is $\rho = 1.0 \cdot 10^3\,\mathrm{kg/m^3}$.

Solution

The air drag on a spherical object moving in a viscous fluid at moderate speed v is given by Stokes' law

$$F_d = 6\pi\eta r v.$$

The terminal speed is reached when the air drag and the buoyant force balance the weight of the sphere and the resulting acceleration of the droplet vanishes,

$$mg - F_d - F_b = 0,$$

or

$$\rho \frac{4\pi}{3} r^3 g = 6\pi\eta r v + \rho' \frac{4\pi}{3} r^3 g,$$

yielding the terminal speed

$$v = \frac{2}{9} \frac{r^2 g}{\eta} (\rho - \rho').$$

Numerically,

$$v = \frac{2}{9} \frac{(1.0 \cdot 10^{-5}\,\mathrm{m})^2 \cdot (9.81\,\mathrm{m \cdot s^{-2}})}{(1.0 \cdot 10^{-5}\,\mathrm{kg \cdot m^{-1} \cdot s^{-1}})}$$

$$\cdot (1.0 \cdot 10^3\,\mathrm{kg \cdot m^{-3}} - 1.29\,\mathrm{kg \cdot m^{-3}}) = 2.2 \cdot 10^{-2}\,\frac{\mathrm{m}}{\mathrm{s}}.$$

3 **(B)** Find the velocity $v(t)$ as a function of time for a raindrop of mass m falling in air in the case in which the friction force is proportional to the square of the drop's velocity, $F_{\mathrm{friction}} = -\epsilon v^2$. Assume that the raindrop starts with zero initial velocity, and neglect the buoyant force, which is numerically irrelevant at larger speeds. Find the time scale over which the terminal speed is approached by knowing that the density of water is $1.0 \cdot 10^3\,\mathrm{kg/m^3}$, the spherical raindrop has radius 1.0 mm, and $\epsilon = C_D \rho' A/2$, where the drag coefficient $C_D = 0.5$, $\rho' = 1.29\,\mathrm{kg/m^3}$ is the air density, and A is the cross section presented to the air.

Solution

Neglecting the buoyant force, the equation of motion of the raindrop is

$$ma = F_{\mathrm{gravity}} + F_{\mathrm{friction}}$$

or

$$m \frac{d^2y}{dt^2} = mg - \epsilon v^2,$$

where g is the acceleration of gravity, which is assumed to be constant over the scale of distances interesting for rainfall, y is a vertical coordinate pointing downward, and $v(t) = dy/dt$. One can write

$$\frac{dv}{dt} + \frac{\epsilon}{m} v^2 - g = 0, \tag{3.15}$$

which is recognized to be a Riccati equation [32] of the form

$$\frac{dv}{dt} + \alpha v^2 + \beta = 0$$

with $\alpha = \epsilon/m$ and $\beta = -g$. The solution is obtained by using the auxiliary variable u defined by

$$v \equiv \frac{1}{\alpha u} \frac{du}{dt}. \tag{3.16}$$

Since

$$\frac{dv}{dt} = \frac{1}{\alpha u} \frac{d^2u}{dt^2} - \frac{1}{\alpha u^2} \left(\frac{du}{dt} \right)^2,$$

substitution of Eq. (3.16) into Eq. (3.15) yields

$$\frac{d^2u}{dt^2} + \alpha \beta \, u = 0$$

with $\alpha \beta = -g\epsilon/m < 0$. The general solution of this equation is

$$u(t) = u_1 \exp \left(\sqrt{|\alpha\beta|}\, t \right) + u_2 \exp \left(-\sqrt{|\alpha\beta|}\, t \right)$$

and the velocity is

$$v(t) = \frac{1}{\alpha u} \frac{du}{dt} = \sqrt{\left| \frac{\beta}{\alpha} \right|} \frac{u_1 \, e^{\sqrt{|\alpha\beta|}\, t} - u_2 \, e^{-\sqrt{|\alpha\beta|}\, t}}{u_1 \, e^{\sqrt{|\alpha\beta|}\, t} + u_2 \, e^{-\sqrt{|\alpha\beta|}\, t}},$$

where u_1 and u_2 are integration constants. The initial condition imposed by the problem is

$$v(0) = 0,$$

which yields the relation between the integration constants $u_1 = u_2$. Hence, the solution for the velocity can be written as

$$v(t) = \sqrt{\frac{gm}{\epsilon}} \tanh \left(\sqrt{\frac{g\epsilon}{m}}\, t \right).$$

The velocity $v(t)$ of the raindrop describes a substantial acceleration stage only during an initial transient and quickly approaches the asymptotic value $\sqrt{gm/\epsilon}$ on the time scale $\gamma = \sqrt{m/(g\epsilon)}$.

The time scale γ can be written as

$$\gamma = \sqrt{\frac{m}{g\epsilon}} = \left[\frac{(4\pi r^3/3)\rho}{g(C_D/2)\rho'\pi r^2}\right]^{1/2} = \left(\frac{8r}{3C_Dg}\frac{\rho}{\rho'}\right)^{1/2}$$

$$= \left[\frac{8 \cdot (1.0 \cdot 10^{-3}\,\text{m})}{3 \cdot 0.5 \cdot (9.81\,\text{m} \cdot \text{s}^{-2})}\frac{1.0 \cdot 10^3\,\text{kg} \cdot \text{m}^{-3}}{1.29\,\text{kg} \cdot \text{m}^{-3}}\right]^{1/2} = 0.65\,\text{s}.$$

4 (A) A spherical raindrop of radius 1.5 mm falls in air with velocity v. In the reference frame in which the drop is at rest, the fluid flow is turbulent. Assuming that the Reynolds number is $N_R = 10$ (the threshold for the onset of turbulence in a fluid flow with obstacles), estimate the vertical velocity of the drop. The dynamic viscosity coefficient of air is $\eta = 1.8 \cdot 10^{-5}\,\text{Pa} \cdot \text{s}$ and its density is $\rho = 1.29\,\text{kg/m}^3$.

Solution
The Reynolds number appropriate for fluid flow in the presence of obstacles is the dimensionless quantity

$$N_R = \frac{\rho v d}{\eta},$$

where d is the transversal size of the obstacle (the diameter of the sphere). Hence,

$$v = \frac{N_R \eta}{\rho d} = \frac{10 \cdot (1.8 \cdot 10^{-5}\,\text{Pa} \cdot \text{s})}{(1.29\,\text{kg} \cdot \text{m}^{-3}) \cdot (3.0 \cdot 10^{-3}\,\text{m})} = 4.7 \cdot 10^{-2}\,\frac{\text{m}}{\text{s}}.$$

5 (A) Discuss the air drag on an object moving rapidly in a fluid for which the buoyant force is negligible compared to the air drag. Compute the terminal velocity of a spherical raindrop with radius of 1 mm assuming that the drag coefficient is $C_D = 0.41$, the air density is $\rho' = 1.29\,\text{kg/m}^3$, and the density of water is $\rho = 1.0 \cdot 10^3\,\text{kg/m}^3$.

Solution
For an object moving rapidly in a fluid, the buoyant force is negligible in comparison to the air drag. The latter is proportional to the square of the velocity of the object relative to the fluid,

$$F_D = b v^2,$$

where the coefficient b is given by

$$b = \frac{1}{2} C_D \rho A;$$

ρ is the density of the fluid and A is the area of the object normal to the flow. The *drag coefficient* C_D depends on the shape of the moving body.

For a spherical raindrop falling with terminal velocity, the weight equals the air drag force,

$$F_D = mg,$$

or

$$\left(\frac{1}{2} C_D \rho'\right) \cdot \left(\pi r^2\right) v^2 = \frac{4\pi}{3} r^3 \rho g,$$

where ρ' is the air density. Therefore,

$$
\begin{aligned}
v &= \left(\frac{8r}{3C_D} \frac{\rho}{\rho'} g\right)^{1/2} \\
&= \left[\frac{8 \cdot (1.0 \cdot 10^{-3}\,\text{m})}{3 \cdot 0.41} \frac{\left(1.0 \cdot 10^3\,\text{kg} \cdot \text{m}^{-3}\right)}{(1.29\,\text{kg/m}^3)} \cdot \left(9.81\,\text{m} \cdot \text{s}^{-2}\right)\right]^{1/2} \\
&= 7.0 \frac{\text{m}}{\text{s}}.
\end{aligned}
$$

6 **(A)** Compute the latent heat of vaporization L_V of water at 10°C, 30°C, and 60°C. Is the temperature dependence of L_V a factor in explaining the abundant precipitation in tropical regions?
Hint: Use the Regnault equation.

Solution
The temperature dependence of the latent heat of vaporization of water is given by the empirical Regnault equation

$$L_V = \left[606.5 - 0.695 \left(\frac{T}{1°\text{C}}\right)\right] \frac{\text{Kcal}}{\text{kg}},$$

an equation valid for temperatures above a certain threshold, $T \geq T_c$. The latent heats at 10°C, 30°C, and 60°C are, respectively,

$$L_V^{(1)} = 600 \frac{\text{Kcal}}{\text{kg}},$$

$$L_V^{(2)} = 586 \, \frac{\text{Kcal}}{\text{kg}},$$

$$L_V^{(3)} = 565 \, \frac{\text{Kcal}}{\text{kg}}.$$

L_V decreases linearly with the temperature (for $T \geq T_c$). In tropical regions a large amount of solar energy is available to evaporate water, and this is the main factor explaining the abundant precipitation. However, the comparatively low value of L_V at the relatively high temperatures found in these regions also facilitates the evaporation of even larger masses of water.

Chapter 4

ELECTROMAGNETIC RADIATION AND RADIOACTIVITY

There is something fascinating about science. One gets such wholesale return of conjecture out of such trifling investments of facts.
—Mark Twain, *Life on the Mississippi*

The electromagnetic force is one of the four fundamental forces, the others being the gravitational force, the strong nuclear force, and the weak interaction. The electromagnetic force dominates on the scale of atoms and molecules, determining the chemical properties of the elements and compounds, the chemical reactions occurring on the planet, and the mechanical, thermal, electrical, and optical properties of materials. Although the electromagnetic interaction has infinite range like gravity, in practice the electromagnetic force is negligible on large scales where gravity dominates instead. This is due to the fact that there are two sources of electric forces—positive and negative charges—which can cancel each other, as opposed to the case of gravity, which is only generated by positive masses. On the larger scales there is no net electric charge.

Electromagnetic radiation is emitted in the form of continuous spectra by accelerated charges and in the form of discrete or band spectra by atoms, molecules, and nuclei, or by electrons in solids. Emission and absorption processes related to discrete spectra are described by quantum mechanics and the emitted radiation carries the fingerprint of the emitting atom or molecule. For this reason spectroscopy is an invaluable tool to identify the presence of even small amounts of different chemicals, and spectroscopic techniques are widely used as tools in envi-

ronmental physics to monitor, for example, the presence and abundance
of pollutants, greenhouse gases, and of gases destroying the ozone layer.

As general references for the material in this chapter we suggest [59,
22] while Ref. [37] is the professional reference on electromagnetism.
References [4, 68] are useful for environmental spectroscopy, while [61,
41, 49, 23] are standard references on quantum mechanics.

4.1   The electromagnetic spectrum

Electromagnetic waves cover a semi-infinite range of frequencies and
wavelengths ranging from low-frequency (long wavelength) radio waves
to the infrared, to the rather narrow optical band that we call "light" in
everyday language (or, more properly, "visible" band), to the ultraviolet, and to the X and γ radiation corresponding to the highest frequencies. Electromagnetic waves travel at the speed of light, reflect, refract,
diffract, and scatter, and these phenomena are frequency-dependent.
The color of objects is explained by selective absorption and scattering.
Many of the following exercises focus on the propagation of electromagnetic waves through an absorbing medium.

1 **(C)** How is the electromagnetic spectrum divided into wavelength
bands?

Solution
At the low-frequency end of the spectrum there are extremely long
waves (usually classified as noise) with wavelengths $\lambda > 10^3$ m; then,
in order of decreasing wavelength, corresponding to increasing frequency and energy, there are
— radio waves with $0.1\,\text{m} < \lambda < 10^3\,\text{m}$
— microwaves with $10^{-4}\,\text{m} < \lambda < 0.1\,\text{m}$
— infrared radiation with $7 \cdot 10^{-7}\,\text{m} < \lambda < 10^{-4}\,\text{m}$
— visible light with $4 \cdot 10^{-7}\,\text{m} < \lambda < 7 \cdot 10^{-7}\,\text{m} = 700\,\text{nm}$
— ultraviolet light with $10^{-8}\,\text{m} < \lambda < 4 \cdot 10^{-7}\,\text{m} = 400\,\text{nm}$
— X-rays with $10^{-11}\,\text{m} < \lambda < 10^{-8}\,\text{m} = 10\,\text{nm} = 100\,\text{Å}$
— γ-rays with $\lambda < 10^{-11}\,\text{m} = 10^{-2}\,\text{nm} = 0.1\,\text{Å}$.

2 **(A)** The energy of electromagnetic radiation is often measured in eV
and the inverse wavelength in cm^{-1}. Show that $1\,\text{cm}^{-1}$ is equivalent
to $1.24 \cdot 10^{-4}$ eV.

Solution
The relation between frequency ν and wavelength λ of electromag-

netic radiation $c = \lambda\nu$ (where c is the speed of light in vacuum), combined with the relation between energy and frequency of a photon $E = h\nu$, yields $E = hc/\lambda$. If $\lambda = 1$ cm,

$$E = \frac{hc}{\lambda} = \frac{(6.625 \cdot 10^{-34}\,\text{J} \cdot \text{s}) \cdot (2.998 \cdot 10^8\,\text{m/s})}{1.00 \cdot 10^{-2}\,\text{m}} = 1.986 \cdot 10^{-23}\,\text{J}$$

$$= 1.24 \cdot 10^{-4}\,\text{eV},$$

where the conversion factor 1 eV$= 1.60 \cdot 10^{-19}$ J is used.

3 **(B, C)** The absorption of electromagnetic radiation by chlorophyll a and chlorophyll b present in green plants peaks at wavelengths $\lambda_a = 6.80 \cdot 10^{-7}$ m and $\lambda_b = 6.44 \cdot 10^{-7}$ m, respectively. In what region of the electromagnetic spectrum do these wavelengths fall? Is this fact related to the color of chlorophyll?

Photosynthesis of glucose ($C_6 H_{12} O_6$) can be summarized in the net reaction

$$6\,CO_2 + 6\,H_2 O \longrightarrow C_6 H_{12} O_6 + 6\,O_2.$$

The total energy needed to make one CO_2 molecule to react is $E = 2.34 \cdot 10^{-18}$ J $= 14.6$ eV. On average, how many photons must be absorbed by a CO_2 molecule in chlorophyll a to react? In chlorophyll b?

Solution
The absorption wavelengths λ_a and λ_b both fall in the red band of the visible spectrum, hence both chlorophyll a and b absorb red light and look green (the complementary color).
The energy of a photon at wavelength λ_a is

$$E_a = h\nu_a = \frac{hc}{\lambda_a} = \frac{(6.625 \cdot 10^{-34}\,\text{J} \cdot \text{s})\,(2.998 \cdot 10^8\,\text{m/s})}{6.80 \cdot 10^{-7}\,\text{m}}$$

$$= 2.92 \cdot 10^{-19}\,\text{J} = 1.83\,\text{eV},$$

while the energy of a photon at wavelength λ_b is

$$E_b = h\nu_b = \frac{hc}{\lambda_b} = \frac{(6.625 \cdot 10^{-34}\,\text{J} \cdot \text{s})\,(2.998 \cdot 10^8\,\text{m/s})}{6.44 \cdot 10^{-7}\,\text{m}}$$

$$= 3.08 \cdot 10^{-19}\,\text{J} = 1.93\,\text{eV}.$$

The average number of photons of frequency ν_a necessary to add up to the energy E is

$$\frac{E}{E_a} = \frac{2.34 \cdot 10^{-18}\,\text{J}}{2.92 \cdot 10^{-19}\,\text{J}} = 8,$$

while for the photons at frequency ν_b we have

$$\frac{E}{E_b} = \frac{2.34 \cdot 10^{-18}\,\mathrm{J}}{3.08 \cdot 10^{-19}\,\mathrm{J}} = 7.60.$$

4.2   Blackbody radiation

The spectrum of electromagnetic radiation emitted by blackbodies was the subject of investigation in theoretical physics and led to the introduction of the quantum of action by Max Planck in 1900, beginning a revolution in physics now called quantum mechanics. For the environmental scientist, a blackbody or a graybody is a convenient model to describe emission or absorption by the Sun, the planet, the atmosphere, the soil, and the oceans.

1 **(A)** What is the total energy per unit area and per unit time radiated over all frequencies by an ideal blackbody at $2.50 \cdot 10^3$ K?

Solution
The total energy radiated by a blackbody per unit time and per unit normal area is given by the Stefan–Boltzmann law,

$$\frac{dE}{dtdS} = \sigma T^4 = \left(5.67 \cdot 10^{-8}\,\frac{\mathrm{W}}{\mathrm{m^2 K^4}}\right)\left(2.50 \cdot 10^3\,\mathrm{K}\right)^4 = 2.21 \cdot 10^6\,\frac{\mathrm{W}}{\mathrm{m^2}}.$$

2 **(A)** What is the wavelength of the peak of the blackbody curve for an object radiating at $3.00 \cdot 10^3$ K?

Solution
Wien's law of displacement $\lambda_{\max} T = b = 2.898 \cdot 10^{-3}\,\mathrm{m} \cdot \mathrm{K}$ yields

$$\lambda_{\max} = \frac{b}{T} = 9.66 \cdot 10^{-7}\,\mathrm{m} = 966\,\mathrm{nm}.$$

3 **(A)** Calculate the luminosity of the Sun, i.e., the total energy per unit time radiated in the whole spectrum of frequencies, knowing that the temperature at the surface of the Sun is $T_\odot = 5800$ K and its radius is $R_\odot = 7.00 \cdot 10^5$ km. Compute the solar constant using the average Earth–Sun distance $1.50 \cdot 10^{11}$ m.

Solution
By treating the Sun as a blackbody, the Stefan–Boltzmann law gives the total energy radiated over the whole spectrum per unit time and per unit of normal area

$$\frac{dE}{dtdS} = \sigma T_\odot^4.$$

The luminosity of the Sun is

$$L_\odot = \frac{dE}{dt} = 4\pi R_\odot^2\, \sigma T_\odot^4,$$

where $4\pi R_\odot^2$ is the surface area of the Sun. Using the given data we obtain

$$L_\odot = 4\pi (7.00 \cdot 10^8\,\text{m})^2 \cdot (5.67 \cdot 10^{-8}\,\text{W} \cdot \text{m}^{-2} \cdot \text{K}^{-4}) \cdot (5800\,\text{K})^4$$

$$= 3.95 \cdot 10^{26}\,\text{W}.$$

The solar constant S (energy falling on the unit normal area per unit time) on the Earth is given by $L_\odot = 4\pi d^2 S$, hence

$$S = \frac{L_\odot}{4\pi d^2} = \frac{3.95 \cdot 10^{26}\,\text{W}}{4\pi (1.50 \cdot 10^{11}\,\text{m})^2} = 1.4 \cdot 10^3\,\frac{\text{W}}{\text{m}^2}.$$

4 **(A)** The Earth radiates approximately as a blackbody at 255 K. Calculate the wavelength at which the blackbody distribution peaks and compare the result with the wavelength at which the emission from the Sun peaks (the Sun can be approximated by a blackbody at 5800 K).

Solution
Wien's law of displacement

$$\lambda_{\max} T = b = 0.2898\ \text{cm} \cdot \text{K}$$

yields

$$\lambda_{\max} = \frac{0.2898\ \text{cm} \cdot \text{K}}{255\ \text{K}} = 1.14 \cdot 10^{-3}\ \text{cm} = 11.4\ \mu\text{m}$$

for the Earth, and

$$\lambda_{\max} = \frac{0.2898\ \text{cm} \cdot \text{K}}{5800\ \text{K}} = 5.00 \cdot 10^{-5}\ \text{cm} = 500\ \text{nm}$$

for the Sun. The emission from the Earth is peaked in the infrared while the emission from the Sun is peaked in the visible band (the Sun is classified as a yellow star).

5 **(A)** The radius of the Sun is $R_\odot = 7.0 \cdot 10^8$ m, the average distance between the Earth and the Sun is the astronomical unit (1 A.U.= $1.5 \cdot 10^{11}$ m), and the solar constant on the Earth, i.e., the radiation energy from the Sun falling on the unit of normal area per unit time, is $S = 1.4 \cdot 10^3$ W·m^{-2}. Assuming the solar spectrum to be that of a blackbody, compute the surface temperature of the Sun.

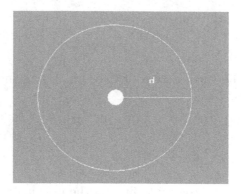

Figure 4.1.   The Sun and a sphere of radius d centered on it.

Solution

The luminosity of the Sun (total energy emitted per unit time over all frequencies) is

$$L_\odot = 4\pi R_\odot^2 F, \tag{4.1}$$

where F is the flux emitted at the surface of the star (energy emitted per unit time and per unit normal area). The Stefan–Boltzmann law gives

$$F = \sigma T_\odot^4, \tag{4.2}$$

where $\sigma = 5.67 \cdot 10^{-8}\,\mathrm{W \cdot m^{-2}\,K^{-4}}$ is the Stefan–Boltzmann constant and $T_\odot$ is the Kelvin temperature of the solar surface. The electromagnetic energy radiated propagates through empty space and its flux S across a sphere centered on the Sun, with radius d equal to the astronomical unit (Fig. 4.1), is given by

$$L_\odot = 4\pi d^2 S. \tag{4.3}$$

By comparing Eqs. (4.1) and (4.3) one obtains, using also Eq. (4.2),

$$
\begin{aligned}
T_\odot &= \left[\frac{S}{\sigma} \left(\frac{d}{R_\odot} \right)^2 \right]^{1/4} \\[2mm]
&= \left[\frac{1.4 \cdot 10^3\,\mathrm{W \cdot m^{-2}}}{5.67 \cdot 10^{-8}\,\mathrm{W \cdot m^{-2} \cdot K^{-4}}} \left(\frac{1.5 \cdot 10^{11}\,\mathrm{m}}{7.0 \cdot 10^8\,\mathrm{m}} \right)^2 \right]^{1/4} \\[2mm]
&= 5800\,\mathrm{K}.
\end{aligned}
$$

6 **(B)** Derive the spectral energy density $u(\lambda, T)$ of a blackbody as a function of wavelength λ, from the Planck distribution $u(\nu, T)$ as a function of frequency ν. Verify Wien's law of displacement qualitatively. How can one verify it quantitatively?

Solution
The Planck distribution for the spectral energy density as a function of frequency is

$$u(\nu, T) = \frac{dE}{dV\, d\nu} = \frac{8\pi h}{c^3} \frac{\nu^3}{e^{\frac{h\nu}{kT}} - 1}.$$

As a function of the wavelength $\lambda = c/\nu$, we have

$$\frac{dE}{dV\, d\lambda} = \frac{dE}{dV\, d\nu} \frac{d\nu}{d\lambda} = -\frac{c}{\lambda^2} u\left(\frac{c}{\lambda}, T\right).$$

By absorbing the negative sign in the definition of the spectral energy density in wavelength, we obtain

$$u(\lambda, T) \equiv -\frac{dE}{dV\, d\lambda} = \frac{c}{\lambda^2} \frac{8\pi hc^3}{\lambda^3 c^3} \frac{1}{e^{\frac{hc}{\lambda kT}} - 1},$$

or

$$u(\lambda, T) = \frac{8\pi hc}{\lambda^5} \frac{1}{e^{\frac{hc}{\lambda kT}} - 1}.$$

To verify Wien's law, note that $du/d\lambda$ vanishes at the maximum of $u(\lambda, T)$:

$$\frac{du}{d\lambda} = \frac{8\pi hc}{\lambda^6} \frac{1}{(e^x - 1)^2} \left[e^x(x - 5) + 5\right] = 0,$$

where

$$x \equiv \frac{hc}{\lambda kT}.$$

The equation $du/d\lambda = 0$ is satisfied if and only if $e^x(5 - x) = 5$. The solutions of this trascendental equation are the intersections between the graphs of the functions $f(x) \equiv e^x(5 - x)$ and $y(x) = 5$. To determine whether such intersections actually exist, we study analytically the shape of the graph of $f(x)$ for $x \geq 0$ (corresponding to $\lambda > 0$). $f(x)$ is everywhere regular, $f(0) = 5$, $f(x) \to -\infty$ as $x \to +\infty$, and

$$\frac{df}{dx} = e^x(4 - x)$$

is positive for $x < 4$, zero for $x = 4$, and negative for $x > 4$. This is sufficient to establish that $f(x)$ has a local maximum that is also an absolute maximum at $x = 4$. The value of this maximum is $f(4) = e^4 \simeq 54.6 > 5$. Hence, there are always only two intersections between the line $y = 5$ and $f(x)$: $x_0 = 0$ and $x = x_1$, with $4 < x_1 < 5$ [note that $f(5) = 0$]. Since the maximum of the blackbody distribution occurs at a particular value x_1 of $x \equiv hc/(\lambda kT)$, Wien's law of displacement

$$\lambda_{\max} T = \frac{hc}{kx_1} \equiv b$$

is qualitatively verified. To verify the law *quantitatively*, i.e., to compute the value of the constant b, we should solve numerically the equation $f(x) = 5$. The result[1] is $x_1 \simeq 4.965$, yielding

$$b = \frac{hc}{kx_1} = \frac{(6.625 \cdot 10^{-34} \,\text{J} \cdot \text{s}) \cdot (3.0 \cdot 10^8 \,\text{m} \cdot \text{s}^{-1})}{(1.38 \cdot 10^{-23} \,\text{J} \cdot \text{K}^{-1}) \cdot 4.965} = 0.29 \,\text{cm} \cdot \text{K}.$$

7 **(B)** Show that the Planck energy distribution

$$u(\nu, T) = \frac{8\pi h\nu^3}{c^3} \frac{1}{e^{h\nu/kT} - 1}$$

leads to the Rayleigh–Jeans law

$$u_{RJ}(\nu, T) = \frac{8\pi\nu^2 kT}{c^3}$$

for low frequencies. Find the asymptotic form of $u(\nu, T)$ at high frequencies and discuss the meaning of "low" and "high" frequencies.

Solution
Low frequency denotes the situation in which the dimensionless ratio of energies $h\nu/kT \ll 1$. In this approximation the Planck distribution reduces to

$$u(\nu, T) = \frac{8\pi h\nu^3}{c^3} \frac{1}{e^{h\nu/KT} - 1} = \frac{8\pi h\nu^3}{c^3} \frac{1}{1 + \frac{h\nu}{kT} + \cdots - 1}$$

$$\simeq \frac{8\pi kT\nu^2}{c^3} \equiv u_{RJ}(\nu, T),$$

where the Taylor expansion of the exponential function $e^x = 1 + x + \cdots$ is used.

---

[1] This numerical solution can be performed by successive approximations by using a simple, nonprogrammable, pocket calculator.

High frequency denotes the opposite limit in which $h\nu/kT \gg 1$. In this limit $\left(e^{h\nu/KT} - 1\right)^{-1} \simeq e^{-h\nu/KT}$ and the Planck distribution reduces to

$$u(\nu, T) = \frac{8\pi h\nu^3}{c^3} e^{-h\nu/KT},$$

a form originally proposed by Wien.

8 **(A)** The filament of a typical incandescent bulb is at a temperature of 3000 K. Approximate the bulb with a blackbody: at what wavelength $\lambda_{\max}$ does then the emitted spectrum peak? In what region of the electromagnetic spectrum does $\lambda_{\max}$ fall? Comment on the efficiency of an incandescent bulb for the purpose of lighting.

Solution
Wien's law of displacement $\lambda_{\max} T = b = 2.898 \cdot 10^{-3}\,\mathrm{m \cdot K}$ yields

$$\lambda_{\max} = \frac{b}{T} = \frac{2.898 \cdot 10^{-3}\,\mathrm{m \cdot K}}{3000\,\mathrm{K}} = 9.66 \cdot 10^{-7}\,\mathrm{m} = 0.966\,\mu\mathrm{m}.$$

This wavelength lies in the *infrared* band of the electromagnetic spectrum, hence most of the energy emitted by the bulb is radiated in the infrared, not in the visible band. An incandescent bulb has a poor efficiency (typically around 5%) for the purpose of lighting.

4.3 Propagation of electromagnetic radiation

The propagation of electromagnetic radiation in a medium is accompanied by absorption, which is differential. Absorption depends on the frequency of the radiation because of the quantum nature of atoms, molecules, and solids that resonate and exchange photons only in correspondence with selected frequencies determined by definite differences between different energy levels of the system.

1 **(A)** *Intensity* of radiation is another name for the flux density of electromagnetic energy. The intensity of solar radiation on the Earth is given by the solar constant S. What is the intensity of solar radiation on Mars, if the average Earth–Sun and Mars–Sun distances are 1 A.U. and 1.53 A.U., respectively? Treat the Sun as a pointlike source.

Solution
The energy emitted by the pointlike Sun propagates radially outward and is distributed over a spherical wavefront. The intensity (or flux

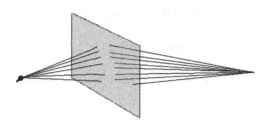

Figure 4.2.   A screen, a lightbulb, and a pointlike source.

density or solar constant) of electromagnetic radiation at distance r from the Sun is $I = L_\odot/\left(4\pi r^2\right)$, where $L_\odot$ is the luminosity (power emitted) of the Sun. At the positions r_E and r_M of the Earth and Mars,

$$I_E = S = \frac{L_\odot}{4\pi r_E^2}, \qquad I_M = \frac{L_\odot}{4\pi r_M^2},$$

and

$$\frac{I_M}{I_E} = \left(\frac{r_E}{r_M}\right)^2$$

so that the intensity (or flux density, or solar constant) at Mars is

$$I_M = S\left(\frac{r_E}{r_M}\right)^2 = S\left(\frac{1\,\text{A.U.}}{1.53\,\text{A.U.}}\right)^2 = 0.43 S.$$

2 **(A)** A 60.0 W lightbulb produces on a very small screen 1.84 m away the same intensity as a pointlike light source of unknown power W_2 at 2.56 m (Fig. 4.2). What is the value of W_2? At what distance from the screen should one place this second source to obtain the intensity of $10.0\,\text{W/m}^2$? Treat the lightbulb as an isotropic point source.

Solution
Assuming for simplicity that the light sources radiate isotropically,

and that the small screen can be approximated by a nearly flat portion of spherical surface centered on the lightbulb, the intensity of light on the screen is

$$I = \frac{W_1}{4\pi d_1^2} = \frac{W_2}{4\pi d_2^2},$$

where W_i and d_i are, respectively, the power and the distance from the screen of the light sources $(i = 1, 2)$. We have

$$W_2 = W_1 \left(\frac{d_2}{d_1}\right)^2 = (60.0 \, \text{W}) \cdot \left(\frac{2.56 \, \text{m}}{1.84 \, \text{m}}\right)^2 = 116 \, \text{W}.$$

When this pointlike source of light is placed at the new distance d' from the screen, the intensity at the screen is $I' = W_2 / \left[4\pi \left(d'\right)^2\right] = 10.0 \, \text{W/m}^2$, yielding

$$d' = \left(\frac{W_2}{4\pi I'}\right)^{1/2} = \left[\frac{116 \, \text{W}}{4\pi \left(10.0 \, \text{W/m}^2\right)}\right]^{1/2} = 0.961 \, \text{m}.$$

3 **(A)** A medium is composed of a substance with molar extinction coefficient $2.5 \cdot 10^5 \, \text{dm}^3 \cdot \text{mol}^{-1} \cdot \text{cm}^{-1}$ at a certain wavelength λ and concentration $2.8 \cdot 10^{-3} \, \text{mol} \cdot \text{dm}^{-3}$. A monochromatic light beam with this wavelength travels through a layer of the substance. What is the distance traveled when the intensity of the light beam is reduced to $1/1000$ of the intensity upon entering the medium? What are the dimensions of the optical density?

Solution
The attenuation of the light beam is given by the Lambert–Beer–Bouguer law

$$I(z) = I_0 \cdot 10^{-\tau},$$

where the *optical density* τ is

$$\tau = \epsilon C z.$$

Here ϵ is the *molar extinction coefficient*, C is the concentration of the chemical, z is the distance traveled by the beam, $I_0 = I(z = 0)$ is the intensity of the beam upon entering the medium, and the attenuation of the beam intensity is exponential. When the beam intensity is reduced to $I/I_0 = 10^{-3}$, then $\tau = 3$ and

$$z = \frac{\tau}{\epsilon C} = \frac{3.0}{\left(2.5 \cdot 10^5 \, \text{dm}^3 \cdot \text{mol}^{-1} \cdot \text{cm}^{-1}\right) \cdot \left(2.8 \cdot 10^{-3} \, \text{mol/dm}^3\right)}$$

$$= 4.3 \cdot 10^{-3} \, \text{cm}.$$

The optical density τ is a pure number and does not carry dimensions.

4 **(A)** The Lambert–Beer–Bouguer law describing the attenuation of the intensity I of electromagnetic radiation due to absorption appears in the literature in two forms:

$$I = I_0 \, e^{-\alpha x},$$

where $I_0 = I(x=0)$ is the intensity upon entering a slab of absorbing material normal to the direction of propagation, x is the distance traveled in the slab, and the *absorption coefficient* α depends on the material and the frequency of radiation. The second form of the Lambert–Beer–Bouguer law is

$$I = I_0 \cdot 10^{-\tau},$$

where $\tau = \epsilon\,C\,x$ is the *optical density*, ϵ is the *molar extinction coefficient*, and C is the concentration of the absorbing substance. What is the relation between α and τ? What are their dimensions?

Solution
By using the formula
$$a^x \equiv e^{x \ln a},$$
we obtain
$$I = I_0 \cdot 10^{-\tau} = I_0 \, e^{-\epsilon C x \ln 10} = I_0 \, e^{-\alpha x},$$
with
$$\alpha = \epsilon\, C \ln 10 \simeq 2.303\, \epsilon\, C.$$

Since the argument of the exponential in the Lambert–Beer–Bouguer law must be dimensionless, the absorption coefficient has the dimensions of the inverse of a length, $[\alpha] = [L^{-1}]$. Similarly, the optical density τ does not carry dimensions. The dimensions of the molar extinction coefficient and of the concentration are

$$[\epsilon] = [\mathrm{dm}^3 \cdot \mathrm{mol}^{-1} \cdot \mathrm{cm}^{-1}]$$
and
$$[C] = [\mathrm{mol/dm}^3].$$

5 **(C)** The attenuation of the intensity I of a beam of electromagnetic radiation in the atmosphere is due to absorption and scattering. Scattering by particles suspended in the atmosphere is described by

$$T = e^{-\beta x},$$

where the *transmittance* $T = I/I_0$ is the ratio between the intensity of the beam I and its intensity I_0 upon entering the atmosphere at $x = 0$, and the coefficient β depends on the radiation wavelength, the concentration of the scatterers, their size r, and the refraction index n. Discuss Rayleigh, Mie, and nonselective scattering according to the size of the scattering particles. How is the radar used for weather monitoring related to scattering?

Solution
When the scattering particles have sizes much smaller than the wavelength of the incident radiation, $r \ll \lambda$, Rayleigh scattering dominates. The coefficient β is strongly λ-dependent,

$$\beta \propto \frac{1}{\lambda^4}.$$

Consequently, the transmittance T is large (close to unity) for long wavelengths and exponentially small for short wavelengths—hence short wavelengths are intensely scattered. This phenomenon is responsible for the blue color of clear, clean skies.

When $r \simeq \lambda$ waves are reflected from different parts of the particle and interfere, Mie scattering dominates and $\beta(\lambda)$ is a complicated function of the wavelength. At shorter wavelengths it is approximately $\beta \propto \lambda^{-2}$.

For $r \gg \lambda$, scattering is nonselective and β is independent of the wavelength, $d\beta/d\lambda = 0$. This is the case, for example, of water droplets in fog and clouds, with size $5 \cdot 10^{-6}$ m $\leq r \leq 10^{-4}$ m, scattering infrared radiation. This scattering phenomenon is used in the weather radar, an important tool that monitors weather and detects suspended droplets and clouds. The speed of an approaching cloud can be measured using the Doppler effect.

6 **(A)** When electromagnetic radiation of frequency ν_1 emitted by the Sun reaches the surface of the Earth after traversing the entire atmosphere radially, its intensity is reduced to 0.35 times its value in outer space because of absorption. The ratio T_1/T_2 of transmittances for waves of frequencies ν_1 and ν_2 is 0.40; what is the value of the absorption coefficient for electromagnetic waves of frequency ν_2? Assume that the thickness of the atmosphere is $1.0 \cdot 10^2$ km.

Solution
The transmittances for radiation at frequencies ν_1 and ν_2 are given

by the Lambert–Beer–Bouguer law

$$T_1 = e^{-\alpha_1 z}, \qquad T_2 = e^{-\alpha_2 z},$$

and their ratio is

$$\frac{T_1}{T_2} = e^{(\alpha_2 - \alpha_1)z},$$

yielding

$$\alpha_2 = \alpha_1 + \frac{1}{z} \ln\left(\frac{T_1}{T_2}\right).$$

The absorption coefficient α_1 is given by $\alpha_1 = -z^{-1} \ln T_1$ and

$$\alpha_2 = \frac{1}{z}\left[\ln\left(\frac{T_1}{T_2}\right) - \ln T_1\right] = \frac{1}{1.0 \cdot 10^5\,\mathrm{m}}\,(\ln 0.40 - \ln 0.35)$$

$$= 1.3 \cdot 10^{-6}\,\mathrm{m}^{-1}.$$

7 **(A)** A beam of monochromatic electromagnetic radiation of initial intensity I_0 propagates through a layer of the atmosphere and is both absorbed and scattered, with absorption and scattering coefficients $\alpha = 0.50 \cdot 10^{-5}\,\mathrm{m}^{-1}$ and $\beta = 2.00 \cdot 10^{-4}\,\mathrm{m}^{-1}$, respectively. What is the beam intensity I (as a fraction of I_0) after it has traveled one kilometer in the air layer?

Solution
Absorption and scattering both contribute to exponential attenuation of the beam intensity, and they occur simultaneously. Without scattering, absorption alone would determine an intensity

$$I_1(z) = I_0\, e^{-\alpha z}.$$

We can take this intensity as the initial value for the beam intensity that is further reduced by scattering by another exponential factor $e^{-\beta z}$, or

$$I(z) = I_1(z)\, e^{-\beta z} = I_0\, e^{-(\alpha+\beta)z}$$

$$= I_0\, e^{-\left(0.50 \cdot 10^{-5}\,\mathrm{m}^{-1} + 2.00 \cdot 10^{-4}\,\mathrm{m}^{-1}\right)(1000\,\mathrm{m})} = 0.815\, I_0.$$

Although this separation is artificial because absorption and scattering occur simultaneously, it may still be useful in order to combine the effects of the two processes.

8 **(A)** The foliage cover of a forest filters sunlight letting only certain wavelengths reach the ground. The attenuation of the intensity of a light beam is due to chlorophyll a, with absorption peaked in the red—the solar energy absorbed is converted into chemical energy due to photosynthesis. What is the molar extinction coefficient (in $dm^3 \cdot mol^{-1} \cdot cm^{-1}$) of chlorophyll for red light, if the leaves are $2.5 \cdot 10^{-2}$ mm thick and the chlorophyll concentration is $C = 8.5 \cdot 10^{-4} mol/dm^3$? Assume that a beam of sunlight, on average, traverses 10 leaves reducing its intensity to 10^{-3} times the value that it has above the foliage cover. What is the value of the extinction coefficient?

Solution

The attenuation of sunlight entering a leaf perpendicularly to its surface is described by the Lambert–Beer–Bouguer law

$$I = I_0 \cdot 10^{-\tau},$$

where the optical density τ is given by $\tau = \epsilon C x$, ϵ is the molar extinction coefficient, C is the chlorophyll concentration, and x is the distance traveled by the beam. Hence,

$$\epsilon = -\left[\lg_{10}\left(\frac{I}{I_0}\right) \right] \frac{1}{Cx} = \frac{-\lg_{10}\left(10^{-3}\right)}{(8.5 \cdot 10^{-4}\, mol/dm^3) \cdot 10 \cdot (2.5 \cdot 10^{-5}\, m)}$$

$$= \frac{3}{(8.5 \cdot 10^{-4}\, mol/dm^3) \cdot 10 \cdot (2.5 \cdot 10^{-3}\, cm)} = 1.41 \cdot 10^5\, \frac{dm^3}{mol \cdot cm}.$$

The Lambert–Beer–Bouguer law can be put in the form

$$T \equiv \frac{I}{I_0} = e^{-\alpha x},$$

where α is the extinction coefficient, or

$$\alpha = -\frac{1}{x} \ln \frac{I}{I_0}.$$

For the average path of a beam of sunlight through the foliage cover, we have

$$\alpha = \frac{-1}{2.5 \cdot 10^{-4}\, m} \ln 10^{-3} = 2.8 \cdot 10^4\, m^{-1}.$$

9 **(B)** In the ionosphere, at altitudes higher than 80 km, the number density of free electrons and ions created by ionizing radiation from the Sun is sufficiently large to affect the propagation of electromagnetic waves. The relation between the angular frequency ω and the

wave vector k (*dispersion relation*) of an electromagnetic wave propagating through the ionospheric plasma is given by

$$\omega^2 = c^2 k^2 + \omega_p^2,$$

where c is the speed of light in vacuum and ω_p is a constant called *plasma frequency*. Compute the phase and group velocities of the electromagnetic wave as functions of the angular frequency ω. Consider a monochromatic plane wave propagating along the positive x-axis and described by the electric field $\vec{E} = \vec{E}_0\, e^{i(kx-\omega t)}$, where $\vec{E}_0$ is a constant vector. Discuss the propagation of the wave for the situations $\omega > \omega_p$ and $\omega < \omega_p$.

The plasma frequency is given by

$$\omega_p = \sqrt{\frac{4\pi e^2 n_e}{m_e \gamma}},$$

where e, m_e, and n_e are, respectively, the electron charge and mass, and the number density of free electrons; $\gamma \equiv \left(1 - v^2/c^2\right)^{-1/2}$ is the Lorentz factor of electrons with speed v. Provide an argument showing that it is harder and harder for radio waves of intermediate frequency to propagate in the ionosphere at higher and higher altitudes.

Solution
The dispersion relation is

$$\omega = ck\sqrt{1 + \left(\frac{\omega_p}{ck}\right)^2},$$

and the corresponding phase velocity is

$$v_p \equiv \frac{\omega}{k} = c\sqrt{1 + \left(\frac{\omega_p}{ck}\right)^2}$$

(see Fig. 4.3), which is larger than the speed of light c. This is not a paradox because the physically relevant velocity is the group velocity

$$v_g \equiv \frac{d\omega}{dk} = \frac{c^2 k}{\sqrt{c^2 k^2 + \omega_p^2}} = c\left[1 + \left(\frac{\omega_p}{ck}\right)^2\right]^{-1/2},$$

which is smaller than c (Fig. 4.4).

For a monochromatic plane wave described by the electric field $\vec{E} = \vec{E}_0\, e^{i(kx-\omega t)}$ the dispersion relation yields

$$k = \frac{1}{c}\sqrt{\omega^2 - \omega_p^2};$$

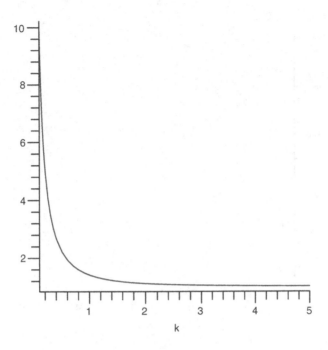

Figure 4.3. The phase velocity in units of the speed of light c.

if $\omega > \omega_p$, the wave vector k is real and the wave propagates almost freely—it is unaffected by the presence of the ionospheric plasma in the limit $\omega \gg \omega_p$, in which $k \approx \omega/c$ as in vacuum. If instead $\omega < \omega_p$, the wave vector k is imaginary, $k = i\,|k|$, and the wave's electric field can be written as

$$\vec{E} = \vec{E}_0\,e^{i(kx - \omega t)} = \vec{E}_0\,e^{-|k|x}\,e^{-i\omega t},$$

in which it is evident that the wave is absorbed by the plasma as it propagates through it. The effective amplitude of the electric field $\left|\vec{E}_0\right| e^{-|k|x}$ decreases exponentially fast with the distance x traveled. Upon traveling the length

$$|k|^{-1} = \frac{c}{\sqrt{\omega^2 - \omega_p^2}}$$

the amplitude is reduced by a factor $1/e$ (one *e-fold*)—long waves cannot propagate through the plasma. High-frequency waves travel

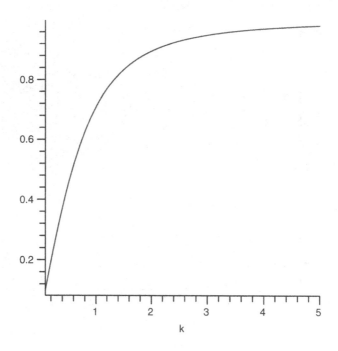

Figure 4.4. The group velocity in units of the speed of light c.

better through the atmosphere than low- or intermediate-frequency waves, a fact well known to radio amateurs.

The plasma frequency (the angular frequency threshold for propagation)

$$\omega_p = \left(\frac{4\pi e^2 n_e}{m_e\, \gamma} \right)^{1/2}$$

is larger and larger at higher elevations due to the larger abundance of electrons freed by ionizing solar radiation in conditions of lower pressure and density (larger n_e). Therefore, it is harder and harder for intermediate frequency electromagnetic waves to propagate at higher altitudes.

4.4   Greenhouse effect and global warming

Global warming is one of the most intensively studied aspects of modern environmental physics. The injection of greenhouse gases, mainly

CO_2, in the atmosphere since the beginning of the Industrial Revolution has created a sort of blanket that allows short-wavelength radiation to reach the surface of the Earth and traps longer-wavelength infrared radiation re-radiated by the Earth, that would normally escape to outer space. The long-term effect is a global warming of the planet. While such a trend is actually being observed, it is not clear whether it is caused by manmade emissions or by natural causes.

1 **(C)** Describe the greenhouse effect in a farm greenhouse. What is the main difference between a greenhouse and the atmosphere?

Solution
In a greenhouse, short wavelength radiation in the visible band (wavelengths 400 nm $< \lambda <$ 700 nm) enters through the glass walls and is absorbed by the plants contained in it. Electromagnetic radiation is then emitted by the plants at shorter wavelengths in the infrared (700 m$< \lambda <$0.1 mm). The glass of the walls is transparent to shorter wavelengths in the visible band and is opaque to the longer wavelengths re-radiated in the infrared by the plants—infrared radiation is therefore trapped inside the greenhouse. The main difference between the Earth's atmosphere and a greenhouse is that infrared radiation escapes from the Earth at night, while it can not escape from the enclosed space of a greenhouse.

2 **(A)** Absorption of ultraviolet light in the atmosphere is described by the Lambert–Beer–Bouguer law. What would happen to the intensity of *UV* radiation on the surface of the Earth if the optical density decreased by one unit?

Solution
The Lambert–Beer–Bouguer law yields the intensity of *UV* radiation reaching the surface of the Earth

$$I = I_0 \cdot 10^{-\tau},$$

where τ is the optical density. If τ decreases by one (τ is dimensionless), the intensity I increases by 10 units:

$$I \longrightarrow I_0 \, 10^{-(\tau-1)} = 10 \, I_0 \, 10^{-\tau}.$$

3 **(A)** The optical density from outer space to a point P in the atmosphere located below the ozone layer is $\tau = 1$. Assume that the ozone concentration drops by 10% and compute the ratio of intensities of UV-B radiation at P before and after the change of the O_3 concentration.

Solution

The optical density to a generic point in the atmosphere is $\tau = \epsilon C z$, where ϵ is the molar extinction coefficient, C is the concentration of ozone responsible for absorbing UV-B radiation, and z is the distance traveled by the UV-B rays. When the concentration of ozone undergoes the change $C \to C' = 0.9\,C$ the optical density $\tau = \epsilon C z$ changes according to $\tau \to \tau' = 0.9\,\tau$. Before the drop in C the intensity of UV-B is given by the Lambert–Beer–Bouguer law, $I = I_0 \cdot 10^{-\tau}$; after C drops, the intensity is $I' = I_0 \cdot 10^{-0.9\,\tau}$. The ratio of intensities is

$$\frac{I'}{I} = \frac{10^{-0.9\,\tau}}{10^{-\tau}} = 10^{0.1\,\tau}.$$

At point P, it is $\tau = 1$ before the change and $I'/I = 10^{0.1} = 1.259$, thus a 10% change in the concentration of atmospheric ozone leads to a 26% change in the intensity of UV-B radiation. Since biomolecules and human tissues are very sensitive to UV-B radiation, a 26% increase in its intensity would have serious consequences on the incidence of skin cancer.

4 **(B)** When the concentration of greenhouse gases in the atmosphere increases, $C_0 \longrightarrow C$, the temperature rises accordingly (*radiative forcing*) to compensate for the decreased flux of infrared radiation leaving the atmosphere. Greenhouse gases are measured by their equivalent CO_2 concentration. Two different models of radiative forcing found in the literature predict the temperature variations

$$\Delta T_1 = \tau \ln \left(\frac{C}{C_0} \right)$$

and

$$\Delta T_2 = \tau \, \frac{\Delta C}{C_0},$$

where $\Delta C \equiv C - C_0$ and $\tau = 6.1$ K. From 1850 to 1990 during the late Industrial Revolution, the value of C rose from 285 ppm to 360 ppm. Compare the predictions of the two models for the corresponding temperature change. Derive the second model from the first. What order of approximation in powers of in $\Delta C/C_0$ is needed in order to reach agreement to two significant figures between the two models?

Solution

In the first model of radiative forcing

$$\Delta T_1 = (6.1\,\text{K}) \ln \left(\frac{360\,\text{ppm}}{285\,\text{ppm}} \right) = 1.4\,\text{K},$$

while in the second model

$$\Delta T_2 = (6.1\,\text{K}) \frac{360\,\text{ppm} - 285\,\text{ppm}}{285\,\text{ppm}} = 1.6\,\text{K}.$$

The disagreement between the predictions of the two models is

$$\frac{\Delta T_2 - \Delta T_1}{\Delta T_2} = 13\%.$$

The second model is nothing but the linear approximation of the first one and is only adequate for small values of $\Delta C/C_0$. By using the series

$$\ln(1+x) = x - \frac{x^2}{2} + \frac{x^3}{3} - \frac{x^4}{4} + \cdots$$

for $|x| < 1$, one computes the corrections to the following orders:

$$\begin{aligned}
\Delta T &= \tau \ln \frac{C}{C_0} = \tau \ln \left(1 + \frac{\Delta C}{C_0}\right) \\
&= \tau \left[\frac{\Delta C}{C_0} - \frac{1}{2}\left(\frac{\Delta C}{C_0}\right)^2 + \frac{1}{3}\left(\frac{\Delta C}{C_0}\right)^3 - \frac{1}{4}\left(\frac{\Delta C}{C_0}\right)^4 + \cdots\right];
\end{aligned}$$

to first order, one recovers the second model of radiative forcing, $\Delta T = \tau \Delta C/C_0$. The leading order correction yields

$$\begin{aligned}
\Delta T &= (6.1\,\text{K}) \left[\left(\frac{360\,\text{ppm} - 285\,\text{ppm}}{285\,\text{ppm}}\right)\right. \\
&\qquad \left. - \frac{1}{2}\left(\frac{360\,\text{ppm} - 285\,\text{ppm}}{285\,\text{ppm}}\right)^2\right] = 1.4\,\text{K}.
\end{aligned}$$

The second-order correction is necessary since $\Delta C/C_0 \simeq 0.263$ is not a very small number and the linear approximation is not accurate. To the relevant accuracy (two significant figures) the second-order approximation agrees with the exact model.

5 **(B)** Assume that the heat flux F radiated by the Earth in space suddenly decreases due to the greenhouse effect, caused by an abrupt increase in the concentration of greenhouse gases in the atmosphere. Since the Earth is in thermal equilibrium with outer space its surface temperature T_s must increase to compensate for the decrease of F, according to the Stefan–Boltzmann law. This *radiative forcing* is described by

$$\Delta T_s \equiv T_s(t) - T_0 = G\,\Delta F,$$

where T_0 is the surface temperature before the change, ΔF is the magnitude of the variation of F, and G is a *gain function*, which in the linear approximation is simply $\partial T_s / \partial (\Delta F)|_{\Delta F = 0}$. Refine this model by taking into account the thermal inertia of the oceans covering 70% of the Earth's surface. Let $c\,m$ be the heat capacity of the top layer of the oceans interested by global warming (m is its mass and c is the specific heat of water); derive a differential equation for the surface temperature $T_s(t)$ and find its solution.

Solution

The heat flux (energy passing per unit time through the unit area normal to the direction of propagation) is $F = dQ/dt\,dS$ and the magnitude of its variation due to the greenhouse effect is ΔF. We obtain

$$\Delta F = \frac{d\,(\Delta Q)}{dt\,dS} + \frac{\Delta T_s}{G},$$

where ΔQ is the difference between the heat lost by the Earth's surface after and before the change in the greenhouse gases concentration. Since $\Delta Q = c\,m\Delta T_s$, it is straightforward to conclude that T_s obeys the differential equation

$$\frac{d\,(\Delta T_s)}{dt} + \frac{\Delta T_s}{\gamma G} = \frac{\Delta F}{\gamma}, \qquad (4.4)$$

where $\gamma \equiv d\,(cm)\,/dS$ is the heat capacity of the oceans per unit area. The general solution of the homogeneous equation associated with Eq. (4.4) is

$$\Delta T_s(t) = A\,\mathrm{e}^{-t/\tau},$$

where $\tau \equiv \gamma G$ is a time scale. A particular solution of the inhomogeneous equation (4.4) is

$$\Delta T_s = G\,\Delta F,$$

and therefore the general solution of Eq. (4.4) is

$$T_s(t) = T_0 + A\,\mathrm{e}^{-t/\tau} + G\,\Delta F.$$

The integration constant A is determined by the initial condition $T_s(0) = T_0$, which yields $A = -G\,\Delta F$, and therefore

$$T_s(t) = T_0 + G\,\Delta F \left(1 - \mathrm{e}^{-t/\tau}\right).$$

The solution goes to its asymptotic value $T_0 + G\,\Delta F$ as $t \to +\infty$. In practice T_s reaches 90% of this value after a time $t = 2.3\,\tau$. The effect of the thermal inertia of the oceans is to introduce a time lag in the global warming of the planet—without the oceans we would have the constant solution $T_0 + G \Delta F$. The time scale τ is estimated to be between 50 and 100 years. In order or magnitude we have

$$\tau = \gamma\, G = \frac{d\,(cm)}{dS}\, G \simeq \frac{cmG}{0.7 \cdot 4\pi R_E^2} = \frac{c\rho \left(0.7 \cdot 4\pi R_E^2\right) hG}{0.7 \cdot 4\pi R_E^2} = c\rho h\, G,$$

where R_E is the Earth's radius and h the depth of the top layer of the oceans. Realistic values are $G = 0.7°\,\mathrm{C} \cdot \mathrm{s} \cdot \mathrm{m}^2/\mathrm{J}$ [4] and $h = 1$ km, yielding

$$\tau \sim \left(4187\, \frac{\mathrm{J}}{\mathrm{kg} \cdot (°\mathrm{C})}\right)\left(1.0 \cdot 10^3\, \frac{\mathrm{kg}}{\mathrm{m}^3}\right)(10^3\,\mathrm{m})\left(0.7\, \frac{°\mathrm{C} \cdot \mathrm{s} \cdot \mathrm{m}^2}{\mathrm{J}}\right)$$

$$\sim\ 3 \cdot 10^9\,\mathrm{s} \sim 93\,\text{years}.$$

4.5   Electromagnetic radiation and human health

The effects of electric fields on biological cells are relatively well understood, while the effects of magnetic fields are not. The effects of electromagnetic fields on complex organisms such as the human body are largely unknown and are the subject of extensive research. Electromagnetic pollution by human activities is the subject of much ongoing research in environmental science.

1 **(C)** What are the various forms of ionizing radiation? What are their nature and origin?

Solution
Ionizing radiation includes UV-B radiation, X-rays, γ-rays, and α-particles. UV-B radiation, X-rays, and γ-rays are electromagnetic waves with increasing frequency ν and energy $E = h\nu$, where h is the Planck constant. The higher the frequency and the energy, the higher the penetrating power of the radiation. As the name says, when ionizing electromagnetic waves propagate through a medium they ionize its atoms or molecules by removing electrons.

- UV-B radiation originates from electronic transitions of electrons that are tightly bound to an atom, with relatively large binding energies.

- X-rays originate from the inner shells of atom—electrons in these inner shells are closer to the nucleus than outer valence electrons and therefore experience the full nuclear charge without being shielded from it, as is the case for the outer electrons. As a consequence, inner electrons are more tightly bound and their excitation involves higher energies and frequencies than those associated with the outer, less tightly bound, electrons.

- γ-rays originate in nuclear radioactivity and in matter–antimatter annihilations.

- α-particles are massive particles composed of two protons and two neutrons, i.e., helium nuclei, and originate in radioactive decay. They also have the ability to ionize the material they propagate through and their penetrating power depends on their energy.

2 **(B)** The absorption of radiation by human tissues can be schematically described as follows. Consider an electromagnetic wave propagating along the positive x-axis and described by the electric field

$$\vec{E} = \vec{E}_0 \, e^{i(kx - \omega t)}$$

for $x \leq 0$, where $\vec{E}_0$ is a constant vector. The wave enters a human tissue modeled by a semi-infinite slab at $x = 0$ and a component is partially transmitted and partially absorbed, while another component is reflected, as described by the electric field

$$\vec{E}(t, \vec{x}) = \vec{E'}_0 \, e^{-x/\delta} \, e^{i(k'x - \omega t)} + \vec{E''}_0 \, e^{i(-kx - \omega t)},$$

for $x > 0$, where $\vec{E'}_0$ and $\vec{E''}_0$ are constants. The new wave vector k' and the *skin depth* δ are given by the equations

$$\left(k'\right)^2 \;=\; \mu\epsilon\omega^2 + \frac{(\mu\sigma\omega)^2}{4\left(k'\right)^2}, \qquad (4.5)$$

$$\delta \;=\; \frac{2k'}{\mu\sigma\omega}, \qquad (4.6)$$

where μ, ϵ, and σ are, respectively, the magnetic permeability, dielectric constant, and conductivity of the material. Discuss Eqs. (4.5) and (4.6) in the limits of an ideal conductor $\sigma \to +\infty$ and of an ideal dielectric $\sigma \to 0$ (in reality, human tissues are neither ideal conductors nor ideal insulators). In both cases, find an expression for the phase velocity of the transmitted wave.

If $\delta = 15$ cm, at what value of x is the electric field amplitude reduced by 90%?

Solution

In the case of an ideal conductor ($\sigma \to +\infty$) the first term in the right-hand side of Eq. (4.5) can be neglected, obtaining

$$k' = \sqrt{\frac{\mu\sigma\omega}{2}},$$

while Eq. (4.6) yields

$$\delta = \sqrt{\frac{2}{\mu\sigma\omega}} \longrightarrow 0.$$

The phase velocity of the damped part of the wave is

$$v = \frac{\omega}{k'} = \sqrt{\frac{2\omega}{\mu\sigma}} \longrightarrow 0$$

as $\sigma \to +\infty$.

In the case of an ideal dielectric ($\sigma \to 0$), Eqs. (4.5) and (4.6) yield

$$k' \simeq \sqrt{\epsilon\mu}\,\omega,$$

$$\delta = \frac{2k'}{\mu\sigma\omega} \simeq \frac{2}{\sigma}\sqrt{\frac{\epsilon}{\mu}};$$

then $\delta \to +\infty$ and the transmitted wave is not absorbed, $e^{-x/\delta} \to 1$. The phase velocity is

$$v' = \frac{\omega}{k'} = \frac{1}{\sqrt{\mu\epsilon}}.$$

The amplitude of the electric field describing the part of the wave that is partially transmitted and partially absorbed is reduced by 90% when $e^{-x/\delta} = 0.1$, or $x = (-\ln 0.1)\,\delta \simeq 2.3\,\delta = 35$ cm.

3 **(A)** What is the strength of the magnetic field 5 m, 10 m, and 50 m away from a lightning bolt during the short time it carries a current of $1.0 \cdot 10^4$ A? For simplicity, approximate the bolt with a straight line. Compare your result with the strength of the geomagnetic field $B_g \simeq 10^{-4}$ T and with the strength of the field below a high voltage power line, $B_0 = 10^{-3}$ T. Is there a point in studying the health effects of manmade electromagnetic fields?

Solution

The magnetic field surrounding a long, thin, straight wire is

$$B = \frac{\mu_0 I}{2\pi r},$$

where μ_0 is the magnetic permeability of vacuum, I is the intensity of the current carried by the conductor, and r is the distance from the wire. The geomagnetic field's average intensity is $B_g \simeq 10^{-4}$ T. At 5 m from the lightning bolt,

$$B = \frac{\left(4\pi \cdot 10^{-7} \,\text{T} \cdot \text{m} \cdot \text{A}^{-1}\right) \cdot \left(1.0 \cdot 10^4 \,\text{A}\right)}{2\pi \cdot (5\,\text{m})} = 4 \cdot 10^{-4} \,\text{T} = 4B_g.$$

At 10 m, $B = 2 \cdot 10^{-4}\,\text{T} = 2B_g$ and, at 50 m from the bolt, $B = 4 \cdot 10^{-5}\,\text{T} = 0.4B_g$.

Even at the very small distance of 10 m from the lightning bolt, the ratio between the magnetic induction due to the bolt and the magnetic induction B_0 below a high voltage power line is

$$\frac{B\,(10\,\text{m})}{B_0} = \frac{2 \cdot 10^{-4}\,\text{T}}{10^{-3}\,\text{T}} = 0.2.$$

Therefore, electromagnetic fields produced by human activities can have much larger intensities than natural electromagnetic fields, even when the latter are at their strongest intensities. In principle there is a point in studying the health effects of manmade electromagnetic fields.

4.6   Environmental spectroscopy

The quantum structure of matter (atoms, molecules, and solids) determines discrete energy levels: quantum systems can only absorb or emit electromagnetic radiation in discrete packets of energy corresponding to the difference between two energy levels—a resonance phenomenon. This fact explains, for example, why absorption of electromagnetic radiation is so highly selective in frequency. Emission spectra by gases consist of spectral lines characteristic of the atom or molecule that constitute a fingerprint useful to identify that atom or molecule. Spectroscopy in environmental applications is used to detect the presence of a molecule naturally present or of a pollutant in the atmosphere, soil, and water. Examples are the monitoring of the concentration of ozone in the stratosphere with remote sensing, the measurement of CO_2 (the main greenhouse gas) abundance in tree rings formed a few centuries ago, or the detection of toxic chemicals in soil and water. Suggested introductory references are [68] and [4].

1 **(C)** Summarize the main spectroscopic techniques useful in environmental science according to the energy levels excited, and give examples of their uses.

Solution

Spectroscopic techniques can be classified according to the energy levels of atoms and molecules that are excited and originate spectra.

- *Electronic transitions* generate spectra in the near-infrared, visible, and ultraviolet regions; transitions involving inner shell electrons produce X-rays. Optical spectroscopy, X-ray emission and absorption, photoelectron, Auger, and *PIXE* (particle-induced X-ray emission) spectroscopy are based on these transitions.

- Rotational and vibrational spectroscopy study transitions between *rotational and vibrational states of molecules*. Raman spectroscopy studies light scattered by molecules and rotational and vibrational transitions.

- *NMR* (nuclear magnetic resonance) and *ESR* (electron spin resonance) spectroscopy study, respectively, transitions between nuclear and electron spin states in the presence of an external magnetic field in interaction with atoms or molecules.

Spectroscopy is a very important scientific tool in many fields; its applications to environmental science include qualitative and quantitative analysis of the composition of the atmosphere, soil, surface waters and groundwater, monitoring the concentration of greenhouse gases in the atmosphere, the variation of concentration of chemicals destroying the ozone layer, and the relative abundance of atmospheric CO_2 in the past (analysis of tree rings and deep ice samples).

4.6.1 Quantum mechanics

It is beyond the scope of this book and outside the interest of most environmental science students to review quantum mechanics through exercises. Here we present a few selected problems that are useful to reinforce the understanding of basic aspects of environmental spectroscopy. Standard references are [23, 41, 49].

1 **(C)** Discuss the relative size in meters of molecules, atoms, and nuclei.

Solution

Molecules have typical sizes of order 10^{-9} m, however heavy, complicated molecules composed of many atoms can be much larger. The size of the hydrogen atom is given by the Bohr radius a_0, the radius at which the probability density for the ground state of the single electron peaks: $a_0 = 0.5 \cdot 10^{-10}$ m. Hence, atoms have typical size of order 10^{-10} m$= 1$ Å. A nucleus has a typical size of 10^{-15} m $= 1$ fm (1 Fermi). Therefore, $l_{atom}/l_{molecule} \approx 10^{-1}$ and $l_{nucleus}/l_{atom} \approx 10^{-5}$.

2 **(A)** Consider a system of two identical quantum particles described by the wave function $\psi(\vec{x}_1, \vec{x}_2)$. Any function $\psi(\vec{x}_1, \vec{x}_2)$ can be decomposed into a symmetric and an antisymmetric part as

$$\psi(\vec{x}_1, \vec{x}_2) = \psi^{(S)}(\vec{x}_1, \vec{x}_2) + \psi^{(A)}(\vec{x}_1, \vec{x}_2),$$

where

$$\psi^{(S)}(\vec{x}_1, \vec{x}_2) \equiv \frac{1}{2}\left[\psi(\vec{x}_1, \vec{x}_2) + \psi(\vec{x}_2, \vec{x}_1)\right],$$

$$\psi^{(A)}(\vec{x}_1, \vec{x}_2) \equiv \frac{1}{2}\left[\psi(\vec{x}_1, \vec{x}_2) - \psi(\vec{x}_2, \vec{x}_1)\right].$$

Prove that this decomposition of ψ into a symmetric and an antisymmetric part is unique.

Consider the *exchange operator* $\hat{P}_{12}$ defined by

$$\hat{P}_{12}\,\psi(\vec{x}_1, \vec{x}_2) \equiv \psi(\vec{x}_2, \vec{x}_1);$$

what are the eigenvalues and the eigenvectors of $\hat{P}_{12}$? What kind of particles do they describe?

Solution
Let

$$\psi(\vec{x}_1, \vec{x}_2) = A(\vec{x}_1, \vec{x}_2) + B(\vec{x}_1, \vec{x}_2)$$

with A symmetric and B antisymmetric be another decomposition of ψ into symmetric and antisymmetric parts. Then

$$\psi(\vec{x}_2, \vec{x}_1) = A(\vec{x}_2, \vec{x}_1) + B(\vec{x}_2, \vec{x}_1)$$

$$= A(\vec{x}_1, \vec{x}_2) - B(\vec{x}_1, \vec{x}_2).$$

By adding and subtracting the expressions of $\psi(\vec{x}_1, \vec{x}_2)$ and $\psi(\vec{x}_2, \vec{x}_1)$ we obtain, respectively,

$$A(\vec{x}_1, \vec{x}_2) = \frac{1}{2}\left[\psi(\vec{x}_1, \vec{x}_2) + \psi(\vec{x}_2, \vec{x}_1)\right] \equiv \psi^{(S)}(\vec{x}_1, \vec{x}_2),$$

$$B(\vec{x}_1, \vec{x}_2) = \frac{1}{2}\left[\psi(\vec{x}_1, \vec{x}_2) - \psi(\vec{x}_2, \vec{x}_1)\right] \equiv \psi^{(A)}(\vec{x}_1, \vec{x}_2),$$

which proves the uniqueness of the decomposition.

The operator $\hat{P}_{12}$ is nilpotent, i.e., $\hat{P}_{12}{}^2 = \hat{I}_d$, where $\hat{I}_d$ is the identity operator, hence the eigenvalues a of $\hat{P}_{12}$ simultaneously satisfy the relations

$$\hat{P}_{12}\,\psi = a\psi$$

and

$$\hat{P}_{12}{}^2\psi = a^2\psi = \psi,$$

which imply that $a^2 = 1$, or that the eigenvalues are $a = \pm 1$. *Symmetric* wavefunctions are associated with the eigenvalue $+1$ and describe *bosons*, while *antisymmetric* wavefunctions are associated with the eigenvalue -1 and describe *fermions*.

3 **(B)** The width of a spectral line due to *lifetime broadening* (or *uncertainty broadening*) is described by the Lorentzian function

$$L(E) = \frac{\hbar/\tau}{(E - E_f)^2 + [\hbar/(2\tau)]^2}, \tag{4.7}$$

where E_f is the energy of the spectral line and τ is the lifetime of the excited state. Study analytically the shape of the Lorentzian curve, sketch its graph, and find its full width at half-maximum (FWHM). Does the area under the Lorentzian curve depend on the lifetime τ?

Solution
We have $L(E) = 2f(x)$, where $x \equiv E$, $x_0 = E_f$, and

$$f(x) = \frac{\alpha}{(x - x_0)^2 + \alpha^2},$$

with $\alpha \equiv \hbar/(2\tau)$. The function $f(x)$ is regular for every value of x and its graph is symmetric about the vertical line $x = x_0$. If one performs the translation $x \to x' = x - x_0$, then $f(x') = f(-x')$. In addition,

$$\lim_{x \to \pm\infty} f(x) = 0.$$

The first derivative of f is

$$f'(x) = \frac{-2\alpha\,(x - x_0)}{[(x - x_0)^2 + \alpha^2]^2}.$$

We have $f'(x) > 0$ for $x < x_0$, $f'(x_0) = 0$, and $f'(x) < 0$ for $x > x_0$, hence $f(x)$ has a local and absolute maximum $f_{\max} = f(x_0) = \alpha^{-1}$ at $x = x_0$ (or $E = E_f$). The graph of the Lorentzian (in units $1/E_f$)

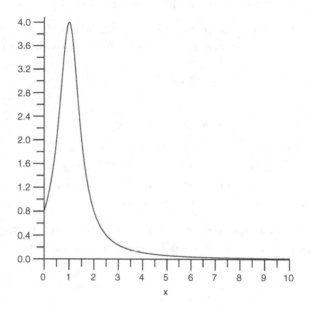

Figure 4.5. A Lorentzian curve.

versus x (in units E_f) is sketched in Fig. 4.5 for $\alpha = 0.5 E_f$.

The half-maximum of f is attained at the values $x_\pm$ of x satisfying the equation

$$\frac{f_{\max}}{2} = \frac{\alpha}{(x - x_0)^2 + \alpha^2},$$

or $(x - x_0)^2 = \alpha^2$, which yields $x_\pm = x_0 \pm \alpha$ corresponding to $E_\pm = E_f \pm \hbar/(2\tau)$. The FWHM is then

$$\Delta E = E_+ - E_- = \hbar/\tau.$$

As α increases, the maximum of the Lorentzian $2 f_{\max} = 2\alpha^{-1}$ decreases and the curve becomes less and less peaked around E_f while the FWHM $\Delta E = 2\alpha$ increases. The total area between the Lorentzian curve and the x-axis is given by

$$\int_{-\infty}^{+\infty} 2 f(x) dx = 2\alpha \int_{-\infty}^{+\infty} dx \frac{1}{(x - x_0)^2 + \alpha^2}$$

$$= 2\alpha \int_{-\infty}^{+\infty} dz \frac{1}{z^2 + \alpha^2}$$

$$= 2\alpha \lim_{M \to +\infty} \left[\frac{1}{\alpha} \operatorname{arctg} \left(\frac{z}{\alpha} \right) \right]_{-M}^{+M} = 2\pi,$$

and therefore does not depend on α nor on the lifetime τ of the excited state.

4.6.2 Vibrational and rotational levels of molecules

For the environmental scientist, rotational and vibrational spectroscopy are very useful to study the presence and abundance of specific molecules — usually pollutants — in water, in the atmosphere, or in samples of material. Like electronic energy levels, the rotational and vibrational energies of a molecule are also quantized. There are very common diatomic molecules that are quite simple and have relatively uncomplicated rotational and vibrational spectra, described by simple formulas for the energy levels that can be used without extensive knowledge of quantum mechanics.

1 **(A)** The CO molecule has a bond strength $k = 1860 \, \text{N·m}^{-1}$ while the masses of C and O are, respectively, $1.99 \cdot 10^{-26}$ kg and $2.66 \cdot 10^{-26}$ kg. Find the frequency $\nu_{(12,11)}$ corresponding to the transition $v = 12 \longrightarrow v = 11$ between vibrational levels of the CO molecule. Is the frequency $\nu_{(9,8)}$ corresponding to the $v = 9 \longrightarrow v = 8$ transition different from $\nu_{(12,11)}$?

Solution
The CO molecule is a linear oscillator and in the harmonic approximation its vibrational energy levels are the energy levels of the one-dimensional quantum mechanical harmonic oscillator (see, e.g., Refs. [61, 41, 49, 23])

$$E_v = \left(v + \frac{1}{2} \right) \hbar\omega \qquad v = 0, 1, 2, 3, \cdots,$$

where $\omega = (k/\mu)^{1/2}$ and μ is the reduced mass of the molecule. The energy difference between adjacent energy levels is

$$\Delta E_{(v+1,v)} = \left[\left(v + 1 + \frac{1}{2} \right) - \left(v + \frac{1}{2} \right) \right] \hbar\omega = \hbar\omega$$

and does not depend on v: the energy levels of the harmonic oscillator are equally spaced. Note that transitions between adjacent energy levels are the only ones allowed by the selection rule $\Delta v = \pm 1$ [41, 49, 23].

The reduced mass of the molecule is

$$\mu = \frac{m_C m_O}{m_C + m_O}$$

$$= \frac{(1.99 \cdot 10^{-26}\,\text{kg}) \cdot (2.66 \cdot 10^{-26}\,\text{kg})}{(1.99 \cdot 10^{-26}\,\text{kg}) + (2.66 \cdot 10^{-26}\,\text{kg})}$$

$$= 1.14 \cdot 10^{-26}\,\text{kg},$$

and the frequency of any $v + 1 \longrightarrow v$ transition is

$$\nu_{(v+1,\,v)} = \frac{1}{2\pi}\sqrt{\frac{k}{\mu}} = \frac{1}{2\pi}\sqrt{\frac{1860\,\text{N} \cdot \text{m}^{-1}}{1.14 \cdot 10^{-26}\,\text{kg}}} = 6.43 \cdot 10^{13}\,\text{Hz}.$$

2 (A) The separation between the carbon and oxygen atoms in the CO molecule is $r = 1.13 \cdot 10^{-10}$ m and the masses of the C and O atoms are, respectively, $1.99 \cdot 10^{-26}$ kg and $2.66 \cdot 10^{-26}$ kg. Compute the energy of the two lowest rotational states of the molecule in Joules and in electronvolts.

The strength of the C-O bond is $k = 1860\,\text{N} \cdot \text{m}^{-1}$. Find the energies of the first two vibrational levels of the molecule in Joules and in electronvolts. Compare the orders of magnitude of the energy of the vibrational and the rotational levels.

Solution
The CO molecule is a linear rotor, and the two-body problem can be reduced to an equivalent one for a single particle with reduced mass μ at distance r from the rotation axis. The reduced mass is

$$\mu = \frac{m_1 m_2}{m_1 + m_2} = \frac{(1.99 \cdot 10^{-26}\,\text{kg}) \cdot (2.66 \cdot 10^{-26}\,\text{kg})}{(1.99 \cdot 10^{-26}\,\text{kg}) + (2.66 \cdot 10^{-26}\,\text{kg})}$$

$$= 1.14 \cdot 10^{-26}\,\text{kg}.$$

The moment of inertia for the reduced problem is

$$\mathcal{I} = \mu r^2 = (1.14 \cdot 10^{-26}\,\text{kg}) \cdot (1.13 \cdot 10^{-10}\,\text{m})^2 = 1.46 \cdot 10^{-46}\,\text{kg} \cdot \text{m}^2.$$

The rotational energy levels are given by

$$E_J = J(J + 1)\,hcB,$$

where $B = \hbar/(4\pi c\mathcal{I})$ is the *rotational constant* and $J = 0, 1, 2, 3, \dots$. Hence,

$$E_J = \frac{\hbar^2}{2\mathcal{I}} J(J+1) = 3.8 \cdot 10^{-23}\,J(J+1)\ \text{Joules} = 2.38 \cdot 10^{-4} J(J+1)\ \text{eV}.$$

The two lowest rotational states corresponding to $J = 0$ and $J = 1$ have energies $E_0 = 0$ and

$$E_1 = 7.60 \cdot 10^{-23} \text{ Joules} = 4.76 \cdot 10^{-4} \text{ eV}.$$

The vibrational energy levels of a diatomic molecule are given by the energy eigenvalues of the one-dimensional quantum harmonic oscillator [61, 41, 49, 23],

$$E_v = \left(v + \frac{1}{2}\right)\hbar\omega \qquad (v = 0, 1, 2, ...),$$

where $\omega = \sqrt{k/\mu}$ is the vibrational angular frequency. For the CO molecule,

$$\omega = \left(\frac{1860 \text{ N m}^{-1}}{1.14 \cdot 10^{-26} \text{ kg}}\right)^{1/2} = 4.04 \cdot 10^{14} \frac{\text{rad}}{\text{s}}.$$

The energy levels are

$$E_v = \left(v + \frac{1}{2}\right) \cdot (4.26 \cdot 10^{-20} \text{ J}) = 0.266 \left(v + \frac{1}{2}\right) \text{ eV}.$$

The energies of the two lowest vibrational states are $E_0 = 0.133$ eV and $E_1 = 0.399$ eV, and the ratio between rotational and vibrational energies is of the order

$$\frac{E_J}{E_v} = \frac{2.38 \cdot 10^{-4} \text{ eV}}{0.266 \text{ eV}} \simeq 10^{-3}.$$

3 **(A)** The bond strength of the H_2 molecule is $k = 5.80 \cdot 10^2 \text{ N} \cdot \text{m}^{-1}$, and the internuclear separation is $d = 8.00 \cdot 10^{-11}$ m. Compute the energy corresponding to the $J = 0 \rightarrow J = 1$ rotational transition and the energy of the $v = 1$ vibrational level.

Solution
The H_2 molecule is a linear rotor and the two-body system is mechanically equivalent, in the center of mass frame, to a single particle of reduced mass $\mu = m_p/2$ (where m_p is the proton mass) at a distance d from the center of force. The moment of inertia of the reduced particle relative to rotations about an axis perpendicular to the H-H axis is

$$\mathcal{I} = \mu d^2 = \frac{m_p d^2}{2}.$$

The rotational energy eigenvalues are given by

$$E_J = \frac{J(J+1)\hbar^2}{2\mathcal{I}} = \frac{J(J+1)\hbar^2}{m_p d^2} \qquad (J = 0, 1, 2, ...).$$

The energy difference corresponding to the $J = 0 \rightarrow J = 1$ transition is

$$\Delta E = E_1 - E_0 = E_1 = \frac{2\hbar^2}{m_p d^2}$$

$$= \frac{2 \cdot (1.054 \cdot 10^{-34}\ \mathrm{J \cdot s})^2}{(1.67 \cdot 10^{-27}\ \mathrm{kg}) \cdot (8.00 \cdot 10^{-11}\ \mathrm{m})^2} = 2.08 \cdot 10^{-21}\ \mathrm{J}.$$

The vibrational energy eigenvalues are given by

$$Ev = \left(v + \frac{1}{2}\right) \hbar\omega \qquad (v = 0, 1, 2, ...),$$

where $\omega = \sqrt{k/\mu} = \sqrt{2k/m_p}$. The energy of the $v = 1$ level is

$$E\,(v = 1) = \frac{3}{2}\hbar\omega = 3\hbar\sqrt{\frac{k}{2m_p}}$$

$$= 3 \cdot (1.054 \cdot 10^{-34}\ \mathrm{J \cdot s}) \cdot \left[\frac{(5.80 \cdot 10^2\ \mathrm{N \cdot m^{-1}})}{2 \cdot (1.67 \cdot 10^{-27}\ \mathrm{kg})}\right]^{1/2}$$

$$= 1.32 \cdot 10^{-19}\ \mathrm{J}.$$

4 **(A)** Find the rotational spectrum of a triatomic molecule with principal moments of inertia $\mathcal{I}_1, \mathcal{I}_2$ and rotational Hamiltonian operator

$$\hat{H}_{\mathrm{rot}} = \frac{\hat{L}_x^2 + \hat{L}_y^2}{2\mathcal{I}_1} + \frac{\hat{L}_z^2}{2\mathcal{I}_2}.$$

Treat the molecule as if it were rigid.

Solution
In the approximation of constant internuclear distances $\mathcal{I}_1$ and $\mathcal{I}_2$ are constant and, since

$$\hat{L}_x^2 + \hat{L}_y^2 = \hat{L}^2 - \hat{L}_z^2,$$

we obtain

$$\hat{H}_{\mathrm{rot}} = \frac{\hat{L}^2}{2\mathcal{I}_1} + \frac{\hat{L}_z^2}{2}\left(\frac{1}{\mathcal{I}_2} - \frac{1}{\mathcal{I}_1}\right).$$

The rotational Hamiltonian is constructed from the operators $\hat{L}^2$ and $\hat{L}_z$ and it commutes with them. Therefore, the energy eigenstates are simultaneous eigenstates of these operators. These eigenstates are the spherical harmonics $Y_{lm}(\theta, \varphi)$, which satisfy

$$\hat{L}^2 Y_{lm} = l(l+1)\hbar^2 Y_{lm},$$

$$\hat{L}_z Y_{lm} = l\, m\hbar\, Y_{lm},$$

where l and m assume the values $l = 0, 1, 2, 3, \ldots$ and $m = -l, -l + 1, \ldots 0, \ldots, l-1, l$. The eigenvalues of the rotational Hamiltonian are then

$$E_{lm} = \frac{\hbar^2}{2}\left[\frac{l(l+1)}{\mathcal{I}_1} + m^2\left(\frac{1}{\mathcal{I}_2} - \frac{1}{\mathcal{I}_1}\right)\right].$$

4.7   Radioactivity

The phenomenon of radioactivity is due to the decay of naturally occurring or artificially produced isotopes. These can originate naturally in the decay of parent nuclei or in the bombardment of atoms by cosmic rays in the atmosphere. One naturally occurring isotope, radon, causes environmental problems: accumulation of radon in gaseous form in buildings has been discovered to be potentially dangerous for humans (*radon problem*).

In addition to natural radioactivity, manmade radioactive isotopes are used in nuclear power generation and its related aspects such as uranium mining and the storage and transport of radioactive waste, nuclear weapons and nuclear testing, and medical applications (radiation therapy is widely used in cancer treatments). Exercises on these aspects require some knowledge of nuclear physics and are beyond the scope of this book. In this section we focus on the main phenomenological features of radioactivity.

1 **(A)** Find an approximate value for the density of the nuclide ^{55}Cs. Does the nuclear density of the elements in the periodic table depend on the atomic mass number A?

Solution
The radius of a nucleus is given by the approximate law

$$r = r_0\, A^{1/3},$$

where $r_0 = 1.2$ fm and A is the mass number of the nuclide. The density of ^{55}Cs is approximately

$$\rho = \frac{m}{V} = \frac{m}{4\pi r^3/3} = \frac{3\,(A\,\text{a.m.u.})}{4\pi r_0^3 A} = \frac{3 \cdot (1.66 \cdot 10^{-27}\,\text{kg})}{4\pi \cdot (1.2 \cdot 10^{-15}\,\text{m})^3}$$

$$= 2.30 \cdot 10^{17}\,\text{kg}.$$

The atomic mass number A cancels out and the value of ρ computed is therefore typical for the nuclear density of most nuclides.

2 **(A)** Radon, a radioactive element, is in the gas phase at room temperature. It originates from the decay of uranium and thorium in rocks and soil and can invade foundations of buildings by seeping out of the ground and accumulating to dangerously high levels (*radon problem*). The isotope $^{222}_{86}$Rn is a decay product of ^{238}U, while $^{220}_{86}$Rn originates from the decay of ^{232}Th. What are the electron, proton, and neutron contents of $^{222}_{86}$Rn and $^{220}_{86}$Rn? How can one minimize the radon problem in buildings?

Solution
To answer this question, we must look at the atomic and mass numbers of the isotopes considered. The atomic number denotes the number of protons in the nucleus, which is also the number of electrons of the neutral (nonionized) isotope. $^{222}_{86}$Rn has 86 proton, 86 electrons, and $(222 - 86) = 136$ neutrons, while $^{220}_{86}$Rn has 86 protons, 86 electrons, and $(220 - 86) = 134$ neutrons.

The radon problem in buildings can be minimized by sealing all cracks in the foundations and walls and by installing adequate ventilation systems to prevent accumulation of the radioactive gas in crawl spaces and in basements. Basements without concrete foundations should be avoided and adequate ventilation provided, especially in winter.

3 **(A, B)** Radon (a naturally occurring radioactive gas) penetrates into the basement of a house by diffusing through a homogeneous slab of compacted soil in its basement. Let d be the thickness of the horizontal slab and describe it, for simplicity, as extending to infinity in the x and y directions; let D be the diffusion coefficient. Let C_B and C_T be the radon concentration at the bottom and the top of the slab, respectively. Assuming a stationary regime (i.e., C does not depend on time), find the dependence $C(z)$ of C on the vertical coordinate z.

Solution 1 (level A)
We have, using Fick's law $F = -\Delta C/\Delta x$,

$$F = -D\,\frac{C_T - C(z)}{d - z}$$

between the top of the slab and level z, and

$$F = -D\,\frac{C(z) - C_B}{z}$$

between level z and the bottom of the slab. By comparing the two expressions of the flux we obtain

$$\frac{C_T - C(z)}{d - z} = \frac{C(z) - C_B}{z},$$

from which

$$C_T z - C(z)z = C(z)d - C(z)z - C_B d + C_B z$$

and

$$C(z) = (C_T - C_B)\,\frac{z}{d} + C_B.$$

Solution 2 (level B)
The diffusion equation

$$\frac{\partial C}{\partial t} = D\,\nabla^2 C \tag{4.8}$$

reduces to

$$\frac{d^2 C}{dx^2} = 0,$$

with the linear solution $C(z) = \alpha z + \beta$. The boundary conditions

$$C(z = 0) = C_B$$

and

$$C(z = d) = C_T$$

yield, respectively, $\beta = C_B$ and $\alpha = (C_T - C_B)/d$ and the solution is

$$C(z) = (C_T - C_B)\,\frac{z}{d} + C_B.$$

Solution 3 (level B)
The flux of radon is $\vec{F} = (0, 0, F_z)$. The continuity equation

$$\frac{\partial C}{\partial t} + \vec{\nabla} \cdot \vec{F} = 0$$

leads to $\partial F_z/\partial z = 0$ (i.e., F does not depend on z) in the stationary regime $\partial C/\partial t = 0$. Fick's law $\vec{F} = -D\vec{\nabla}C$ for the flux density $(0, 0, F)$ then yields $\partial F/\partial z = 0$ and

$$\frac{\partial C}{\partial z} = \frac{dC}{dz} = -\frac{F}{D} = \text{constant.}$$

Integration then gives

$$C(z) = -\frac{F}{D}z + C_0.$$

The boundary conditions yield, as in Solution 2,

$$C(z) = (C_T - C_B)\frac{z}{d} + C_B$$

with $F = (C_B - C_T)\,D/d > 0$ and $dC/dz = -F/D < 0$.

4 **(A)** A radioactive nucleus $_Z^A$X is observed to emit an α-particle and a γ-ray, followed by two β-particles and two γ-rays. What are the mass and atomic number of the final nucleus?

Solution
An α-particle (helium nucleus) consists of two protons and two neutrons and carries away four units of mass, that is, the mass number of the nucleus decreases by four. Two protons are emitted in this decay and therefore the atomic number decreases by two. The final nucleus is $_{Z-2}^{A-4}$X.

5 **(A)** Discuss how the principles of conservation of electric charge and baryon number apply to the nuclear reaction

$$C^{14} \longrightarrow N^{14} + e^-.$$

Solution
All isotopes of carbon have 6 protons and C^{14} is no exception; nitrogen has 7 protons and in the nuclear reaction considered, a nucleus of C^{14} acquires a positive charge by emitting an electron (β-particle). On the left-hand side of the nuclear reaction the C^{14} nucleus has electric charge $+6e$, while on the right-hand side the N^{14} nucleus has charge $+7e$ and the electron has charge $-e$, with total electric charge $+6e$ on the right-hand side. Electric charge is conserved in this nuclear process (charge $+6e$ before the decay and charge $+6e$ after the decay).

Protons and neutrons each carry baryon number +1, while the electron is assigned baryon number 0. On the left-hand side of the reaction there are 14 nucleons in C^{14}, 6 protons and 8 neutrons, for a total baryon number 14. On the right-hand side there are again 14 nucleons in N^{14} (7 protons and 7 neutrons) with total baryonic number 14 (the electron does not contribute to the baryonic number). Also, the baryonic number is conserved in this nuclear reaction.

6 **(B)** Consider a radioactive sample with a half-life of 10.5 days, write the differential equation expressing the number of atoms $N(t)$ present in the sample as a function of time, and solve it. Express the disintegration rate r after 15 days in units of the initial disintegration rate r_0.

Solution
The number of atoms decaying per unit time is proportional to the number N of atoms present in the sample at that particular instant of time, or

$$-\frac{dN(t)}{dt} = \alpha N(t),$$

where $\tau \equiv \alpha^{-1} \ln 2$ is the *half-life* of the nuclide. The solution of this elementary differential equation is

$$N(t) = N_0 \, e^{-\alpha t},$$

where $N_0 = N(0)$ is the initial number of atoms. By using the fact that $\alpha = \tau^{-1} \ln 2$, one also has

$$N(t) = N_0 \, 2^{-t/\tau}.$$

The disintegration rate at time t is

$$r(t) \equiv \frac{dN}{dt} = -\alpha N(t),$$

and the ratio between the disintegration rate at $t = 15$ days and the initial disintegration rate is

$$\frac{r(t)}{r_0} = \frac{-\alpha N_0 \, e^{-\alpha t}}{-\alpha N_0} = e^{-\alpha t} = \exp\left(-\frac{15\,\text{days}}{10.5\,\text{days}} \ln 2\right) = 0.37,$$

or $r\,(15\,\text{days}) = 0.37\,r_0$.

7 **(B)** The radioactive isotope C^{14} is present in the atmosphere together with the much more abundant isotope C^{12}. The ratio of the abundances of isotopes C^{14} and C^{12} in the atmosphere is constant, having reached its equilibrium value many millions of years ago. C^{14}

decays with a half-life $\tau_{14} = 5730$ years. A living organism constantly absorbs C^{14} and C^{12} in the atmospheric ratio

$$r_0 \equiv \frac{\text{number of } C^{14} \text{ nuclides}}{\text{number of } C^{12} \text{ nuclides}} = 1.6 \cdot 10^{-10}.$$

When the organism dies, C^{14} is not absorbed anymore, it decays and disappears as an isotope, whereas C^{12} remains and the ratio r decreases with time. Derive a formula[2] that helps an archaeologist dating the remains that she has just uncovered at time t from the measured ratio $r(t) = 8.5 \cdot 10^{-12}$.

Solution
At the time t_0 when the organism died, the ratio between the number N_{14} of C^{14} isotopes and the number N_{12} of C^{12} isotopes was

$$r_0 = \left. \frac{N_{14}}{N_{12}} \right|_{t_0} = 1.6 \cdot 10^{-10}.$$

Since N_{12} is constant while $N_{14}(t) = N_{14}(t_0) \, 2^{-\frac{(t-t_0)}{\tau_{14}}}$, the ratio r at time t is

$$r(t) = r_0 \, 2^{-\frac{(t-t_0)}{\tau_{14}}} = r_0 \, e^{-\frac{(t-t_0)}{\tau_{14}} \ln 2}$$

and, by taking the logarithm of both sides,

$$t - t_0 = -\frac{\tau_{14}}{\ln 2} \ln \left[\frac{r(t)}{r_0} \right] = -\frac{5730 \, \text{years}}{\ln 2} \ln \left(\frac{8.5 \cdot 10^{-12}}{1.6 \cdot 10^{-10}} \right)$$

$$= 24000 \, \text{years}.$$

8 **(B)** A radioactive substance has decay constant r and is produced at a constant rate (number of particles created per second) α. Find the number $N(t)$ of nuclides of that nuclear species present at time t if at the initial time $t = 0$ this number is $N(0) = N_0$. How many nuclides will be present as $t \to +\infty$? Give a practical meaning to this limit.

Solution
The rate of change of the number of nuclides with time is the creation rate α (input) minus the decay rate (output) $rN(t)$. Therefore,

$$\frac{dN}{dt} = \alpha - rN(t). \tag{4.9}$$

[2]This is the basis of the method of radiocarbon dating discovered by W. F. Libby, winner of the Nobel Prize for chemistry in 1960.

The general solution of this first-order linear, nonhomogeneous ODE is the sum the general solution of the complementary equation

$$dN/dt = -rN(t)$$

(which is $C\mathrm{e}^{-rt}$, where C is an integration constant) and of a particular solution N_p of the nonhomogeneous equation. By using the method of variation of parameters we look for a particular solution of the form $N_p(t) = u(t)\,\mathrm{e}^{-rt}$. Substitution into Eq. (4.9) yields

$$\frac{du}{dt} = \alpha\,\mathrm{e}^{rt}$$

and

$$u(t) = \frac{\alpha}{r}\,\mathrm{e}^{rt};$$

hence,

$$N_p(t) = \frac{\alpha}{r}.$$

The general solution of the inhomogeneous equation is therefore

$$N(t) = \frac{\alpha}{r} + C\mathrm{e}^{-rt}.$$

By imposing the initial condition $N(0) = N_0$ one determines that $C = N_0 - \alpha/r$ and

$$N(t) = \frac{\alpha}{r} + \left(N_0 - \frac{\alpha}{r}\right)\mathrm{e}^{-rt}.$$

This solution is the sum of a transient that dies off exponentially fast and of the steady-state α/r. At late times $t \to +\infty$ we have $N(t \to +\infty) \approx \alpha/r$. In practice, because of the exponentially fast decay, the late time limit $t \to +\infty$ can be taken as $t \approx$ a few time scales r^{-1}.

The steady-state solution can also be found directly by looking for equilibrium solutions and setting $dN/dt = 0$. The fact that this equilibrium state is approached irrespective of the initial condition N_0 means that the steady-state is stable. This guess is confirmed by a perturbation analysis. Let

$$N(t) = \frac{\alpha}{r} + \delta N(t),$$

where δN is a perturbation. By substituting this expression into Eq. (4.9) for $N(t)$, we find the evolution equation for the perturbation δN:

$$\frac{d\,(\delta N)}{dt} = -r\delta N,$$

which has the general solution $\delta N(t) = \delta_0\, e^{-rt}$ (with $\delta_0 = \delta N(t=0)$). Perturbations of any sign or amplitude decay exponentially fast and the steady-state is asymptotically stable.

9 **(A)** The accident at the Chernobyl nuclear power plant in April 1986 released a large amount of ^{137}Cs with half-life $\tau = 30$ years. When will the number of atoms of this nuclide[3] be reduced by 90% of its original value?

Solution
The number of ^{137}Cs atoms present at time t after the disaster is

$$N(t) = N_0\, 2^{-t/\tau} = N_0\, e^{-\frac{t\ln 2}{\tau}};$$

it is reduced by 90% of its original value N_0 when $N(t) = 0.1\, N_0$, or

$$t = -\frac{\ln 0.10}{\ln 2}\, \tau = 3.3\, \tau \simeq 100\,\text{years}.$$

Ninety percent of the ^{137}Cs atoms will have disappeared due to radioactive decay by the year 2086.

10 **(B)** Some amounts of Sr^{90} were injected into the environment as fallout during atmospheric nuclear tests and, because it is chemically similar to calcium, it tends to be assimilated by organisms in the same way and ends up in bone tissues. The half-life of Sr^{90} is 28 years. When will the number of Sr^{90} nuclides released in a 1960 test be reduced to 10% of its initial value? To 1% of it?

Solution
The number of Sr^{90} nuclei present at time t is

$$N(t) = N_0\, 2^{-t/\tau} = N_0\, e^{-\frac{t\ln 2}{\tau}}$$

and the time t at which the number of nuclei present is a fraction N/N_0 of its original value is obtained by taking the logarithm of both sides of this equation, obtaining

$$t = -\frac{\tau}{\ln 2}\, \ln \frac{N}{N_0}.$$

$N(t)$ will be reduced to 10% of its original value N_0 at the time

$$t_{0.1} = -\frac{28\,\text{years}}{\ln 2}\, \ln 0.1 = 93\,\text{years},$$

---

[3]Cesium is chemically similar to potassium and it tends to be absorbed by living organisms in the same way.

i.e., in the year 2053. Reduction to 1% of the original value will occur in the time

$$t_{0.01} = -\frac{28 \, \text{years}}{\ln 2} \ln 0.01 = 186 \, \text{years},$$

i.e., in the year 2146.

Chapter 5

ENERGY AND THE ENVIRONMENT

Make everything as simple as possible, but no simpler.

—Albert Einstein

The purposes of studying energy and energy transport in environmental physics are manifold. The planet receives energy from the Sun and emits energy in space, part of which is reflected back or trapped in the atmosphere because of the greenhouse effect. The planet is mostly in an equilibrium configuration in which the energy balance determines the average temperature of the Earth.

The heavy use of fossil fuels as a source of energy to face an increasing demand in the post–Industrial Revolution world during the last two centuries has led to the depletion of fossil fuel reserves and to major problems related to the emission of pollutants. Another issue is the emission of greenhouse gases and the related climate change. The conservation of energy and the need to reduce the use of fossil fuels and minimize the associated pollution problems call for energy conservation practices and the search for alternative energy sources. The conversion of energy from one form to another is unavoidably related to the environment. Even a relatively clean method of power generation such as hydroelectric generation impacts the environment (e.g., the damming of rivers).

Energy transfer is studied to attempt to reduce the need for energy, e.g., by minimizing the loss of heat from a building through the understanding of heat conduction and new insulating materials and the rational design and location of windows. Other motivations for the study of energy transfer include the need to make the transfer of thermal energy from one part of a machine to another more efficient, thus improving overall efficiency and reducing the consumption of fossil fuels; or transporting electric power from a power station to the users in a way that minimizes energy losses and the impact of power lines on the human and natural environment.

References recommended for the general aspects of energy production, transport and consumption in relation to the environment are [4, 47, 35, 16].

5.1   Mechanical energy

Various aspects of mechanics are of interest in the physical world and specifically to the environmental physicist. Here we focus on the storage and transport of mechanical energy, on the conversion of eolic[1] energy into mechanical or electrical energy, and on energy losses in vehicles.

5.1.1   Storage and transport

Understanding how, and how much, energy can be stored in a flywheel clarifies the limitations of "green" vehicles designed with the intention of limiting the consumption of gasoline or diesel fuel by storing kinetic energy in a flywheel instead of dissipating it as heat in the vehicle's brakes.

1 **(C)** In cars the continuous start-and-stop driving of rush hour wastes energy that is transformed into heat generated by the brakes, and the same is true for urban public transport vehicles at all times. Flywheels could be used to store rotational kinetic energy in the vehicle for later use. How much rotational kinetic energy is stored in a flywheel? How can you maximize it for practical purposes?

Solution
The rotational kinetic energy stored in a flywheel is

$$K = \frac{1}{2} \mathcal{I} \omega^2,$$

where $\mathcal{I}$ is the moment of inertia of the flywheel with respect to its rotation axis and ω is the angular velocity. The moment of inertia is $\mathcal{I} = \alpha M R^2$, where M is the mass of the flywheel, R is its size, and α is a dimensionless coefficient depending on the flywheel's geometry and density distribution. To maximize K one either maximizes $\mathcal{I}$ or ω,

---

[1]From the name of Eolus, the keeper of the winds in Greek mythology.

or both. Since it is not practical to increase the mass of the flywheel beyond certain limits in a vehicle, one maximizes $\mathcal{I}$ by appropriately designing the mass distribution (e.g., a thin ring has a moment of inertia that is twice that of a cylinder with equal mass and radius). In practice, we try to increase ω by using rapidly spinning flywheels. The practical upper limit on ω is set by the tensile strength of the material used, which should not deform or break under the centrifugal stress.

2 **(B)** Compute the moment of inertia of a homogeneous flywheel with the shape of a hollow cylinder of inner radius R_1, outer radius R_2, and length L with respect to the symmetry axis.

Solution
Since the material composing the flywheel is homogeneous, the mass density ρ does not depend on the position. The moment of inertia is

$$\mathcal{I} = \int \int \int_V d^3 \vec{x} \, \rho \, d^2 (\vec{x}),$$

where $d(\vec{x})$ is the orthogonal distance of the point $\vec{x}$ to the rotation axis. In cylindrical coordinates (r, φ, z),

$$
\begin{aligned}
\mathcal{I} &= \int_{R_1}^{R_2} dr \int_0^{2\pi} d\varphi \int_0^L dz \, r \rho r^2 = \rho \int_{R_1}^{R_2} dr \, r^3 2\pi L \\
&= \frac{1}{2} \rho \left(R_2^4 - R_1^4 \right) \pi L.
\end{aligned}
$$

The mass of the flywheel is

$$M = \int \int \int_V d^3 \vec{x} \, \rho = \int_{R_1}^{R_2} dr \int_0^{2\pi} d\varphi \int_0^L dz \, r \, \rho = \pi L \rho \left(R_2^2 - R_1^2 \right).$$

Hence,

$$\mathcal{I} = \frac{\pi L}{2} \rho \left(R_2^2 - R_1^2 \right) \left(R_2^2 + R_1^2 \right) = \frac{1}{2} M \left(R_1^2 + R_2^2 \right). \qquad (5.1)$$

Given equal mass, the moment of inertia does not depend on the length L of the cylinder, and Eq. (5.1) also applies to a thin cylindrical ring. In the limit $R_1 \to 0$ one recovers the expression of the moment of inertia $M R^2 / 2$ of a solid cylinder or disk.

3 **(A)** A flywheel made of homogeneous steel (density $\rho = 8.0 \cdot 10^3$ kg · m^{-3}) has the shape of a hollow cylinder of inner radius $r_1 = 40$ cm,

outer radius $r_2 = 60$ cm, and length $L = 110$ cm. If the flywheel spins with angular velocity $\omega = 14\,\text{rad}\cdot\text{s}^{-1}$, how much kinetic energy is stored in it?

The flywheel is installed on a vehicle powered by a 20 HP ($\simeq 1.49\cdot 10^4$ W) engine with 22% efficiency. How long will the vehicle run when the engine is replaced by the flywheel rotating with angular velocity $\omega = 14\,\text{rad}\cdot\text{s}^{-1}$?

Solution
The moment of inertia of the flywheel is (cf. previous problem or a table of moments of inertia)

$$I = \frac{1}{2} M \left(r_1^2 + r_2^2 \right),$$

where the mass of the flywheel is $M = \rho V = \rho L \pi \left(r_2^2 - r_1^2 \right)$. The rotational kinetic energy stored in the flywheel is

$$
\begin{aligned}
K &= \frac{1}{2} I \omega^2 = \frac{\pi}{4} \rho L \left(r_2^2 - r_1^2 \right) \left(r_2^2 + r_1^2 \right) \omega^2 \\
&= \frac{\pi}{4} \rho L \left(r_2^4 - r_1^4 \right) \omega^2 = \frac{\pi}{4} \left(8.0 \cdot 10^3 \,\text{kg} \cdot \text{m}^{-3} \right) \cdot (1.10\,\text{m}) \\
&\quad \cdot \left[(0.6\,\text{m})^4 - (0.4\,\text{m})^4 \right] \cdot \left(14\,\text{rad}\cdot\text{s}^{-1} \right)^2 = 1.4 \cdot 10^5 \,\text{J}.
\end{aligned}
$$

The power generated by the vehicle's engine is energy divided by time, $W = E/t$; if the engine is replaced by the flywheel, the vehicle will run on the stored rotational energy of the flywheel for the time

$$t = \frac{E}{W} = \frac{1.4 \cdot 10^5 \,\text{J}}{0.22 \cdot (1.49 \cdot 10^4 \,\text{W})} = 43 \,\text{s},$$

a very short time indeed. We need to increase the kinetic energy stored in the flywheel by a significant amount if it is going to be useful. Since it is not practical to increase the mass or the size of the flywheel beyond certain limits for use in a vehicle, we must increase the angular velocity ω instead. This has led to the search for new materials (e.g., fiber composites) capable of withstanding the large centrifugal stresses arising during fast rotation.

4 **(A)** In order to reduce pollution and save energy the small town of Cleverville uses a trolleybus running on the energy stored in a large cylinder of mass $m = 800$ kg, radius $R = 1$ m, spinning at the angular velocity $\omega_0 = 5$ revolutions per second at full speed.
How much kinetic energy is stored in the flywheel rotating at full speed? The power needed to run the bus is 25 HP on average. How

long will the trolleybus run on the kinetic energy stored in the flywheel alone?

Solution
The kinetic energy stored in a cylinder rotating around its axis of symmetry is

$$T = \frac{1}{2} \mathcal{I} \omega^2,$$

where $\mathcal{I} = mR^2/2$ is the moment of inertia of the cylinder. At full speed,

$$T = \frac{1}{4} \left(8 \cdot 10^2 \, \text{kg}\right) \cdot (1 \, \text{m})^2 \cdot \left(2\pi \cdot 5 \, \text{rad} \cdot \text{s}^{-1}\right)^2 = 2 \cdot 10^5 \, \text{J}.$$

The power used by the bus is $W = 25$ HP$= 25 \cdot 746$ W$= 1.87 \cdot 10^4$ W and, since power is the rate at which energy is used, $W = E/t$, the bus will run on the kinetic energy of the flywheel for

$$t = \frac{E}{W} = \frac{2 \cdot 10^5 \, \text{J}}{1.87 \cdot 10^4 \, \text{W}} = 11 \, \text{s}.$$

5 **(A)** Compute the moment of inertia of a hoop of radius R and mass M about its axis of symmetry (Fig. 5.1).

Solution
By definition the moment of inertia of a body is

$$\mathcal{I} = \sum_i m_i r_i^2, \tag{5.2}$$

where m_i is the mass of the ith particle composing the body, r_i is its perpendicular distance to its rotation axis, and the sum is extended over all particles composing the body. In a hoop all particles are at the same distance R from the rotation axis and

$$\mathcal{I} = \left(\sum_i m_i\right) R^2 = MR^2, \tag{5.3}$$

where $M = \sum_i m_i$ is the total mass of the hoop.

6 **(B)** Compute the moment of inertia of a thin homogeneous rod about an axis passing through its center and perpendicular to its length. Let l be the length of the cylindrical rod and R its radius and express your result in terms of l, R, and the mass m of the rod.

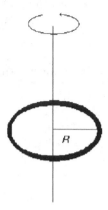

Figure 5.1. A hoop and its rotation axis.

Solution

It is appropriate to use cylindrical coordinates (r, θ, z) adapted to the axis of symmetry, that is taken as the z-axis. Then, the required moment of inertia is

$$\mathcal{I} = \int \int \int d^3 \vec{x}\, \rho\, d^2\,(\vec{x})\,,$$

where ρ is the density of the material (which is constant, since the rod is homogeneous) and $d\,(\vec{x})$ is the distance of a generic point $\vec{x}$ of the rod from the rotation axis. We have

$$\mathcal{I} = \rho \int_0^R dr \int_0^{2\pi} d\varphi \int_{-l/2}^{l/2} dz\, r z^2 = 2\pi \rho \int_0^R dr\, r \int_{-l/2}^{l/2} dz\, z^2$$

$$= 2\pi \rho \left[\frac{r^2}{2}\right]_0^R \cdot \left[\frac{z^3}{3}\right]_{-l/2}^{l/2} = \frac{\pi}{12} \rho R^2 l^3.$$

The mass of the rod is

$$m = \rho V = \rho \cdot (\pi R^2 l),$$

where V is the volume of the rod, and the moment of inertia can be written as

$$\mathcal{I} = \frac{\pi}{12} \rho R^2 l^3 = \left(\rho \pi R^2 l\right) \frac{l^2}{12} = \frac{m l^2}{12}.$$

5.1.2   Transportation and vehicles

Vehicles are a major source of pollution as is especially evident at rush hour in a big city. Limiting the emission of pollutants, or controlling their chemical nature and relative abundances, can have a significant impact on air pollution and fossil fuel consumption.

1 **(A, B)** Assume that the forces of air drag F_d and rolling resistance F_r acting on a car moving at constant speed v are constant. Find the total power dissipated against these forces.

Solution 1 (level A)
The dissipated power is equal to the work done per unit time against the friction force,
$$W = \frac{F \cdot d}{t},$$
where d is the displacement and t is the time during which the constant friction force $F = F_d + F_r$ acts. Since $d/t = v$, the constant speed of the car, we have
$$W = (F_d + F_r)\, v.$$

Solution 2 (level B)
The dissipated power W is the rate at which energy E is spent doing work against the friction forces,
$$W = \frac{dE}{dt} = \frac{d}{dt}\left(\int_0^{x(t)} dx\, F\right) = \frac{d}{dt}\left(\int_0^t Fv\, dt\right) = Fv = (F_d + F_r)\, v,$$
where $v = dx/dt$ is the velocity of the car. This expression is valid even when the friction forces and the velocity v are not constant.

2 **(A)** A modern car with mass $m = 1000$ kg, air drag coefficient $C_d = 0.3$, and frontal area $A = 1.89$ m^2 starts from rest. What is the dominant source of power loss at low and at high speeds? At what speed (in km/h) is the power dissipated against air drag equal to the power dissipated against rolling resistance? Is this speed different from the speed at which the air drag force equals the rolling resistance? The density of air is $\rho = 1.2\,\text{kg/m}^3$ and the rolling friction coefficient for a paved road is $C_r = 0.01$.

Solution
The air drag is given by
$$F_d = \frac{C_d}{2}\rho A v^2,$$

where v is the car's speed, and the rolling resistance is given by

$$F_r = C_r mg,$$

where g is the acceleration of gravity. The power dissipated against friction is $W = W_d + W_r$, where

$$W_d = \frac{C_d}{2} \rho A v^3, \qquad W_r = C_r mg v.$$

The ratio between the power dissipated against air drag and the power dissipated against rolling resistances is

$$\frac{W_d}{W_r} = \frac{C_d \rho A}{2 C_r mg} v^2,$$

proportional to the square of the car's velocity. Rolling resistances dominate at small speeds while air drag dominates at higher speed. The dissipated powers W_d and W_r are equal when

$$\frac{C_d}{2} \rho A v^2 = C_r mg,$$

or

$$v = \left(\frac{2 C_r}{C_d} \frac{mg}{\rho A} \right)^{1/2} = \left(\frac{2 \cdot 0.01}{0.3} \frac{(1.0 \cdot 10^3 \, \text{kg}) \cdot (9.81 \, \text{m} \cdot \text{s}^{-2})}{(1.2 \, \text{kg} \cdot \text{m}^{-3}) \cdot (1.89 \, \text{m}^2)} \right)^{1/2}$$

$$= 17 \, \frac{\text{m}}{\text{s}} = 61 \, \frac{\text{km}}{\text{h}}.$$

The condition that $F_d = F_r$ is exactly the same as $W_d = W_r$ and yields the same threshold speed.

5.1.3   Eolic energy

There is a fundamental limit (the *Betz limit*) on the efficiency of windmills converting the kinetic energy of the wind into mechanical work. In general, the potential and the limitations of renewable energy sources should be understood before advocating them as a panacea for the problems created by an ever-increasing demand for energy. This does not mean that we should shy away from renewable energy sources: they are worth tapping into where the conditions are favorable and they will become more and more competitive with fossil fuels as the price of the latter increases.

1 **(A)** Estimate the maximum power output of a windmill with a rotor of radius $r = 4$ m in a uniform constant wind with a speed of 10 m/s

operating with an efficiency $\eta = 40\%$. How does your result depend on the wind speed? Is it realistic to replace a 1 GW nuclear or coal-fired power station with a wind farm?

Hint: Consider the kinetic energy of a horizontal cylinder of air of radius r with one end against the windmill blades.

Solution

The windmill converts the kinetic energy of air impinging on it into electric energy. Consider a horizontal cylinder of air of radius r with one end on the windmill rotor (Fig. 5.2). During the time t the air that pushes the rotor travels a distance $L = vt$, where v is the (constant and uniform) wind speed. The kinetic energy of the air contained in the cylinder of length L and radius r is

$$\frac{1}{2} m v^2 = \frac{1}{2} \rho_{\text{air}} \left(L \pi r^2 \right) v^2 = \frac{1}{2} \rho_{\text{air}} \pi r^2 v^3 \, t,$$

where $m = \rho_{\text{air}} \pi r^2 L$ is the mass of air contained in the cylinder of volume $V = \pi r^2 L$. The power output of a windmill with efficiency η is

$$W = \frac{dE_{\text{generated}}}{dt} = \eta \frac{d \left(m v^2 / 2 \right)}{dt} = \frac{\eta \pi}{2} \rho_{\text{air}} r^2 v^3,$$

and it depends on the *third power* of the wind speed v. Numerically,

$$W = \frac{0.40 \pi}{2} \left(1.2 \, \frac{\text{kg}}{\text{m}^3} \right) (4\text{m})^2 \left(10 \, \frac{\text{m}}{\text{s}} \right)^3 = 10^4 \, \text{W}.$$

In order to replace a 1 GW power station with a wind farm one would need

$$\frac{1 \, \text{GW}}{10^4 \, \text{W}} = 10^5$$

windmills, assuming for simplicity that the presence of a windmill does not reduce the efficiency of nearby windmills (a questionable assumption). Such an option is clearly unviable and windmills are only appropriate for the small-scale production of eolic energy. The dependence of the power output on v^3 makes windmills useful only in regions traversed by dominant winds with appreciable speed.

2 **(B)** a) What is the maximum amount of power that can be theoretically generated by a domestic windmill with horizontal rotor (the blades have radius $R = 1.5$ m) in horizontal wind with average speed $v = 3.1$ m/s? The density of air at 20°C is 1.2 kg/m³. Is this power significant for household purposes?

Figure 5.2. A cylinder of air of radius R and length vdt.

b) Now assume that the windmill is upgraded to a new design with rotor blades of 5-m radius, mounted on a tower that allows the blades to spin and catch wind at the higher speed of 4.0 m/s. What maximum power can theoretically be generated?

Solution

The maximum possible efficiency of a windmill with horizontal rotor is the *Betz limit* $\eta_{\text{Betz}} = 16/27 = 59\%$. The air impinging on the rotor blades horizontally in the time interval dt traverses a horizontal cylinder of radius R and length vdt (Fig. 5.2) and carries the kinetic energy

$$dE = \frac{1}{2}\left(dm\right)v^2 = \frac{1}{2}\rho\left(dV\right)v^2 = \frac{1}{2}\rho\left(\pi R^2 vdt\right)v^2$$

and the power

$$\frac{dE}{dt} = \frac{\pi}{2}\rho R^2 v^3.$$

The wind is not stopped by the rotor and only a fraction η_{Betz} of this power can theoretically be converted into electric power (in practice, the real efficiency is lower than the Betz limit), or

$$\eta_{Betz} \frac{dE}{dt} = \frac{\pi \eta_{Betz}}{2} \rho R^2 v^3$$

$$= \frac{\pi \cdot 0.59}{2} \left(1.2 \frac{kg}{m^3}\right) (1.5\,m)^2 \left(3.1 \frac{m}{s}\right)^3 = 75\,W,$$

an amount too small even for household needs.

b) For the improved and enlarged version of the windmill it is instead

$$\eta_{Betz} \frac{dE}{dt} = \frac{\pi \eta_{Betz}}{2} \rho R^2 v^3$$

$$= \frac{\pi \cdot 0.59}{2} \left(1.2 \frac{kg}{m^3}\right) (5.0\,m)^2 \left(4.0 \frac{m}{s}\right)^3 = 1.8 \cdot 10^3\,W,$$

or 24 times the previous result. This is not surprising considering that doubling the size of the rotor quadruples the extracted energy and doubling the wind speed would multiply the generated energy by eight times. This new number suggests that it is worth investing in a larger windmill.

3 **(B)** It can be shown by physical considerations that the theoretical efficiency of a windmill with horizontal rotor in horizontal wind is given by

$$\eta = \frac{\left(v_{in} + v_{out}\right)\left(v_{in}^2 - v_{out}^2\right)}{2v_{in}^3},$$

where v_{in} and v_{out} are, respectively, the wind velocities upstream and downstream the turbine. Find the maximum possible value of η (called the *Betz limit*).

Hint: Use the variable $x \equiv v_{out}/v_{in}$.

Solution
Inspection of the expression for the windmill efficiency η suggests using the new variable $x \equiv v_{out}/v_{in}$, eliminating one of the two variables [this is possible because of the special form of $\eta\left(v_{in}, v_{out}\right)$]. The efficiency becomes

$$\eta(x) = \frac{1}{2}\left(1 + x\right)\left(1 - x^2\right)$$

with $0 \leq x \leq 1$, and we are looking for a maximum of $\eta(x)$ in this interval. Because $\eta(x)$ is a polynomial, it is continuous with all its derivatives of all orders. Furthermore, $\eta(0) = 1/2$ and $\eta(1) = 0$. The first derivative is

$$\frac{d\eta}{dx} = -\frac{1}{2}\left(3x^2 + 2x - 1\right) = -\frac{3}{2}(x+1)\left(x - \frac{1}{3}\right).$$

The factor $x + 1$ is positive in this interval, hence the sign of $\eta(x)$ is the opposite of the sign of $(x - 1/3)$ and

$$\frac{d\eta}{dx} > 0 \quad \text{if} \quad 0 \leq x < \frac{1}{3},$$

$$\left.\frac{d\eta}{dx}\right|_{x=1/3} = 0,$$

$$\frac{d\eta}{dx} < 0 \quad \text{if} \quad \frac{1}{3} < x \leq 1.$$

The function $\eta(x)$ is therefore strictly increasing between 0 and 1/3, has horizontal tangent at $x = 1/3$, and is strictly decreasing between 1/3 and 1: since η is continuous this is sufficient to conclude that there is a local maximum at $x = 1/3$. Since $\eta(1/3) = 16/27 \simeq 0.59 > \eta(0) = 0.5$, this is also an absolute maximum in $[0, 1]$. The Betz limit (maximum theoretical efficiency) of the horizontal windmill is therefore

$$\eta_{\text{Betz}} = \frac{16}{27} = 59\%.$$

4 **(B)** Compute the maximum energy that can theoretically be extracted in one day from a continuous horizontal wind by a windmill with a horizontal rotor equipped with blades of radius $R = 1.5$ m. The air density is $\rho = 1.25\,\text{kg/m}^3$ and the probability that the wind has speed v is given by the two-parameter Weibull distribution function[2]

$$f(v) = \frac{\beta}{\alpha}\left(\frac{v}{\alpha}\right)^{\beta-1} e^{-(v/\alpha)^\beta},$$

where, at the given location, the scale parameter has the value $\alpha = 5.1$ m/s and the shape parameter is $\beta = 1.5$.

Hint: Use the Weibull distribution to compute averages of the quantities in your formulas that depend on v.

---

[2] The Weibull distribution is widely used in reliability analysis, especially in the prediction of failures in the metallurgical industry.

Solution

The kinetic energy dE impinging on the rotor during the time interval dt is carried by the air that traverses a horizontal cylinder of radius R, length vdt, volume $dV = \pi R^2 v dt$, and mass $dm = \rho dV$,

$$dE = \frac{1}{2}(dm)\,v^2 = \frac{1}{2}\rho\,(dV)\,v^2 = \frac{\pi}{2}\rho R^2 v^3 dt,$$

and the maximum power that can theoretically be extracted by this air mass is

$$\eta_{\text{Betz}}\,\frac{dE}{dt} = \frac{\pi\eta_{\text{Betz}}}{2}\rho R^2 v^3,$$

where the Betz limit $\eta_{\text{Betz}} = 16/27 = 59\%$ is the maximum theoretical efficiency of such a windmill. We now have to replace the factor v^3 with its statistical average $\overline{v^3}$ taken with the Weibull distribution. We have

$$\overline{v^3} \equiv \int_0^{+\infty} dv\, f(v) v^3 = \int_0^{+\infty} dv\, \frac{\beta}{\alpha}\left(\frac{v}{\alpha}\right)^{\beta-1} e^{-(v/\alpha)^\beta} v^3$$

$$= \alpha^3 \beta \int_0^{+\infty} d\left(\frac{v}{\alpha}\right)\left(\frac{v}{\alpha}\right)^{\beta-1} e^{-(v/\alpha)^\beta}\left(\frac{v}{\alpha}\right)^3.$$

By using the auxiliary variable $z \equiv v/\alpha$,

$$\overline{v^3} = \alpha^3 \beta \int_0^{+\infty} dz\, z^{\beta+2}\, e^{-z^\beta},$$

and changing variables again to $\zeta \equiv z^\beta$, we have $\beta z^{\beta+2} dz = \zeta^{3/\beta}\, d\zeta$, which yields, using a table of integrals [21],

$$\overline{v^3} = \alpha^3 \int_0^{+\infty} d\zeta\, \zeta^{3/\beta}\, e^{-\zeta} = \alpha^3\, \Gamma\left(\frac{3}{\beta}+1\right),$$

where Γ is the gamma function. Therefore, the maximum theoretical power that can be extracted from the windmill with the given parameters is

$$W = \frac{\pi\eta_{\text{Betz}}}{2}\rho R^2 \overline{v^3} = \frac{\pi\eta_{\text{Betz}}}{2}\rho R^2 \alpha^3\, \Gamma\left(\frac{3}{\beta}+1\right)$$

$$= \frac{\pi \cdot 0.59}{2}\left(1.25\,\frac{\text{kg}}{\text{m}^3}\right)(1.5\,\text{m})^2\left(5.1\,\frac{\text{m}}{\text{s}}\right)^3 \Gamma(3) = 690\,\text{W}.$$

In one day of continuous operation the energy extracted from this wind is

$$E = (690\,\text{W})\,(24 \cdot 3600\,\text{s}) = 6 \cdot 10^7\,\text{J}.$$

This is only the theoretical upper limit based on the Betz efficiency: the real efficiency could be much lower.

5.2   Heat transfer

There are three mechanisms of heat transfer: *conduction, convection,* and *radiation.* In many applications of environmental science also *latent heat transfer* should be added to this list: it occurs when water vapor is removed from the surface of a body of water (ocean), from soil, or from a sweating human body, and the latent heat of evaporation is carried away (evapotranspiration is reviewed in Chapter 7 together with phase transitions). The most common model for a radiating body is a blackbody or a graybody—these are discussed in Chapter 4. Convection is associated with fluid motions—for the environmental scientist, moving air or water—and is intimately connected with fluid dynamics. The focus of this section is on conduction and convection.

Heat conduction is described by the heat or diffusion equation, which is also the tool to describe problems such as the spread of pollutants and the formation of plumes in air, water, or groundwater, the infiltration of rain into a soil, the transport of groundwater in an aquifer, percolation of rain through a snowpack, diffusion of salts in sea water, and many other environmental physical problems. Therefore, the student spending time and effort on heat conduction will become acquainted with a mathematical formalism useful in many other fields of environmental physics. Direct applications of heat conduction include the study of how heat penetrates through a soil, minimizing heat losses from buildings in colder climates, or helping to reduce the energy demand for air conditioning in warmer climates. In engineering instead, one often wants to maximize heat transfer, e.g., between different parts of a machine, or to cool a power plant (which in turn leads to thermal pollution), or other devices.[3]

Temperature affects biological systems and biochemical reactions. As a rule of thumb, a 10°C increase in temperature doubles the reaction rate (with the exception of exothermic reactions) and this increases the rate of cell subdivision and plant growth. In addition, temperature affects evaporation and humidity in the atmosphere.

1 (C) Discuss the three mechanisms of heat transfer.

Solution
Heat is transferred by conduction, convection, and radiation. In *conduction* the kinetic energy of particles that are relatively free to move, e.g., free electrons in a metal, is transferred to other particles through

---

[3]Think, for example, of the black or silvered fins that are usually found on the back of a stereo amplifier or similar electronic devices.

scatterings resulting in a net heat flux. In *convection*, macroscopic motions of a fluid (liquid or gas) transport heat. In *radiation*, electromagnetic waves in the infrared band carry away energy from the heat source without transport of matter. Electromagnetic waves can propagate through vacuum, and therefore radiation does not require the presence of a medium through which heat passes. For example, radiation from the Sun reaches the surface of the planets propagating almost entirely through empty space.

2 **(A)** Calculate the heat lost per unit time through a glass window of surface area $2.6\,\text{m}^2$ and the heat resistance of the window. The thickness of the glass pane is $d = 5.0$ mm, its thermal conductivity is $k = 1.2\,\text{W}\cdot\text{m}^{-1}\cdot\text{K}$, the temperature of the room is $18°C$, and the temperature outside is $-2°C$. How can the heat loss be minimized?

Solution
According to Fourier's law, the temperature difference between outside and inside is

$$T_o - T_i = -\frac{q''}{k}\,d,$$

where q'' is the heat flux. The heat lost per unit time is

$$q = q''\Lambda = -k\frac{(T_o - T_i)\,A}{d},$$

where A is the surface area of the window. Hence,

$$q = \left(-1.2\,\frac{\text{W}}{\text{m}\cdot\text{K}}\right)\frac{(-20\,\text{K})}{(5.0\cdot 10^{-3}\,\text{m})}\cdot(2.6\,\text{m}^2) = 1.25\cdot 10^4\,\text{W}.$$

The heat resistance is

$$R = \frac{\Delta T}{q} = \frac{20\,\text{K}}{1.25\cdot 10^4\,\text{W}} = 1.6\cdot 10^{-3}\,\frac{\text{K}}{\text{W}}.$$

Alternatively, we could calculate the heat resistance by using its expression for a slab of material

$$R = \frac{d}{kA},$$

and obtain the same result.

The heat loss can be minimized by using a double-pane glass instead of single pane because the air trapped between the two layers of glass acts as an insulator. In addition, a shutter or other insulating surface can be placed in front of the glass and a curtain on the inside. At the stage of designing the building, the window area can be minimized.

3 **(A)** The thermal equivalent of a certain refrigerator is a styrofoam box ($k = 5.0 \cdot 10^{-2}\,\mathrm{W} \cdot \mathrm{m}^{-1} \cdot \mathrm{K}^{-1}$) with inner surface area $4.0\,\mathrm{m}^2$ and thickness $d = 1.0$ cm. At what rate is heat being removed from the interior of the refrigerator, given that it is 20°C below the outside temperature and that the motor is running 11% of the time?

Solution
Were the motor of the refrigerator running 100% of the time, the heat flux would have magnitude

$$q'' = k\frac{\Delta T}{d}$$

and the heat removed per unit time would be

$$q = k\frac{\Delta T}{d}\,A.$$

Given that the motor only runs 11% of the time, the actual heat removed from the interior of the refrigerator when the motor is operating is

$$q \;=\; k\frac{\Delta T}{d}A\frac{1}{0.11} = \frac{\left(5.0 \cdot 10^{-2}\,\mathrm{W} \cdot \mathrm{m}^{-1} \cdot \mathrm{K}^{-1}\right) \cdot \left(4.0\,\mathrm{m}^2\right) \cdot (20\,\mathrm{K})}{0.11 \cdot \left(1.0 \cdot 10^{-2}\,\mathrm{m}\right)}$$

$$\;=\; 3.6\,\mathrm{kW}.$$

4 **(B)** Integrate the heat equation for the temperature $T(t, \vec{x})$ in a material

$$\rho\,c_p\frac{\partial T}{\partial t} = \vec{\nabla} \cdot (k\,\vec{\nabla}T) + \dot{q}$$

over a fixed volume of space V and describe the heat balance for V. ρ, c_p and k are, respectively, the density, specific heat at constant pressure, and thermal conductivity of the material. The source term $\dot{q}$ denotes the heat generated inside the material per unit volume and per unit time.

Solution
Let the closed surface S be the boundary of the volume V (Fig. 5.3). By integrating the heat equation over V we obtain

$$\iiint_V dV\,\rho\,c_p\frac{\partial T}{\partial t} = \iiint_V dV\,\vec{\nabla} \cdot (k\vec{\nabla}T) + \iiint_V dV\,\dot{q}.$$

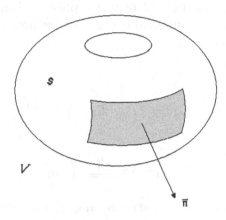

Figure 5.3.   Heat flow from V through S.

By applying Gauss' law and exchanging the operators of integration and differentiation with respect to time in the left-hand side, we obtain

$$\frac{d}{dt}\left(\int\int\int_V dV\, \rho\, c_P\, T\right) = \int\int_S k\vec{\nabla}T\cdot\vec{n}\, dS + \int\int\int_V dV\dot{q}, \quad (5.4)$$

where $\vec{n}$ is the unit normal to S. The use of Fourier's law expressing the heat flux density (heat flowing through the unit of normal area per unit time), $\vec{q''} = -k\,\vec{\nabla}T$, gives

$$\int\int\int_V dV\dot{q} = \frac{d}{dt}\left(\int\int\int_V dV\rho c_P T\right) + \int\int_S \vec{q''}\cdot\vec{n}\, dS.$$

The last equation describes the heat balance for the volume V and the terms appearing in it have the following physical interpretation:
$\int\int\int_V dV\dot{q}$ is the heat generated in V per unit time;
$\int\int\int_V dV\, \rho\, c_P\, T$ is the heat absorbed by the material in V (which increases its temperature) per unit time, and $\frac{d}{dt}\left(\int\int\int_V dV\rho c_P T\right)$ is its rate of variation;
$-\int\int_S k\vec{\nabla}T\cdot\vec{n}\, dS = \int\int_S \vec{q''}\cdot\vec{n}\, dS$ is the amount of heat escaping the volume V through the surface S per unit time. By adopting the usual convention on the sign of the normal to a surface, $\vec{n}$ points outwards

and $\int \int_S k\vec{\nabla} T \cdot \vec{n}\, dS$ is then the heat leaving V. The physical meaning of the integral form (5.4) of the heat equation is that the total heat generated in V (per unit of time) equals the heat absorbed by the medium in V (in the unit of time) plus the heat escaping V (in the unit of time). The physical content of the heat equation is simply an energy balance.

5 **(B)** Verify that

$$T(t, x) = T_1 + (T_0 - T_1)\,\mathrm{erf}\left(\frac{x}{2\sqrt{at}}\right), \qquad (5.5)$$

where

$$\mathrm{erf}(s) \equiv \frac{2}{\sqrt{\pi}} \int_0^s d\xi\, e^{-\xi^2},$$

is the *error function*, is the solution of the one-dimensional heat equation

$$\frac{\partial T}{\partial t} = a\,\frac{\partial^2 T}{\partial x^2} \qquad (5.6)$$

in $[\,0, +\infty\,)$ with the boundary condition at $x = 0$

$$T(t, 0) = T_1 \qquad (t \geq 0)\,,$$

the boundary condition $\lim_{x \to +\infty} T(t, x)$ finite, and the initial condition

$$T(0, x) = T_0 \qquad (x > 0)$$

corresponding to a sudden temperature change at $t = 0$.

Solution
It is straightforward to check that the boundary and the initial conditions are verified by (5.5) by using the property of the error function $\mathrm{erf}(0) = 0$ and the fact that the integral of the Gaussian is

$$\int_0^{+\infty} ds\, e^{-s^2} = \frac{\sqrt{\pi}}{2}\,,$$

which yields

$$\lim_{t \to 0^+} \mathrm{erf}\,(s(t)) = \lim_{s \to +\infty} \mathrm{erf}\,(s) = 1.$$

To check that (5.5) satisfies Eq. (5.6), we compute $\partial T/\partial t$ using

$$\frac{d\,[\mathrm{erf}(s)]}{ds} = \frac{2}{\sqrt{\pi}}\,e^{-s^2}:$$

$$\frac{\partial T}{\partial t} = (T_0 - T_1)\,\frac{2}{\sqrt{\pi}}\,\mathrm{e}^{-\left(\frac{x}{2\sqrt{at}}\right)^2}\,\frac{1}{2\sqrt{a}}\frac{-x}{2t\sqrt{t}} = -\frac{(T_0 - T_1)}{2\sqrt{\pi a}}\,x\,t^{-3/2}\,\mathrm{e}^{-\frac{x^2}{4at}}.$$

On the other hand,

$$\frac{\partial T}{\partial x} = \frac{(T_0 - T_1)}{\sqrt{\pi a t}}\,\mathrm{e}^{-\frac{x^2}{4at}},$$

$$\frac{\partial^2 T}{\partial x^2} = \frac{(T_0 - T_1)}{\sqrt{\pi a t}}\frac{-2x}{4at}\,\mathrm{e}^{-\frac{x^2}{4at}}.$$

Therefore,

$$a\,\frac{\partial^2 T}{\partial x^2} = \frac{-x(T_0 - T_1)}{2\sqrt{\pi a}\,t^{3/2}}\,\mathrm{e}^{-\frac{x^2}{4at}} = \frac{\partial T}{\partial t}.$$

6 **(B)** Verify that the solution of the heat equation in one dimension for daily or seasonal temperature changes, modeled by the boundary conditions

$$T(t,0) = \bar{T} + T_0\cos(\omega t),$$

$$T \to \bar{T} \quad \text{as} \quad x \to +\infty,$$

is

$$T(t,x) = \bar{T} + T_0\,\mathrm{e}^{-Ax}\cos(kx - \omega t),$$

where $\bar{T}$ and T_0 are constants and

$$k = \frac{\omega}{v}, \qquad A = \left(\frac{\omega}{2a}\right)^{1/2}, \qquad v = (2a\omega)^{1/2}.$$

The constants $x_e \equiv A^{-1}$ and $\tau \equiv (1\,\mathrm{m})/v$ are called *damping depth* and *delay time*, respectively.

Solution
We have

$$\frac{\partial T}{\partial t} = T_0\,\omega\,\mathrm{e}^{-Ax}\sin(kx - \omega t),$$

$$\frac{\partial T}{\partial x} = -AT_0\mathrm{e}^{-Ax}\cos(kx - \omega t) - kT_0\mathrm{e}^{-Ax}\sin(kx - \omega t),$$

and

$$\frac{\partial^2 T}{\partial x^2} = T_0\mathrm{e}^{-Ax}\left[A^2\cos(kx - \omega t) + 2Ak\sin(kx - \omega t)\right.$$

$$\left. -k^2\cos(kx - \omega t)\right].$$

Hence,

$$a\frac{\partial^2 T}{\partial x^2} = aT_0 e^{-Ax}\left[\left(A^2 - k^2\right)\cos\left(kx - \omega t\right) + 2Ak\sin\left(kx - \omega t\right)\right]$$

$$= \frac{\partial T}{\partial t}$$

provided that

$$A^2 - k^2 = 0, \qquad 2Aka = \omega.$$

This system of equations has the solution

$$A = k = \sqrt{\frac{\omega}{2a}},$$

and the phase velocity of the damped temperature waves is

$$v = \frac{\omega}{k} = \sqrt{2\,a\,\omega}.$$

The boundary conditions are satisfied since

$$T(t, x = 0) = \bar{T} + T_0 \cos\left(\omega t\right),$$

and

$$T \to \bar{T} \qquad \text{as} \qquad x \to +\infty$$

because $e^{-Ax} \to 0$ as $x \to +\infty$.

7 **(A)** Compute the damping depth and the delay time for a semi-infinite stone wall undergoing daily temperature variations. The Fourier coefficient of the stone used in the wall is $a = 5.00 \cdot 10^{-7}\,\mathrm{m^2/s}$.

Solution
The damping depth is

$$x_e = \sqrt{\frac{2a}{\omega}} = \left[\frac{2 \cdot \left(5.00 \cdot 10^{-7}\,\mathrm{m^2/s}\right)}{2\pi \cdot (24 \cdot 3600\,\mathrm{s})^{-1}}\right]^{1/2} = 0.117\,\mathrm{m}.$$

The delay time is

$$\tau = \frac{(1\,\mathrm{m})}{v} = \frac{(1\,\mathrm{m})}{\sqrt{2a\omega}}$$

$$= \frac{(1\,\mathrm{m})}{\left[2 \cdot \left(5.00 \cdot 10^{-7}\,\mathrm{m^2 \cdot s^{-1}}\right) \cdot 2\pi \cdot (24 \cdot 3600\,\mathrm{s})^{-1}\right]^{1/2}}$$

$$= 1.17 \cdot 10^5\,\mathrm{s} \simeq 33\,\mathrm{hours}.$$

When $x = x_e$, the damping depth, the amplitude of the original temperature wave $T_0 \cos(\omega t)$ at $x = 0$ is reduced by one e-fold, $T_0/e = 0.37\, T_0$.

8 **(B)** Verify that

$$T(t,x) = \frac{\Theta}{2\sqrt{\pi a t}}\, e^{-\frac{x^2}{4at}}, \tag{5.7}$$

where Θ is a constant, is the solution of the one-dimensional heat equation

$$\frac{\partial T}{\partial t} = a\, \frac{\partial^2 T}{\partial x^2} \tag{5.8}$$

with boundary condition

$$T(t,x) \to 0 \qquad \text{as } |x| \to +\infty \tag{5.9}$$

and with initial condition

$$T(0,x) = \Theta\, \delta(x), \tag{5.10}$$

where $\delta(x)$ is the Dirac delta distribution [17]. Consider the solution at an arbitrarily small value of $t > 0$ and an arbitrarily large value of x and discuss its implications for causality. Discuss the graphical interpretation of the integral

$$I(t) = \int_{-\infty}^{+\infty} dx\, T(t,x)$$

for this solution in the limit $t \to 0^+$.

Solution
We have

$$\frac{\partial T}{\partial t} = \frac{\Theta}{4\sqrt{\pi a}\; t^{3/2}}\, e^{-\frac{x^2}{4at}} \left(\frac{x^2}{2at} - 1 \right),$$

$$\frac{\partial T}{\partial x} = -\frac{\Theta x}{4\sqrt{\pi a^3\, t^3}}\, e^{-\frac{x^2}{4at}},$$

and

$$\frac{\partial^2 T}{\partial x^2} = \frac{\Theta}{4\sqrt{\pi a^3\, t^3}}\, e^{-\frac{x^2}{4at}} \left(\frac{x^2}{2at} - 1 \right).$$

Hence $a\, \partial^2 T/\partial x^2 = \partial T/\partial t$ and the heat equation is satisfied.

The boundary condition (5.9) is easily verified due to the fact that $\lim_{|x| \to +\infty} e^{-\alpha^2 x^2} = 0$. To verify the initial condition (5.10), introduce the quantity $\alpha \equiv (2\sqrt{at})^{-1}$; then $\alpha \to +\infty$ as $t \to 0^+$. By using the Dirac delta representation [17]

$$\delta(x) = \frac{1}{\sqrt{\pi}} \lim_{\alpha \to +\infty} \alpha\, e^{-\alpha^2 x^2},$$

we verify that

$$\lim_{t \to 0^+} T(t, x) = \lim_{\alpha \to +\infty} \frac{\Theta}{\sqrt{\pi}} \alpha\, e^{-\alpha^2 x^2} = \Theta\, \delta(x).$$

The initial condition corresponds to releasing a certain amount of thermal energy at $x = 0$ at the initial time $t = 0$. At any later instant of time t and position x, with t arbitrarily small and x arbitrarily large, the temperature $T(t, x)$ is strictly positive—this fact implies that Eq. (5.8) describes *instantaneous* propagation of heat (*heat paradox*). This feature of the heat equation does not constitute a problem for most practical applications because the relativistic corrections that would be needed to take into account the finite velocity of heat propagation are usually negligible. Thus, Eq. (5.8) is an excellent approximation for everyday situations.

The limit $t \to +\infty$ has a graphical interpretation: as $t \to +\infty$, the Gaussian function (5.7) is more and more peaked on $x = 0$. The area between the x-axis and the graph of the Gaussian, given by the integral

$$A = \frac{\Theta}{2\sqrt{\pi at}} \int_{-\infty}^{+\infty} dx\, e^{-x^2/4at} = \frac{\Theta}{\sqrt{\pi at}} \int_{0}^{+\infty} dx\, e^{-x^2/4at} = \Theta,$$

does not depend on time. As $t \to 0^+$ we have $A \to \Theta \int_{-\infty}^{+\infty} dx\, \delta(x) = \Theta$. The total thermal energy content of space does not depend on time. It is initially concentrated at $x = 0$, spreads with infinite velocity,

$$\lim_{t \to 0^+} \frac{\partial T}{\partial t}\Big|_{x=0} = -\infty,$$

and at later times is distributed over the entire x-axis with $\partial T/\partial t \to 0$ as $t \to +\infty$.

9 **(B)** Consider a hollow cylinder with inner radius R_1, outer radius R_2, and length L (with $L \gg R_1, L \gg R_2$), made of a homogeneous and isotropic material with thermal conductivity k (Fig. 5.4). The inner surface of the cylinder ($r = R_1$) is heated and kept at constant temperature T_i with a thermostat, while the outer surface ($r = R_2$) is ventilated, keeping it at constant temperature $T_S < T_i$. No heat

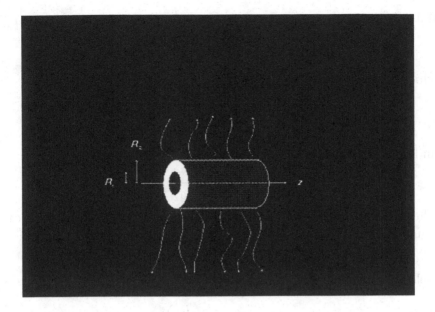

Figure 5.4. The hollow cylinder in steady state.

is generated or retained by the material composing the hollow cylinder. Assuming *steady state*, compute the temperature T of the hollow cylinder as a function of radius r. Compute also the heat current density q'' (heat flowing per unit time through the unit of area normal to the flux).

Solution
We use cylindrical coordinates (r, φ, z) with the axis of the cylinder lying along the z-axis. In steady state, the temperature T does not depend on time, and the heat equation

$$\frac{\partial T}{\partial t} = a \, \nabla^2 T$$

reduces to

$$\nabla^2 T = \frac{1}{r}\frac{\partial}{\partial r}\left(r\,\frac{\partial T}{\partial r}\right) + \frac{1}{r^2}\frac{\partial^2 T}{\partial \varphi^2} + \frac{\partial^2 T}{\partial z^2} = 0,$$

where the expression of the Laplace operator in cylindrical coordinates is used (see Appendix C). Because of the cylindrical symmetry and neglecting end effects, T is independent of φ and z and the heat equation is further simplified, becoming the ordinary differential equation

$$\frac{1}{r}\frac{d}{dr}\left(r\frac{dT}{dr}\right) = 0,$$

which is immediately integrated to

$$\frac{dT}{dr} = \frac{C_1}{r}$$

and again to

$$T(r) = C\ln\left(\frac{r}{r_0}\right),$$

where C_1, C, and r_0 are integration constants. The boundary conditions

$$T(r = R_1) = T_i, \qquad T(r = R_2) = T_S$$

yield

$$T_i = C\ln\left(\frac{R_1}{r_0}\right), \tag{5.11}$$

$$T_S = C\ln\left(\frac{R_2}{r_0}\right). \tag{5.12}$$

By subtracting Eq. (5.11) from Eq. (5.12), we obtain

$$C = \frac{T_S - T_i}{\ln(R_2/R_1)},$$

while the addition of Eqs. (5.11) and (5.12) yields

$$T_i + T_S = C\ln\left(\frac{R_1 R_2}{r_0^2}\right) = \frac{T_S - T_i}{\ln(R_2/R_1)}\ln\left(\frac{R_1 R_2}{r_0^2}\right)$$

and

$$\frac{T_i + T_S}{T_i - T_S}\ln\left(\frac{R_2}{R_1}\right) = \ln\left(\frac{r_0^2}{R_1 R_2}\right);$$

finally,

$$r_0 = \left\{R_1 R_2 \exp\left[\frac{T_i + T_S}{T_i - T_S}\ln\left(\frac{R_2}{R_1}\right)\right]\right\}^{1/2}.$$

The solution of the heat equation can thus be expressed as

$$T(r) = \frac{T_S - T_i}{\ln(R_2/R_1)}\ln\left(\frac{r}{r_0}\right),$$

while the heat current density is

$$q''(r) = -k\frac{dT}{dr} = -k\frac{T_S - T_i}{\ln(R_2/R_1)}\frac{1}{r}.$$

The heat flux through the side wall of a cylinder of radius r (with $R_1 \le r \le R_2$) and length L is

$$q(r) = 2\pi r L q''(r) = -2\pi r L k \frac{T_S - T_i}{\ln(R_2/R_1)}\frac{1}{r} = -2\pi L k \frac{T_S - T_i}{\ln(R_2/R_1)},$$

and does not depend on r, as it should be, since heat cannot stop in the material (there are no sources or sinks of heat inside the cylinder).

10 **(A)** In cold weather you exit an overheated building without gloves and you put your sweaty hand on an iron railing. By knowing that your body temperature is $36.8°C$, the outside temperature is $-15°C$, the values of the contact coefficient for the human skin and for iron are $1200\,\mathrm{kg \cdot s^{-5/2} \cdot K}$ and $1.7 \cdot 10^4\,\mathrm{kg \cdot s^{-5/2} \cdot K}$, respectively, find the contact temperature of the skin.

Solution
The contact temperature of the hand's skin is

$$T_c = \frac{b_1 T_1 + b_2 T_2}{b_1 + b_2}$$

$$= \left\{ \left(1200\,\mathrm{kg \cdot s^{-5/2} \cdot K} \cdot 36.8°C \right) \right.$$

$$\left. + \left[1.7 \cdot 10^4\,\mathrm{kg \cdot s^{-5/2} \cdot K} \cdot (-15°C) \right] \right\}$$

$$\cdot \left[(1200 + 17000)\,\mathrm{kg \cdot s^{-5/2} \cdot K} \right]^{-1} = -12°C.$$

Moisture on your hand from sweating will freeze upon touching the railing.

11 **(B, C)** Does the heat equation

$$\frac{\partial T}{\partial t} = a\,\nabla^2 T$$

describe reversible or irreversible processes? Provide an answer based on the mathematical form of the equation and interpret it from the physical point of view.

Solution

The heat equation describes *irreversible* processes. In fact, under the time reflection $t \longrightarrow -t$ the heat equation changes into

$$\frac{\partial T}{\partial t} = -a \, \nabla^2 T,$$

and therefore the heat equation is not invariant under time reversal and it describes *irreversible* processes. Heat conduction is an irreversible phenomenon. Heat flows spontaneously from regions at higher temperature to regions at lower temperatures. An example is the solution of the heat equation for initial conditions corresponding to thermal energy initially concentrated in a hot spot: the solution is a Gaussian describing heat spreading over all space. The temperature of the initial hot spot decreases monotonically while the temperature at every other point increases and the thermal energy diffuses but never concentrates again spontaneously in a hot spot.

By contrast the basic equation of classical mechanics, Newton's second law of motion

$$m \frac{d^2 \vec{x}}{dt^2} = \vec{F},$$

where $\vec{F}$ is the force acting on a point particle of mass m and position $\vec{x}(t)$, describes reversible phenomena. The time inversion $t \longrightarrow -t$ leaves this equation unchanged because

$$\frac{d}{d\,(-t)} = -\frac{d}{dt},$$

and

$$\frac{d^2}{d\,(-t)^2} = \frac{d^2}{dt^2}.$$

12 **(B)** Derive the heat equation for a homogeneous isotropic medium from a variational principle and provide a Lagrangian and a Hamiltonian formulation of heat diffusion.

Hint: Promote the temperature T to the role of a complex variable and consider T and T^* as Lagrangian fields.

Solution

We proceed in analogy with the Lagrangian and Hamiltonian formalism for the Schrödinger field [51]. Once T is allowed to assume complex values, the heat equation in a homogeneous isotropic medium

$$\frac{\partial T}{\partial t} = a \, \nabla^2 T$$

is formally the same as the Schrödinger equation for a free quantum particle represented by the complex Schrödinger field ψ,

$$i\hbar \frac{\partial \psi}{\partial t} = -\frac{\hbar^2}{2m} \nabla^2 \psi,$$

with the important difference that the coefficients of the heat equation are real. Therefore, we consider the Lagrangian density

$$\mathcal{L} = -a\vec{\nabla}T \cdot \vec{\nabla}T^* - \frac{1}{2}\left(T^*\dot{T} - T\dot{T}^*\right)$$

and the variational principle [20, 51]

$$\delta S = 0$$

for the action

$$S = \int d^3\vec{x}\, \mathcal{L}(t, \vec{x}).$$

The Euler–Lagrange equations equivalent to the variational principle [20, 51]

$$\frac{\partial}{\partial t}\frac{\partial \mathcal{L}}{\partial \dot{T}} + \frac{\partial}{\partial x^i}\frac{\partial \mathcal{L}}{\partial(\partial_i T)} - \frac{\partial \mathcal{L}}{\partial T} = 0,$$

$$\frac{\partial}{\partial t}\frac{\partial \mathcal{L}}{\partial \dot{T}^*} + \frac{\partial}{\partial x^i}\frac{\partial \mathcal{L}}{\partial(\partial_i T^*)} - \frac{\partial \mathcal{L}}{\partial T^*} = 0,$$

then yield

$$\frac{\partial T^*}{\partial t} = -a\,\nabla^2 T^*, \tag{5.13}$$

$$\frac{\partial T}{\partial t} = a\,\nabla^2 T, \tag{5.14}$$

using

$$\frac{\partial \mathcal{L}}{\partial T} = \frac{1}{2}\dot{T}^*, \qquad \frac{\partial \mathcal{L}}{\partial T^*} = -\frac{1}{2}\dot{T},$$

$$\frac{\partial \mathcal{L}}{\partial(\partial_i T)} = -a\,\partial_i T^*, \qquad \frac{\partial \mathcal{L}}{\partial(\partial_i T^*)} = -a\,\partial_i T.$$

Equation (5.13) tells us that $T^*(t) - T(-t)$ is simply the time reverse of the solution $T(t)$ since it satisfies the same equation with t changed into $-t$. The Hamiltonian formulation is obtained by defining the canonical momenta

$$\pi_T \equiv \frac{\partial \mathcal{L}}{\partial \dot{T}} = -\frac{1}{2} T^*,$$

$$\pi_{T^*} \equiv \frac{\partial \mathcal{L}}{\partial \dot{T}^*} = \frac{1}{2} T,$$

from which one reads the Hamiltonian density

$$\mathcal{H} = \pi_T \dot{T} + \pi_{T^*} \dot{T}^* - \mathcal{L} = a \vec{\nabla} T \cdot \vec{\nabla} T^*$$

and the Hamiltonian

$$H = \int d^3\vec{x}\, \mathcal{H} = a \int d^3\vec{x}\, \vec{\nabla} T \cdot \vec{\nabla} T^*.$$

The Hamilton equations

$$\dot{T} = \frac{\partial \mathcal{H}}{\partial \pi_T}, \qquad \dot{T}^* = \frac{\partial \mathcal{H}}{\partial \pi_{T^*}},$$

$$\dot{\pi}_T = -\frac{\partial \mathcal{H}}{\partial T}, \qquad \dot{\pi}_{T^*} = -\frac{\partial \mathcal{H}}{\partial T^*},$$

are equivalent to the Euler–Lagrange equations and yield again Eqs. (5.14) and (5.13).

13 **(B)** The loss of heat from the surface of a body at temperature T to its surroundings at temperature T_∞ can be modeled by *Newton's law of cooling*[4]

$$\frac{dT}{dt} = -\alpha\,(T - T_\infty), \qquad (5.15)$$

where α is a constant coefficient.

a) Solve this equation by assuming that the temperature of the surroundings is not altered by the heat supplied by the cooling body and that at the time $t = 0$ the temperature of the body surface is T_0. Interpret your result physically. What is the late time state of this physical system?

b) Solve the ODE (5.15) by allowing the temperature of the surroundings to vary as heat is supplied by the cooling body. At the initial

---

[4]The nomenclature is a little ambiguous as this name also denotes the law for the heat flux for convective cooling, $q'' = h\,(T - T_\infty)$, from which Eq. (5.15) can be derived.

time $t = 0$ the body and its surroundings have respective temperatures T_0 and $T_\infty^{(0)}$.

Hint: Consider the heat capacities C_{body} and C_{surr} of the body and the surroundings as known quantities.

Solution

a) By using the variable $\theta \equiv T - T_\infty$, we rewrite Eq. (5.15) as

$$\frac{1}{\theta}\frac{d\theta}{dt} = \frac{d}{dt}(\ln \theta) = -\alpha,$$

and it can be immediately integrated to give

$$\theta(t) = \theta_0 \, e^{-\alpha t},$$

where θ_0 is an integration constant to be determined. By imposing the initial condition $T(0) = T_0$, one finds $T_0 - T_\infty = \theta_0$ and

$$T(t) = T_\infty + (T_0 - T_\infty) \, e^{-\alpha t}.$$

At late times (formally, as $t \to +\infty$) the transient $(T_0 - T_\infty)\, e^{-\alpha t}$ decays and the steady-state solution is

$$T\,(t \to +\infty) \approx T_\infty,$$

i.e., the body loses heat (if $T_0 > T_\infty$) and reaches thermal equilibrium with the surroundings at temperature T_∞. Note that the final temperature T_∞ is independent of the initial condition T_0, i.e., beginning with *any* temperature within the realm of validity of Eq. (5.15) one ends with the same final temperature.

b) If the temperature of the surroundings is changed by the heat supplied by the cooling body we have two unknown functions $T(t)$ and $T_\infty(t)$. In order to solve the problem we need a second equation, which is supplied by the calorimetric heat balance for the cooling body and its surroundings:

$$C_{body}\,(T - T_0) + C_{surr}\left(T_\infty - T_\infty^{(0)}\right) = 0,$$

where C_{body} and C_{surr} are the heat capacities of the body and the surroundings (assumed to be independent of the temperature for the process considered). Thus, calorimetry allows us to eliminate the temperature $T_\infty(t)$ by writing it as the function of $T(t)$

$$T_\infty - T_\infty^{(0)} = -\frac{C_{body}}{C_{surr}}\,(T - T_0) \equiv -\gamma\,(T - T_0) \qquad (5.16)$$

and to reduce the problem to the single ODE

$$\frac{dT}{dt} = -\alpha \left[T - T_\infty^{(0)} + \gamma \left(T - T_0 \right) \right],$$

where $\gamma \equiv C_{\text{body}} / C_{\text{surr}}$. By using the variable $\theta \equiv T - T_\infty^{(0)}$ and writing

$$T - T_0 = T - T_\infty^{(0)} + T_\infty^{(0)} - T_0 = \theta - \theta_0,$$

we rewrite Eq. (5.15) as

$$\frac{d\theta}{dt} = -\alpha \left[\theta + \gamma \left(\theta - \theta_0 \right) \right] = -\alpha \left(1 + \gamma \right) \left(\theta - \frac{\gamma}{1+\gamma} \theta_0 \right)$$

or

$$\frac{1}{\theta - \frac{\gamma \theta_0}{1+\gamma}} \frac{d\theta}{dt} = \frac{d}{dt} \left[\ln \left| \frac{\theta - \frac{\gamma \theta_0}{1+\gamma}}{\theta_1} \right| \right] = -\alpha \left(1 + \gamma \right),$$

which is immediately integrated to

$$\theta(t) = \theta_1 \, e^{-\alpha(\gamma+1)t} + \frac{\gamma \theta_0}{1+\gamma},$$

where θ_1 is an integration constant. By imposing $\theta(0) = \theta_0$, we find $\theta_1 = \theta_0 \left(\gamma + 1 \right)^{-1}$ and

$$T(t) = T_\infty^{(0)} + \frac{\theta_0}{\gamma + 1} \left[\gamma + e^{-\alpha(\gamma+1)t} \right]$$

and using the expression (5.16) of $T_\infty(t)$ in terms of $T(t)$,

$$
T(t) = \left[\frac{C_{\text{body}} T_0 + C_{\text{surr}} T_\infty^{(0)}}{C_{\text{body}} + C_{\text{surr}}} \right]
$$
$$
+ \left[\frac{C_{\text{surr}} \left(T_0 - T_\infty^{(0)} \right)}{C_{\text{body}} + C_{\text{surr}}} \right] e^{-\alpha \left(1 + \frac{C_{\text{body}}}{C_{\text{surr}}} \right) t}. \quad (5.17)
$$

The late time steady-state solution is the constant

$$T \left(t \to +\infty \right) = T_\infty \left(t \to +\infty \right) \approx \frac{C_{\text{body}} T_0 + C_{\text{surr}} T_\infty^{(0)}}{C_{\text{body}} + C_{\text{surr}}},$$

a weighted average of the temperatures of the body and the surrounding air with the respective heat capacities as weights. In the limit in which the heat capacity of the surroundings tends to infinity, the solution of part a) is recovered and T_∞ stays constant. Physically, it is clear that this state of thermal equilibrium is stable. This is confirmed by the mathematics by considering this steady-state constant solution rewritten in terms of the variable θ, which we call

$$\theta_* = \frac{\gamma \theta_0}{\gamma + 1}.$$

$\theta = \theta_*$ is an exact solution and is stable. In fact, assume that the initial datum is $\theta_0 > \theta_*$ (corresponding to the surface of the body initially hotter than the surrounding air, $T_0 > T_\infty^{(0)}$). Then it is $\theta(t) > \theta_*$ at all times t. For any solution $\theta(t)$, the curve representing $\theta(t)$ in the (t, θ) plane never crosses the straight line $\theta = \theta_*$, or else the uniqueness theorems for the solutions of ODEs would be violated— hence it is $\theta(t) > \theta_*$. Then

$$\frac{d\theta}{dt} = -\alpha (1 + \gamma)(\theta - \theta_*) < 0,$$

$$\frac{d^2\theta}{dt^2} = -\alpha (1 + \gamma)\frac{d\theta}{dt} > 0,$$

and the solution is monotonically decreasing with upward-facing concavity. It must approach asymptotically its lower bound θ_*, which it cannot cross.

Similarly if $\theta_0 < \theta_*$ initially, then $d\theta/dt > 0$ and $d^2\theta/dt^2 < 0$, and again $\theta(t)$ approaches θ_* as $t \to +\infty$. If $\theta_0 = \theta_*$, then the solution stays constant forever (an unphysical fine-tuned situation). The final equilibrium state is stable, as expected.

14 **(A)** What is the heat flux between the ground surface at 15.6°C and the surrounding air at 5.2°C due to free convection? Measurements have provided the empirical value of 9.5 W·m^{-2}·K^{-1} for the convection coefficient.

Solution
The heat flux due to free convection is proportional to the temperature difference between air and ground according to Newton's law of cooling

$$q'' = h(T_g - T_\infty) = \left(9.5 \, \frac{\text{W}}{\text{m}^2 \cdot \text{K}}\right)(15.6 - 5.2)\,\text{K} = 99 \, \frac{\text{W}}{\text{m}^2}.$$

15 **(A)** The *Bergmann rule* of biology states that in warm-blooded, wide-ranging animal species, members living in colder climates tend to be

larger than animals of the same species living in warmer climates. Explain this rule on the basis of your knowledge of heat transfer.

Solution

In cold climates or cold waters it is essential to preserve body heat. The total heat produced by an animal is proportional to the amount of metabolizing tissue, hence to the body mass. Let us model the animal as a sphere[5] of radius R; then the body mass is $m = 4\pi\rho R^3/3$, where ρ is the average density, while the heat dissipated by convection and radiation is given by the heat flux densities $q_1'' = h\left(T - T_\infty\right)$ and $q_2'' = \epsilon\sigma\left(T^4 - T_\infty^4\right)$ (here T is the body surface temperature, T_∞ is the temperature of the surroundings, h is the convection coefficient, σ is the Stefan–Boltzmann constant, and ϵ is the average graybody factor). If also conduction is an important factor of heat loss, Fourier's law gives the heat flux density $\vec{q''}_3 = -k\vec{\nabla}T$. In all these cases the heat lost per unit time is proportional to the surface area A of the cooling body, $q_i = Aq_i''$. For the spherical animal $A = 4\pi R^2$, and the ratio

$$\frac{\text{heat lost per unit time}}{\text{heat generated per unit time}} \propto \frac{4\pi R^2}{4\pi\rho R^3/3} \approx \frac{1}{R}$$

is inversely proportional to the size of the animal. It is expected that natural selection favors larger individuals within the same species.

5.3   Thermodynamics

From the point of view of energy demand and its environmental consequences, thermodynamics places fundamental limits on the efficiency of processes converting thermal energy into mechanical and then electrical energy. These limits are summarized in the second law of thermodynamics. From a more general point of view, thermodynamics regulates the exchanges of heat and mechanical work between different subsystems of the environment or between organisms and their habitat. A review of equilibrium and nonequilibrium thermodynamics is beyond the scope of this book—Refs. [9, 15, 62, 74] are suggested as general references and [8, 45, 52] for biological applications.

1 **(B, C)** Define heat capacity, specific heat, latent heat of fusion, and latent heat of vaporization.

---

[5]This is in the spirit of the spherical cow of Ref. [27] and of many jokes on physicists.

Solution

The *heat capacity* C of a body is the amount of heat Q necessary to rise its temperature by one degree Celsius: it is proportional to the mass m of the body and it depends on the nature of the material composing the body. In formulas,

$$C \equiv \frac{dQ}{dT}.$$

where dQ is the infinitesimal amount of heat exchanged with the body and dT is the consequent variation of its temperature. In principle, heat can be supplied or removed at constant volume or at constant pressure, obtaining two different heat capacities $C_V \equiv (dQ/dT)_V$ and $C_P \equiv (dQ/dT)_P$. It is $C_P > C_V$ because when the volume is allowed to change in a process at constant pressure part of the heat energy supplied goes into mechanical work ($dW = PdV$ for an infinitesimal change in volume dV). In practice processes of interest for the environmental scientist occur at constant atmospheric pressure and C_P is the relevant heat capacity.

The *specific heat* (capacity) c is the heat capacity of the unit mass of the material composing the body—for a small temperature change ΔT we have

$$Q = cm\,\Delta T,$$

where $C = cm$ is the heat capacity. This formula holds if there are no phase changes and if the temperature change ΔT is small. For larger temperature changes $T_i \longrightarrow T_f$ the heat exchanged is given by the integral

$$Q = \int_{T_i}^{T_f} c\,m\,dT,$$

where in principle $c = c(T)$ depends on the temperature.

The *latent heat* of fusion L_f of a substance is the heat necessary to melt a unit of mass of it. The heat necessary to melt a mass m of the same substance is

$$Q = L_f\, m;$$

this thermal energy is given back by the substance when it solidifies.

The *latent heat of vaporization* L_v of a liquid is the heat necessary to evaporate a unit of mass of it. The heat required to evaporate a mass m of the same substance is

$$Q = L_v\, m;$$

this energy is removed from the vapor when it liquefies.

Note that there is no temperature variation in the formulas for phase transitions. The temperature stays constant during a phase change of a chemically pure substance.

2 **(A)** In a heat engine hot steam is injected at 180°C and is expelled at 50°C. Compute the maximum possible theoretical efficiency of the engine. Given that 35% of the mechanical work produced is dissipated against internal friction, what is the real maximum efficiency possible?

Solution
The maximum possible theoretical efficiency is given by the Carnot factor

$$\eta_{max} = 1 - \frac{T_C}{T_H} = 1 - \frac{323 \text{ K}}{453 \text{ K}} = 0.29 = 29\%.$$

The work extracted is

$$W_{out} = (1 - 0.35) \cdot 0.29 \, Q_{in} = 0.19 \, Q_{in},$$

and the real maximum efficiency possible is[6]

$$\eta_{real} = \frac{W_{out}}{Q_{in}} = 0.19 = 19\%.$$

3 **(A)** A power station uses hot steam at 650°C as the hot reservoir and water from a nearby river at 16°C as a coolant. What is the maximum efficiency of the plant that is theoretically possible according to thermodynamics?

Solution
The maximum theoretical efficiency allowed by the laws of thermodynamics is given by the Carnot factor

$$\eta_{max} = 1 - \frac{T_C}{T_H},$$

where T_C and T_H are, respectively, the Kelvin temperatures of the cold and hot reservoirs. Therefore[7],

$$\eta_{max} = 1 - \frac{289 \text{ K}}{923 \text{ K}} = 0.687 = 68.7\%.$$

---

[6]A common mistake is to express temperatures in the Celsius scale instead of the correct Kelvin scale. This incorrect procedure would lead to a maximum theoretical efficiency of 72%, a significant deviation from the true value.
[7]The incorrect procedure of expressing temperatures in the Celsius scale instead of the correct Kelvin scale would lead to a maximum theoretical efficiency of 97.5%, very different from the actual theoretical value.

4 **(A)** An experimental heat engine in a laboratory uses a cycle extremely close to a Carnot cycle with 40% efficiency, and the cold reservoir is at 21°C. Find the temperature of the hot reservoir. Why is such an engine useless for most practical purposes?

Solution
The efficiency of the Carnot cycle is given by the Carnot factor

$$\eta = 1 - \frac{T_C}{T_H},$$

and therefore the temperature of the hot reservoir is

$$T_H = \frac{T_C}{1 - \eta} = \frac{294\,\text{K}}{1 - 0.4} = 490\,\text{K}.$$

A reversible Carnot engine is useless for most practical purposes because the time required for the engine to perform work in reversible and quasistatic conditions is extremely long and the power output is extremely small. This disadvantage arises from the need to keep the transformations reversible at every point along the cycle. Alternatively, an operation on reasonable time scales would require the temperature difference $T_H - T_C$ to be infinitesimal and the amount of work extracted would be negligible.

5 **(A)** A thermal field in Volcanoland has a steam temperature of 276°C. A small thermal power station operates using this steam as the hot reservoir and water at 16.5°C as the cold reservoir. What is the maximum theoretical efficiency of this plant?

Solution
The maximum theoretical efficiency of the power station is given by the Carnot factor

$$\eta = 1 - \frac{T_C}{T_H} = 1 - \frac{2.90 \cdot 10^2\,\text{K}}{5.49 \cdot 10^2\,\text{K}} = 0.472 = 47.2\,\%.$$

6 **(A)** The average efficiency of the muscles in the human body is approximately 20% and it can almost double in a trained athlete. Assume that the average temperature of the environment in which the body operates is 15°C. Can the human body be described as a heat engine? If not, what kind of thermodynamic processes describe it?

Solution
Were the human body correctly described as a heat engine, its efficiency would have as upper limit the Carnot factor

$$\eta_{\max} = 1 - \frac{T_C}{T_H} = 1 - \frac{288\,\mathrm{K}}{309\,\mathrm{K}} = 6.8 \cdot 10^{-2} = 6.8\,\%,$$

where the hot reservoir would be the body itself at a temperature
of approximately 36°C= 309 K. Since its efficiency is much larger
than the Carnot factor, the human body cannot be described as a
heat engine—it is better described as a *isothermal* machine at the
constant temperature of approximately 36°C.

7 **(A)** The exhausts of a gas turbine, which has efficiency $\eta_A = 0.32$,
are used to produce the steam necessary to run a second turbine that
has efficiency $\eta_B = 0.42$. What is the overall efficiency of the system?

Solution
Let Q_A be the heat entering the gas turbine, Q_B the heat leaving it,
and Q_C the heat leaving the steam turbine. The efficiency of the gas
turbine is

$$\eta_A = \frac{Q_A - Q_B}{Q_A} = 1 - \frac{Q_B}{Q_A},$$

which yields

$$Q_B = Q_A\,(1 - \eta_A).$$

The efficiency of the steam turbine is

$$\eta_B = \frac{Q_B - Q_C}{Q_B} = 1 - \frac{Q_C}{Q_B},$$

which yields

$$Q_C = Q_B\,(1 - \eta_B) = Q_A\,(1 - \eta_A)\,(1 - \eta_B).$$

The overall efficiency of the system is

$$\eta_{\text{total}} = \frac{Q_A - Q_C}{Q_A} = \frac{Q_A - Q_B}{Q_A} + \frac{Q_B - Q_C}{Q_A}$$

$$= \eta_A + \frac{Q_A\,(1 - \eta_A) - Q_A\,(1 - \eta_A)\,(1 - \eta_B)}{Q_A}$$

$$= \eta_A + (1 - \eta_A)\,[1 - (1 - \eta_B)]$$

$$= \eta_A + \eta_B - \eta_A\eta_B.$$

Numerically,

$$\eta_{total} = 0.32 + 0.42 - 0.32 \cdot 0.42 = 0.61.$$

5.4 Electricity

Electricity is the main form in which energy for human use is transported over large distances through power lines. It is used to run motors, for lightning, for heating, in industrial processes, and in many devices including computers. Power generation plants convert thermal or mechanical energy into more readily usable and transportable electric energy.

1 **(A)** An underground parking lot has its lights on twenty-four hours a day. Ten 60 W incandescent bulbs are replaced by 10 20 W fluorescent bulbs providing the same amount of light. How much energy is saved in one year of operation?

Solution
Since power W is energy divided by time, the energy saved per second is $10 \cdot (60 - 20)$ W·s $= 400$ J, and the energy saved in one year is

$$E = Wt = 10 \cdot (40 \text{ W}) \cdot (365 \text{ days}) \cdot (24 \text{ hours}) = 3500 \text{ kWh}.$$

2 **(A)** What is the cost of operating a 4.50 W electrical alarm clock for a year if your electric company charges 9.75 cents/kWh?

Solution
The energy used in one year of operation by the alarm clock is

$$E = (4.50 \text{ W}) \cdot (365 \text{ days}) \cdot (24 \text{ hours}) = 3.94 \cdot 10^4 \text{ Wh} = 39.4 \text{ kWh}.$$

The cost of a year of operation is $(9.75 \text{ cents/kWh}) \cdot (39.4 \text{ kWh}) = 3.84 \cdot 10^2$ cents $= \$ 3.84$.

3 **(A)** A power line at 350 kV distributes 100 MW of power; what current is carried in it, if losses are neglected?

Now taking into account heat generation due to the Joule effect, find the amount of heat generated per unit time by the power line if it has a resistance of 10 Ω. What fraction of the power is lost as heat?
Hint: For simplicity, treat the line as a direct current circuit.

Solution

Neglecting heat losses, the power transported by the power line is $W = VI$ and the intensity I of the current is

$$I = W/V = \frac{1.0 \cdot 10^8 \text{ W}}{3.5 \cdot 10^5 \text{ V}} = 2.86 \cdot 10^2 \text{ A} \simeq 2.9 \cdot 10^2 \text{ A}.$$

The power dissipated into heat is

$$W_{\text{lost}} = I^2 R = (286 \text{ A})^2 \cdot (10 \, \Omega) = 8.2 \cdot 10^5 \text{ W}.$$

The fraction of power lost is

$$\frac{W_{\text{lost}}}{W} = \frac{8.2 \cdot 10^5 \text{ W}}{10^8 \text{ W}} = 8.2 \cdot 10^{-3}.$$

4 **(A)** The Niagara Falls are approximately 52 m high with a flow rate of $6.0 \cdot 10^3 \text{ m}^3/\text{s}$. If all the gravitational potential energy of the water in the falls could be converted into electrical energy, what power would be available? Take the density of water to be $\rho = 1.0 \cdot 10^3 \text{ kg/m}^3$. Five hydroelectric stations at Niagara Falls produce a total of 1800 MW; what fraction of the total potential energy of water is converted into electricity?

Solution

The generated power W would be the rate at which gravitational potential energy is converted into electrical energy,

$$W = \frac{E}{t}.$$

Since the gravitational potential energy of a mass m of water occupying the volume V is $E = mgh = \rho V gh$, we have

$$
\begin{aligned}
W &= \rho \frac{V}{t} gh \\[2mm]
&= \left(1.0 \cdot 10^3 \text{ kg} \cdot \text{m}^{-3}\right) \left(6.0 \cdot 10^3 \text{ m}^3 \cdot \text{s}^{-1}\right) \left(9.81 \text{ m} \cdot \text{s}^{-2}\right) (52 \text{ m}) \\[2mm]
&= 3.1 \cdot 10^9 \text{ W} = 3.1 \text{ GW}.
\end{aligned}
$$

The fraction of gravitational potential energy converted into electricity is

$$\eta = \frac{1.8 \cdot 10^9 \text{ W}}{3.1 \cdot 10^9 \text{ W}} = 0.58.$$

5 **(A)** Compute the maximum flux density of kinetic energy, available for conversion into electric energy by a turbine, which is carried by a fluid of density ρ moving with speed v.

Hint: For simplicity, assume laminar flow and consider a portion of stream tube traversed by the fluid in the infinitesimal time dt.

Solution
Assuming laminar flow, in the time dt the fluid covers the distance $dL = vdt$ and defines a portion of stream tube of length dL and cross-sectional area A, which can be assumed to be constant along the stream tube because the latter has infinitesimal length. The mass contained in this portion of stream tube is $dm = \rho dV = \rho A v dt$ and its kinetic energy is $dE = (dm)v^2/2 = \rho A v^3 dt/2$. The power (energy per unit time) carried by the fluid per unit area, which is in principle available for conversion into electric power, is

$$\frac{W}{A} = \frac{1}{A}\frac{dE}{dt} = \frac{1}{2}\rho v^3,$$

proportional to the *cube* of the fluid speed.

6 **(A)** A power plant converts the energy of tides into electric energy. If the tidal range is $R = 5.2$ m, what is the maximum power per unit of area of the collecting basin that can be extracted? The density of sea water is $\rho = 1.03 \cdot 10^3 \, \mathrm{kg/m^3}$.

Solution
Assuming that the collecting basin has a depth equal to the tidal range R and that the water is discharged onto a turbine at low tide, the gravitational potential energy of the water in the collecting basin is mgR, where $m = \rho V$ is the mass of sea water, of density ρ, in the basin of volume V. Therefore, if A is the area of the collecting basin, the total energy converted into electric energy is

$$E = mgh = \rho A R g R$$

or

$$\frac{E}{A} = \rho g R^2 = \left(1.03 \cdot 10^3 \, \frac{\mathrm{kg}}{\mathrm{m^3}}\right)\left(9.81 \, \frac{\mathrm{m}}{\mathrm{s^2}}\right)(5.2\,\mathrm{m})^2 = 2.7 \cdot 10^5 \, \frac{\mathrm{J}}{\mathrm{m^2}}.$$

Since there is a tide every 12 hours, the power available per square meter of the collecting basin is

$$W = \frac{E}{At} = \frac{2.7 \cdot 10^3 \, \mathrm{kg/m^2}}{12 \cdot 3600\,\mathrm{s}} = 6.3 \, \frac{\mathrm{W}}{\mathrm{m^2}}.$$

This modest number suggests that the collecting basin should have a large area, which would impact a large portion of shoreline or tidal flats, if present.

Chapter 6

FLUID MECHANICS

You may have inner tranquillity, but you can't escape surface tension.
—V. Louise Roth

Liquids and gases, whether in static equilibrium or in motion, are essential constituents of the natural environment, and various fluids are also essential for the functioning of biological organisms. Earth sciences such as atmospheric science, meteorology, dynamical oceanography, and hydrology rely on the understanding of fluid mechanics. While liquids are only slightly compressible, gases are easily compressed and this accounts for many of their different physical properties. It is convenient to separate the study of fluids from that of liquids and gases. Applications to the oceans and the atmosphere are presented in Chapter 3.

Any standard first-year physics textbook will suffice as a reference for the problems labeled **A** or **C** in this chapter—a fluid dynamics book (e.g., Refs. [1, 50, 40, 57]) is needed for the rest of the material in Section 1, and a book on gas laws, thermodynamics, and kinetic theory (e.g., [62, 9, 38, 10]) is useful for Section 2.

6.1 Liquids

The main liquid of interest for the environmental scientist is water, but various hydrocarbons are also important for the petroleum industry and its related aspects, e.g., oil exploration, pipelines, and environmental clean-up.

1 **(C)** Summarize the unusual physical and chemical properties of water.

Solution

Due to the hydrogen bonds, the boiling point of water (100°C at 1 atm) is rather high and the difference between melting point and boiling point is also rather large (100°C), making it possible for water to exist in the liquid state and allowing for the existence of oceans, lakes, and rivers.

Water expands when it turns into ice unlike most common substances; ice does not sink to the bottom of lakes and oceans, but floats on top. In addition, the density of water is maximum at 4°C and water at this temperature sinks, allowing life to continue under water. The expansion of water when it freezes, with large stresses applied to the walls of a container confining it, constitutes an important agent of erosion when the temperature oscillates around 0°C.

The specific heat of water $(4187 \, \text{J} \cdot \text{kg}^{-1} \cdot (^\circ \text{C})^{-1})$ is unusually high, about five times that of most common rocks and soils. As a consequence oceans and lakes mitigate climates with their thermal inertia, reduce temperature excursions, and create local winds and sea and mountain breezes. The high specific heat also makes water an excellent coolant medium for industrial applications.

The latent heat of vaporization of water (2258 J/kg) is very large in comparison to most substances, implying that a large amount of latent heat can be stored in the water vapor evaporated at the tropics and residing in the lower atmosphere. This energy can be transported (an important item in the global distribution of solar energy around the planet), fuel tropical hurricanes, or be released when water vapor condenses, causing precipitation and continuing the hydrologic cycle.

The surface tension of water is also relatively high, a fact that has various applications, e.g., to soil physics, surface waves on bodies of water, and the rising of sap in the xylem of a tree.

Finally, water molecules are polar, making water a very good solvent for many substances and a good medium to transport nutrients and waste products in plant and animal organisms. Most biochemical reactions occur in water solutions.

6.1.1   Fluid statics

Very often we see water at rest or in slow motion and it is important for the environmental scientist to have a clear understanding of the properties of static fluids. For example, most of the sea water in the oceans far away from major oceanic currents can be considered at rest or in very slow motion and the motion of groundwater is also very slow.

The static state is the result of an equilibrium that has been achieved through previous motion. Understanding hydrostatics helps to model bodies of water, capillary phenomena in the soil or plant and animal tissues, buoyancy phenomena in the atmosphere, or the floating of icebergs.

1 **(A)** A river has average width 46 m, has an average depth of 21 m, and is 156 km long. What is the mass of water in it?

Solution
The mass of the river is the product of the density of water $\rho \simeq 1.0 \cdot 10^3$ kg/m^3 and of the volume of the river,

$$M = \rho V = \rho S h$$

$$= (1.0 \cdot 10^3 \text{ kg} \cdot \text{m}^{-3}) \cdot (46 \text{ m}) \cdot (21 \text{ m}) \cdot (1.56 \cdot 10^5 \text{ m})$$

$$= 1.5 \cdot 10^{11} \text{ kg}.$$

2 **(A)** A dam creates an artificial lake with surface area $S = 112$ km^2 and average depth $h = 63$ m. What is the mass of the water in the lake? At the beginning of summer the average temperature of the lake was 8°C, while at the end of the summer it is 15.5°C. How much heat was absorbed by the lake during the summer?

Solution
The mass of the lake is the product of the density of water $\rho = 1.00 \cdot 10^3$ kg/m^3 and of the volume of the lake,

$$M = \rho V = \rho S h = (1.0 \cdot 10^3 \text{ kg/m}^3) \cdot (1.12 \cdot 10^8 \text{ m}^2) \cdot (63 \text{ m})$$

$$= 7.1 \cdot 10^{12} \text{ kg}.$$

The heat absorbed by the lake during summer is given by

$$Q = cm\,\Delta T = \left(1.0 \frac{\text{Kcal}}{\text{kg} \cdot {}^\circ\text{C}}\right) \cdot (7.1 \cdot 10^{12} \text{ kg}) \cdot (7.5^\circ\text{C})$$

$$= 5.3 \cdot 10^{13} \text{ Kcal} = 2.2 \cdot 10^{17} \text{ J}.$$

3 **(A)** The Columbia Plateau is a large basaltic lava flow over an area shared by four states of the present northwestern USA. It covers a surface of area $A = 5.76 \cdot 10^5$ km^2 and has an average depth $h = 150$ m.

Compute the mass of basalt that flowed forming the plateau and the pressure that it exerts on the underlying rock. The density of the basalt is $\rho = 2.2 \cdot 10^3 \, \text{kg/m}^3$.

Solution
The mass of the Columbia Plateau is given by its average density times its volume,

$$m \;=\; \rho V = \rho A h = \left(2.2 \cdot 10^3 \, \frac{\text{kg}}{\text{m}^3} \right) \cdot \left(5.76 \cdot 10^{11} \, \text{m}^2 \right) \cdot (150 \, \text{m})$$

$$= \; 1.9 \cdot 10^{17} \, \text{kg}.$$

The pressure exerted by this mass on the underlying rock is the ratio of its weight to its area,

$$P = \frac{mg}{A} = \rho g h = \frac{\left(1.9 \cdot 10^{17} \, \text{kg}\right) \cdot \left(9.81 \, \text{m} \cdot \text{s}^{-2}\right)}{5.76 \cdot 10^{11} \, \text{m}^2} = 3.2 \cdot 10^6 \, \text{Pa}.$$

4 (A) What is the fraction of the volume that is submerged for an iceberg? The densities of water and ice are, respectively, $1.00 \cdot 10^3 \, \text{kg/m}^3$ and $0.917 \cdot 10^3 \, \text{kg/m}^3$. Is there a simple formula giving the height of the part of the iceberg that sticks out of the water?

Solution
According to Archimede's principle, the upward-directed buoyant force F_B on the iceberg balances the weight of the ice,

$$F_B = m_{\text{water}} \, g = m_{\text{ice}} \, g,$$

where m_{water} is the mass of the volume of water displaced. At equilibrium $m_{\text{water}} = m_{\text{ice}}$ and we have, for the corresponding densities and volumes,

$$\rho_{\text{water}} V_{\text{displaced}} = \rho_{\text{ice}} V_{\text{iceberg}}.$$

The fraction of the volume of the iceberg that is submerged is then

$$\frac{V_{\text{displaced}}}{V_{\text{iceberg}}} = \frac{\rho_{\text{ice}}}{\rho_{\text{water}}} = \frac{0.917 \cdot 10^3 \, \text{kg} \cdot \text{m}^{-3}}{1.00 \cdot 10^3 \, \text{kg} \cdot \text{m}^{-3}} = 0.917,$$

or 91.7%. In practice, even more of the volume of the iceberg sticks out of the water because saltwater in the ocean has a higher density than the freshwater composing the iceberg, and because the ice contains air bubbles that lower its density.

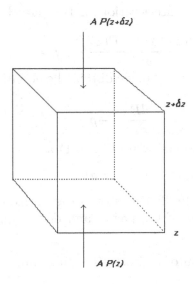

Figure 6.1. A parcel of fluid of infinitesimal volume $\delta V = A\delta z$.

There is no simple formula giving the height of the part of the iceberg sticking out of the water: it depends on the shape of the iceberg.

5 **(B)** Derive the equation of hydrostatic equilibrium in a fluid.

Solution
Consider a vertical axis z pointing upward and a parcel of fluid with the shape of a vertical parallelepiped with volume V and horizontal faces of area A at z and $z + \delta z$ (Fig. 6.1). The only forces acting in the vertical direction are due to the pressures on the horizontal faces,

$-AP(z+\delta z)$ (directed downward) and $AP(z)$ (directed upward), and the weight $-mg$ of the parcel pointing downward (g is the acceleration of gravity). Since the mass of the parcel is $m = \rho V = \rho\, A\, \delta z$, where ρ is the density of the fluid, the balance of forces in the vertical direction yields

$$A\left[P(z) - P(z + \delta z)\right] - \rho g A \delta z = ma,$$

where a is the vertical acceleration of the parcel. In equilibrium, $a = 0$ and

$$\frac{P(z + \delta z) - P(z)}{\delta z} + \rho g = 0.$$

By taking the limit as $\delta z \to 0$, we obtain the equation of hydrostatic equilibrium

$$\frac{dP}{dz} = -\rho g. \tag{6.1}$$

Using differentials,[1] we can write Eq. (6.1) as

$$dP = -\rho g\, dz. \tag{6.2}$$

The forces acting on the fluid parcel in the horizontal direction balance on opposite sides of the parcel and do not provide additional mathematical relations.

6 **(B)** State the equation of hydrostatic equilibrium

a) under water, where $P(z)$ is the pressure and z is the depth;

b) in the atmosphere, where $P(z)$ is the pressure and z is the elevation.
Solve the equation in both cases.
Hint: In case b) substitute the actual temperature with an *average* temperature that is independent of elevation.

Solution
a) The equation of hydrostatic equilibrium is

$$dP = \rho g\, dz,$$

where z is the water's depth. Since water is almost incompressible, one can assume that ρ is independent of z and the equation of hydrostatic equilibrium is immediately solved, yielding

$$P(z) = \rho g\, z.$$

---
[1] More rigorously, using differential forms, or 1-forms.

b) The equation of hydrostatic equilibrium is

$$dP = -\rho g \, dz,$$

where z is the elevation. In this case, the air density depends on z, and the ideal gas law in the form

$$P = \rho \frac{kT}{m},$$

where k is the Boltzmann constant, T is the temperature, and m is the average mass of the air particles, yields

$$\rho(z) = \frac{mP(z)}{kT}$$

and

$$\frac{dP}{dz} = -\rho g = -\frac{mgP(z)}{kT}.$$

By replacing the temperature T with an average, z-independent temperature $\bar{T}$, we obtain the ordinary differential equation

$$\frac{1}{P} \frac{dP}{dz} = \frac{d\left(\ln P(z)\right)}{dz} = -\frac{mg}{k\bar{T}}$$

that has the exponential solution (Fig. 6.2)

$$P(z) = P_0 \exp\left(-\frac{z}{H}\right),$$

where

$$H \equiv \frac{k\bar{T}}{mg}$$

is a length scale characteristic of the atmosphere and $P_0 = P(z = 0)$.

7 **(B)** Prove that in an incompressible fluid in hydrostatic equilibrium the pressure P satisfies the Laplace equation $\nabla^2 P = 0$.

Solution
The equation of hydrostatic equilibrium is

$$\vec{\nabla} P - \rho \vec{g} = 0,$$

where $\vec{g} = -\vec{\nabla}\Phi$ is the opposite of the gradient of the gravitational potential Φ. By taking the divergence of this equation, we obtain

$$\nabla^2 P - \left(\vec{\nabla}\rho\right) \cdot \vec{g} - \rho \vec{\nabla} \cdot \vec{g} = 0.$$

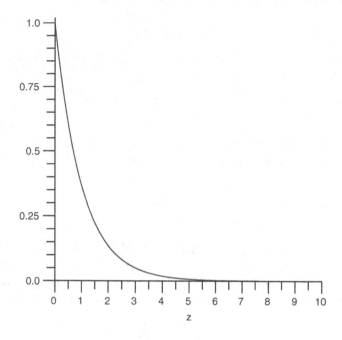

Figure 6.2. The pressure (atm) versus elevation (km) for $H = 1$ km.

Since the acceleration of gravity $\vec{g}$ can be considered as constant on regions small compared to the radius of the Earth and the fluid is incompressible ($\rho =$ constant), both $\vec{\nabla}\rho$ and $\vec{\nabla} \cdot \vec{g}$ vanish and we are left with $\nabla^2 P = 0$.

8 **(A)** Compute the percent change in volume of ocean water at a depth of 6000 m due to the weight of the water above. The bulk modulus and the average density of sea water are, respectively, $B = 2.2 \cdot 10^9$ N·m^{-2} and $\rho = 1.03 \cdot 10^3$ kg/m^3.

Solution
The fractional change of volume is given by

$$\frac{\delta V}{V} = \frac{1}{B}\frac{F}{A}, \qquad (6.3)$$

where B is the bulk modulus and F is the force per unit normal area A. The pressure at depth h is approximately given by

$$P = \rho g h = (1.03 \cdot 10^3 \, \text{kg/m}^3) \cdot (9.81 \, \text{m/s}^2) \cdot (6000 \, \text{m}) = 6.06 \cdot 10^7 \, \text{Pa},$$

where g is the acceleration of gravity. Here we neglect the compressibility of water in the computation of the pressure, which amounts to a second-order error in the final formula for $\delta V/V$. Therefore,

$$\frac{\delta V}{V} = \frac{P}{B} = \frac{6.06 \cdot 10^7 \, \text{Pa}}{2.2 \cdot 10^9 \, \text{N} \cdot \text{m}^{-2}} = 2.8 \cdot 10^{-2} = 2.8\,\%.$$

6.1.2 Capillarity and surface tension

Capillarity and surface tension are important concepts in understanding the adsorption of thin films of water at the interface of pores and solid particles in a soil. Capillarity is also important to understand the horizontal transport of liquids (especially water) and the vertical transport of liquids over short distances which is counteracted by gravity. This has applications, e.g., to heat pipes in machines and to some extent to plant tissues. Surface tension also regulates the physics of membranes forming the interface between water and air, of bubbles and drops and is relevant, e.g., for the detergent industry which is responsible for some water pollution and for eutrophication, or the ability of certain insects to walk on the surface of a pond.

1 **(B)** Compute the difference between the internal and the external pressure in a soap bubble by using the fact that the surface tension represents the energy per unit area. Consider an infinitesimal dilation during which the bubble radius increases from r to $r + dr$. Repeat the exercise for a spherical bubble of air in water.

Solution
The difference between the internal pressure and the external atmospheric pressure is $P_i - P_0$. The total force normal to the surface of the bubble is $(P_i - P_0) \, 4\pi r^2$, and the infinitesimal work done by this force during the dilation $r \longrightarrow r + dr$ is $4\pi r^2 \, (P_i - P_0) \, dr$. This work must equal the energy change of the bubble γdA, where A is the surface area. Since a soap bubble has *two* surfaces of contact between the liquid and air, each of area $4\pi r^2$, it is $\gamma \, dA = \gamma \, d \left(8\pi r^2 \right) = 16\pi \gamma r dr$. Therefore, the energy balance is

$$4\pi r^2 \, (P_i - P_0) \, dr = 16\pi \gamma \, r dr,$$

and the gauge pressure inside the bubble is given by

$$P_i - P_0 = \frac{4\gamma}{r}.$$

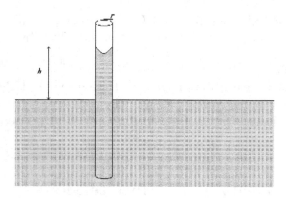

Figure 6.3. A capillary tube.

Smaller bubbles have larger internal pressure.

For a bubble of air in water there is only one surface of contact between liquid and gas. Therefore,

$$4\pi r^2 (P_i - P_0) \, dr = 8\pi\gamma \, rdr,$$

which yields the difference between the air pressure inside and the water pressure outside the bubble:

$$P_i - P_0 = \frac{2\gamma}{r}.$$

2 **(B)** Derive a formula for the height h reached by a liquid of surface tension γ_S and contact angle θ in a capillary tube of given radius r, using only dimensional analysis (Fig. 6.3).

Solution

The height h reached by a liquid in a capillary tube will depend on the nature of the liquid, the radius r of the tube, and the gravitational acceleration g. The physical variables characterizing the liquid and playing a role in this problem will be the density ρ, the surface tension γ_S, and the contact angle θ of the liquid. However, because the contact angle does not carry dimensions, the dependence of h on θ will be missed. The dimensions of the quantities involved are

$$[h] \quad = \quad [r] = [L],$$

$$[\gamma_S] \quad = \quad [MT^{-2}],$$

$$[g] = [LT^{-2}],$$

$$[\rho] = [L^{-3}M],$$

and by setting

$$h = A\,\gamma_S^\alpha\,\rho^\beta\,g^\gamma\,r^\delta,$$

where A is a dimensionless coefficient, we obtain the dimensional equation

$$[L] = [M^\alpha T^{-2\alpha}]\left[L^{-3\beta}M^\beta\right][L^\gamma T^{-2\gamma}]\left[L^\delta\right]$$

$$= \left[L^{-3\beta+\gamma+\delta}T^{-2\alpha-2\gamma}M^{\alpha+\beta}\right],$$

which yields

$$-3\beta + \gamma + \delta = 1,$$

$$-2\alpha - 2\gamma = 0,$$

$$\alpha + \beta = 0.$$

This linear system of three equations for four unknowns is underdetermined and has the solution

$$\alpha \text{ arbitrary}, \qquad \beta = \gamma = -\alpha, \qquad \delta = 1 - 2\alpha,$$

which yields only

$$h = A\left(\frac{\gamma_S}{\rho g}\right)^\alpha r^{1-2\alpha},$$

leaving α to be determined by experiment, e.g., by studying the dependence of h on the radius of the capillary tube and establishing that $h \propto r^{-1}$ we would obtain the formula $h = A\gamma/(\rho g r)$. However, this is not what is required, and we can refine the dimensional analysis and solve the problem as follows below.

The improvement consists in distinguishing between lengths in the vertical direction L_z and lengths in the transverse or horizontal direction, L_r. Then the relevant physical quantities have dimensions

$$[h] = [L_z],$$

$$[r] = [L_r],$$

$$[\gamma_S] = \left[\frac{\text{Force}}{L}\right] = \left[\frac{L_z T^{-2} M}{L_r}\right],$$

$$[g] = \left[L_z T^{-2}\right],$$

$$[\rho] = \left[L_z^{-1} L_r^{-2} M\right].$$

This yields the dimensional equation

$$[L_z] = \left[L_z^\alpha L_r^{-\alpha} T^{-2\alpha} M^\alpha\right] \left[L_z^{-\beta} L_r^{-2\beta} M^\beta\right] \left[L_z^\gamma T^{-2\gamma}\right] \left[L_r^\delta\right]$$

$$= \left[L_z^{\alpha - \beta + \gamma} L_r^{-\alpha - 2\beta + \delta} T^{-2\alpha - 2\gamma} M^{\alpha + \beta}\right],$$

and the associated linear system

$$\alpha - \beta + \gamma = 1,$$

$$-2\alpha - 2\gamma = 0,$$

$$-\alpha - 2\beta + \delta = 0,$$

$$\alpha + \beta = 0,$$

which now consists of four independent equations for the four un-knowns and is completely determined. The solution is

$$\alpha = 1, \qquad \beta = \gamma = \delta = -1,$$

and

$$h = A\frac{\gamma_S}{\rho g r}.$$

This solution to the problem requires only theoretical considerations and no experiment and constitutes an example of how we can solve a problem that is otherwise unsolvable with the method of dimensions by enlarging the number of quantities treated as independent.

3 **(A)** How high can capillarity lift sap through the tubes with radius $r = 0.02$ mm forming the xylem of a tree? Can capillarity explain how sap reaches the top of a giant sequoia 80 m high? The sap density is $\rho = 1.05 \cdot 10^3$ kg/m^3 and its surface tension is $\gamma = 7.28 \cdot 10^{-2}$ N/m. Assume that the contact angle is zero.

Solution

The height above the free surface of a liquid reached in a capillary tube (Fig. 6.3) is

$$h = \frac{2\gamma \cos \theta}{\rho g r},$$

where g is the acceleration of gravity and θ is the contact angle— hence, the maximum height that sap can reach due to capillarity is

$$h = \frac{2 \cdot \left(7.28 \cdot 10^{-2}\,\mathrm{N/m}\right)}{\left(1.05 \cdot 10^3\,\mathrm{kg/m^3}\right) \cdot \left(9.81\,\mathrm{m/s^2}\right) \cdot \left(2 \cdot 10^{-5}\,\mathrm{m}\right)} = 0.71\,\mathrm{m}.$$

Obviously, capillarity alone cannot be responsible for rising sap even in trees of modest height, let alone giant sequoias. It is a molecular mechanism that explains the rising of sap in trees. Water is lifted upward due to the transpiration from the leaves and the subsequent motion of the water below that is drawn up in small tubular structures due to the relatively high surface tension of this liquid. The work done in raising the water is provided by the latent heat released by evaporation from the leaves on top of the tree.

6.1.3   Fluid dynamics

A fluid in motion is a physical system with an infinite number of degrees of freedom and is described by partial differential equations. The basic equations of fluid dynamics are the Navier–Stokes equations, which are nonlinear and include dissipative terms describing the viscosity of the fluid. Because of these features the study of fluid dynamics is very complicated, few analytical solutions are available, and approximations must be made whenever possible. Nonlinearity and dissipation are responsible for the occurrence of *turbulence*, a regime in which the dynamics becomes chaotic and predictability breaks down. Turbulence, chaotic dynamics, and fractals are the subjects of a recent branch of science that, overall, is still poorly understood—we recommend [70, 11] as references. Here we focus on the basic ideas of fluid dynamics.

1 **(B)** Elementary physics textbooks discuss the *continuity equation* for incompressible fluids in a pipe (or, more generally, in a stream tube)

$$S\, v = \text{constant}, \tag{6.4}$$

where S is the cross-sectional area of the pipe (or stream tube) and v is the velocity of the fluid, normal to S. How does Eq. (6.4) relate to the partial differential equation of fluid dynamics

$$\frac{\partial \rho}{\partial t} + \vec{\nabla} \cdot (\rho \vec{v}) = 0 \tag{6.5}$$

that bears the same name?

Solution
Equation (6.4) is in fact a special form of Eq. (6.5). An incompressible fluid has ρ =constant and

$$\partial\rho/\partial t = \partial\rho/\partial x^i = 0 \qquad (i = 1, 2, 3);$$

then Eq. (6.5) yields $\vec{\nabla} \cdot \vec{v} = 0$. Consider a volume V delimited by a pipe (or by a stream tube) and two cross sections normal to it—the velocity field v is tangent to the pipe (or to the streamlines composing the walls of the tube). By applying Gauss' law to the volume V, we obtain

$$0 = \int\int\int_V dV \, \vec{\nabla} \cdot (\rho\vec{v}) = \rho \int\int_S \vec{v} \cdot \vec{n} \, dS = -S_1 v_1 + S_2 v_2,$$

where S_1 is the area of the cross section where the fluid enters the volume V, with unit normal $\vec{n}_1$ pointing outside the pipe in the direction opposite to $\vec{v}$ (this explains the negative sign of this first term). S_2 is the area of the cross section on the right of the volume V; then the law

$$S_1 \, v_1 = S_2 \, v_2$$

is recovered in the special case of incompressible fluids.

2 **(B)** a) Integrate the continuity equation

$$\frac{\partial\rho}{\partial t} + \vec{\nabla} \cdot (\rho\vec{v}) = 0$$

over a fixed volume of space V that has a closed surface S as its boundary and discuss the physical meaning of this integral form of the continuity equation.

b) How would the continuity equation change if the mass $\dot{q}(\vec{x}, t)$ per unit time and per unit volume was injected or removed at points $\vec{x}$ internal to V?

Hint: Add the corresponding source term to the integral form of the continuity equation studied in a) and then derive the differential form of the equation.

Solution
a) Integration of the continuity equation over a fixed volume of space V (Fig. 5.3) gives

$$\frac{d}{dt}\left(\int\int\int_V dV \, \rho\right) + \int\int\int_V dV \, \vec{\nabla} \cdot (\rho\vec{v}) = 0.$$

By applying Gauss' law to the second integral on the left-hand side, we obtain

$$\frac{d}{dt}\left(\int\int\int_V dV\,\rho\right) = -\int\int_S dS\,\vec{n}\cdot(\rho\vec{v}).\qquad(6.6)$$

The terms appearing in the integral equation (6.6) have the following physical interpretation:

$\int\int\int_V dV\,\rho$ is the mass contained in V and $\frac{d}{dt}\left(\int\int\int_V dV\,\rho\right)$ is its rate of change;

$-\int\int_S dS\,\vec{n}\cdot(\rho\vec{v})$ is the mass flux across the surface S. Since the sign of the *outer* normal is taken as positive, a positive flux represents mass leaving V through S. The integral form (6.6) of the continuity equation states that the rate of change of mass contained in V is equal to minus the mass flux through the boundary S of V. In other words, the physical content of the continuity equation is mass conservation.

b) Let us consider now the case in which the amount $\dot{q}$ of mass per unit volume is injected (if $\dot{q} > 0$) or removed (if $\dot{q} < 0$) from V in the unit of time. In this case the integral equation is modified to

$$\frac{d}{dt}\left(\int\int\int_V dV\,\rho\right) = -\int\int_S dS\,\vec{n}\cdot(\rho\vec{v}) + \int\int\int_V dV\,\dot{q}.$$

Due to the arbitrariness of the integration volume it must be

$$\frac{\partial\rho}{\partial t} + \vec{\nabla}\cdot(\rho\vec{v}) = \dot{q},$$

and $\dot{q}$ represents a source (if $\dot{q} > 0$) or sink (if $\dot{q} < 0$) term. In this case the mass of the fluid is not conserved because there are external sources or sinks adding or removing mass. In the language of thermodynamics, the fluid is not a closed system but exchanges particles and hence mass with its surroundings.

3 **(B)** Show that incompressible fluids have solenoidal velocity field[2] $\vec{v}$, i.e., that $\vec{\nabla}\cdot\vec{v} = 0$.

---

[2]This fact is used in oceanography, where sea water is most often treated as an incompressible fluid and the equation $\vec{\nabla}\cdot\vec{v} = 0$ is used to estimate small vertical velocities from the measurement of horizontal ones.

Solution
An incompressible fluid has uniform and constant density ρ, i.e., $\partial\rho/\partial t = 0$ and $\partial\rho/\partial x^i = 0$ ($i = 1, 2, 3$). The continuity equation

$$\frac{\partial\rho}{\partial t} + \vec{\nabla}\cdot(\rho\vec{v}) = 0$$

then yields $\rho\vec{\nabla}\cdot\vec{v} = 0$, and $\vec{\nabla}\cdot\vec{v} = 0$.

4 **(B, C)** Write the partial differential equations describing the dynamics of a nonviscous and of a viscous fluid in terms of the mass density ρ, pressure P, velocity field $\vec{v}$, and force per unit volume $\vec{f}$.

Solution
Fluid dynamics is described by three partial differential equations expressing the conservation of mass, momentum, and energy.

Mass conservation is expressed by the *continuity equation*

$$\frac{\partial\rho}{\partial t} + \vec{\nabla}\cdot(\rho\vec{v}) = 0.$$

The second equation is Newton's second law of motion adapted to a fluid. For a nonviscous fluid it reads

$$\vec{\nabla}P + \rho\frac{d\vec{v}}{dt} = \vec{f}.$$

For a viscous fluid we have to add terms that describe dissipation. This is done by introducing the stress tensor $\vec{\vec{\sigma}}$ and generalizing the previous equation to

$$\vec{\nabla}P + \rho\frac{d\vec{v}}{dt} = \vec{f} + 2\vec{\nabla}\cdot\left(\eta\vec{\vec{\sigma}}\right),$$

where η is the dynamic viscosity coefficient.

The partial differential equation expressing the conservation of energy is

$$\frac{\partial}{\partial t}\left(\frac{1}{2}\rho v^2 + u\right) + \vec{\nabla}\cdot\left[\left(\frac{1}{2}\rho v^2 + u + P\right)\vec{v}\right] = \vec{f}\cdot\vec{v} - \vec{\nabla}\cdot\vec{q''},$$

where u is the internal energy per unit volume of the fluid and $\vec{q''}$ is the heat flux density describing conduction and possibly radiation in the fluid.

5 **(B)** Consider the velocity field $\vec{v} = \vec{v}(t, \vec{x})$ in a fluid, and the tensor field $v_{ij} \equiv \partial v_i / \partial x^j$. Decompose v_{ij} into a symmetric part θ_{ij}, and an antisymmetric part ω_{ij} (*vorticity* tensor), and prove that the decomposition is unique. Further decompose θ_{ij} as

$$\theta_{ij} = \sigma_{ij} + \frac{\theta}{3}\delta_{ij},$$

where δ_{ij} is the Kronecker delta, σ_{ij} is the *shear* tensor, $(\theta/3)\,\delta_{ij}$ is the *expansion* tensor, and $\theta = \vec{\nabla} \cdot \vec{v}$. Consider at time t a sphere of fluid particles at coordinates x^i and follow their trajectories for a short time interval δt. At time $t + \delta t$ the sphere is deformed into another surface (if δt is sufficiently small, the deviations from spherical shape will be small). Derive the geometrical meaning of the shear, vorticity, and expansion tensors by considering the time evolution map

$$x^i(t) \longrightarrow x^i(t + \delta t)$$

for small time intervals δt.

Solution
One can write

$$v_{ij} = \frac{v_{ij} + v_{ji}}{2} + \frac{v_{ij} - v_{ji}}{2} \equiv \theta_{ij} + \omega_{ij},$$

where θ_{ij} is symmetric ($\theta_{ij} = \theta_{ji}$) and the vorticity tensor ω_{ij} is antisymmetric ($\omega_{ij} = -\omega_{ji}$). This decomposition is unique; in fact, let

$$v_{ij} = A_{ij} + B_{ij}$$

with A_{ij} symmetric and B_{ij} antisymmetric be another decomposition of v_{ij} into symmetric and antisymmetric parts. Then

$$v_{ji} = A_{ji} + B_{ji} = A_{ij} - B_{ij}.$$

By adding and subtracting the last two equations, we obtain, respectively,

$$A_{ij} = \frac{1}{2}\left(v_{ij} + v_{ji}\right) \equiv \theta_{ij},$$

$$B_{ij} = \frac{1}{2}\left(v_{ij} - v_{ji}\right) \equiv \omega_{ij},$$

which proves the uniqueness of the decomposition.
We can further decompose θ_{ij} as follows:

$$\theta_{ij} \equiv \frac{v_{ij} + v_{ji}}{2} = \frac{v_{ij} + v_{ji}}{2} - \frac{\theta}{3}\delta_{ij} + \frac{\theta}{3}\delta_{ij} \equiv \sigma_{ij} + \frac{\theta}{3}\delta_{ij},$$

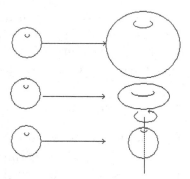

Figure 6.4. The action of expansion, shear, and vorticity.

where

$$\theta = \text{Tr}\,(v_{ij}) \equiv \sum_{i=1}^{3} \frac{\partial v^i}{\partial x^i} = \vec{\nabla} \cdot \vec{v}$$

is the trace of the tensor θ_{ij}, and

$$\sigma_{ij} = v_{ij} + v_{ji} - \frac{\theta}{3}\delta_{ij}$$

is the *shear tensor*. It follows from this definition that the shear tensor is traceless, $\text{Tr}(\sigma_{ij}) = 0$.

Let us consider now a sphere of fluid particles with positions x^i at the time t; at a later time $t + \delta t$ the particles occupy positions

$$x^i(t + \delta t) = x^i + v^i \delta t + \dots$$

(see Fig. 6.4). The Jacobian matrix of the time evolution operator $x^i(t) \longrightarrow x^i(t + \delta t)$ is

$$J_{ij} = \frac{\partial x'^{\,i}}{\partial x^j} = \delta_{ij} + \frac{\partial v^i}{\partial x^j}\,\delta t + \dots = \delta_{ij} + \left[\frac{\theta}{3}\delta_{ij} + \sigma_{ij} + \omega_{ij}\right]\delta t + \dots.$$

The Jacobian is the determinant $J \equiv \det(J_{ij})$. The action of the transformation is then decomposed into the action of three independent transformations described by the expansion, shear, and vorticity tensors, respectively. The *expansion tensor* $(\theta/3)\,\delta_{ij}$, which is diagonal, does not select a preferred direction—it changes the volume of the sphere without altering its shape or orientation. If σ_{ij} and ω_{ij} were zero, we would have

$$J_{ij} = \delta_{ij}\left(1 + \delta t\,\frac{\vec{\nabla} \cdot \vec{v}}{3}\right),$$

corresponding to a change of scale (expansion of the sphere if $\vec{\nabla}\cdot\vec{v} > 0$ or contraction if $\vec{\nabla}\cdot\vec{v} < 0$), and $J = \left(1 + \delta t\, \vec{\nabla}\cdot\vec{v}/3\right)^3$. Since the volume occupied by the particles that previously were forming the sphere is

$$V = \int d^3\vec{x}' = \int d^3\vec{x}\, J = \left(1 + \delta t\, \frac{\vec{\nabla}\cdot\vec{v}}{3}\right)^3 V_{\text{sphere}}$$

$$\simeq \left(1 + \delta t\, \vec{\nabla}\cdot\vec{v}\right) V_{\text{sphere}},$$

then $3\theta = \vec{\nabla}\cdot\vec{v}$ describes the rate $\dfrac{\delta V/V_{\text{sphere}}}{\delta t}$ at which the volume changes.

The symmetric shear tensor σ_{ij} changes the sphere into an ellipsoid without rotating it or changing its volume. The directions along which deformations occur are given by the eigendirections of σ_{ij}, and its eigenvalues give the amount of deformations along these directions. The antisymmetric vorticity tensor w_{ij} rotates the sphere without changing its volume or deforming it.

6 **(B)** Show that by assuming that the stress tensor has the form $\sigma_{ij} = \partial v_i/\partial x^j$, we can write the Navier–Stokes equations

$$\rho\frac{dv_i}{dt} = -\frac{\partial P}{\partial x^i} + 2\sum_{j=1}^{3}\frac{\partial}{\partial x^j}\left(\eta\sigma_{ij}\right) + F_i$$

as

$$\frac{\partial v_i}{\partial t} = -\vec{v}\cdot\vec{\nabla}v_i - \frac{1}{\rho}\frac{\partial P}{\partial x^i} + 2\nu\sum_{j=1}^{3}\frac{\partial^2 v_i}{\partial x^i \partial x^j} + f_i.$$

Here $\rho, P, \vec{v}, \eta, \nu = \eta/\rho, F_i$ and f_i are the density, pressure, velocity, dynamic and kinematic viscosity, external force per unit volume, and external force per unit mass, respectively.

Turbulence occurs when the nonlinear terms, which amplify small velocity perturbations, are larger than the friction terms that tend to smooth out velocity differences between different layers of fluid. Estimate the ratio of the nonlinear and the viscosity term. Does your result have a familiar form?

Solution
By using the fact that

$$\frac{dv_i}{dt} = \frac{\partial v_i}{\partial t} + \sum_{j=1}^{3} \frac{\partial v_i}{\partial x^j} \frac{dx^j}{dt} = \frac{\partial v_i}{\partial t} + \sum_{j=1}^{3} \frac{\partial v_i}{\partial x^j} v_j = \frac{\partial v_i}{\partial t} + \vec{v} \cdot \vec{\nabla} v_i$$

and dividing the Navier–Stokes equations by the density ρ, we obtain

$$\frac{\partial v_i}{\partial t} = -\vec{v} \cdot \vec{\nabla} v_i - \frac{1}{\rho} \frac{\partial P}{\partial x^i} + 2\nu \sum_{j=1}^{3} \frac{\partial^2 v_i}{\partial x^i \partial x^j} + f_i$$

by using also the fact that $\nu \equiv \eta/\rho$ and $f_i = F_i/\rho$.

In order of magnitude, the required ratio is

$$\frac{\vec{v} \cdot \vec{\nabla} v_i}{\nu \sum_{j=1}^{3} \frac{\partial}{\partial x^j} \frac{\partial v_i}{\partial x^i}} \approx \frac{v\,v/L}{\nu\,v/L^2} = \frac{vL}{\nu},$$

where v is a typical value of the velocity and L is a typical length scale over which v varies. The ratio is the *Reynolds number*, usually introduced as ratio between inertial and viscous forces. If $Re < 2000$, the fluid flow is usually laminar, while if $Re > 3000$, it is turbulent.

7 **(A)** The flow rate of oil in a pipeline is $\mathcal{F} = 10.0\,\mathrm{m}^3/\mathrm{s}$, a section of the pipeline between two pumping stations is 10.0 km long, the pressure difference between its ends is $2.5 \cdot 10^4$ Pa, and the dynamic viscosity coefficient of the oil is $\eta = 0.1$ Pa·s at 10°C. What are the radius of the pipe and the power needed to pump the oil?

Solution
According to Poiseuille's law, the flow rate of a viscous fluid in a pipe of radius r and length L is

$$\mathcal{F} = \frac{(P_2 - P_1)\,\pi r^4}{8\eta L};$$

hence,

$$r = \left[\frac{8\mathcal{F}\,\eta\,L}{\pi\,(P_2 - P_1)}\right]^{1/4}$$

$$= \left[\frac{(10.0\,\mathrm{m}^3/\mathrm{s}) \cdot 8 \cdot (0.1\,\mathrm{Pa} \cdot \mathrm{s}) \cdot (1.0 \cdot 10^4\,\mathrm{m})}{\pi \cdot 2.5 \cdot 10^4\,\mathrm{Pa}}\right]^{1/4} = 1.0\,\mathrm{m}.$$

The power needed to pump the oil is

$$W = \frac{8\eta L}{\pi r^4}\, \mathcal{F}^2 = \frac{8 \cdot (0.1\,\mathrm{Pa \cdot s}) \cdot (10^4\,\mathrm{m})}{\pi\,(1\,\mathrm{m})^4}\,\left(10\,\mathrm{m^3/s}\right)^2 = 2.55 \cdot 10^2\,\mathrm{kW}.$$

8 **(B)** Consider the flow of viscous oil in a pipeline as described by Poiseuille's law. The pressure gradient dP/dz is independent of r, φ, and z, where (r, φ, z) are cylindrical coordinates adapted to the pipe symmetry (the z-axis is the axis of the cylindrical pipe).

Let $\mathcal{F}$ be the flow rate of oil, i.e., the volume of fluid passing per unit time through a normal cross section of the pipe; consider two cross sections at z_1 and z_2, with $z_2 - z_1 = L$. Show that the flow rate can be represented, in an electrical analogy with Ohm's law, by the formula

$$P_1 - P_2 = \mathcal{R}\,\mathcal{F}.$$

Find an expression for the fluid-dynamical "resistance" $\mathcal{R}$ in terms of the parameters of the pipe and fluid. Show that the effect of two consecutive sections of the pipeline is obtained, using the electrical analogy, by treating them as resistors in a series connections.

Solution
Poiseuille's law for the flow rate is

$$\mathcal{F} = \frac{dV}{dt} = \frac{\pi G R^4}{8\eta},$$

where $G = -dP/dz$, R is the pipe radius, and η is the oil dynamic viscosity coefficient. Since dP/dz is independent of z, $P(z) = P_0 - Gz$ and

$$P_1 \equiv P(z_1) = P_0 - Gz_1,$$

$$P_2 \equiv P(z_2) = P_0 - Gz_2,$$

and $G = (P_1 - P_2)/L$ is the (constant) pressure gradient. We have

$$\mathcal{F} = \frac{\pi R^4}{8\eta} \frac{(P_1 - P_2)}{L} \equiv \frac{P_1 - P_2}{\mathcal{R}},$$

where $\mathcal{R} = 8\eta L/(\pi R^4)$ is the fluid-dynamical analogue of the electrical resistance in Ohm's law and $8\eta/\pi$ is the analogue of the electrical resistivity.

Two consecutive sections of the pipeline with equal diameters and lengths L_1 and L_2 give the total resistance

$$\mathcal{R} = 8\eta L/(\pi R^4) = \frac{8\eta L_1}{\pi R^4} + \frac{8\eta L_2}{\pi R^4} = \mathcal{R}_1 + \mathcal{R}_2,$$

i.e., the two such hydrodynamical resistances add like resistors in a series connection.

9 **(A)** In the hydrodynamics of viscous fluids it can be proven that flows having the same geometry and Reynolds numbers are dynamically similar, i.e., the two flows are identical when scaled in the appropriate way. Consider water flowing in an irrigation pipe of radius R with speed v: if we quadruple the radius of the pipe, at what speed will the flow be exactly the same as before?
Hint: Consider the Reynolds number.

Solution
The Reynolds number is $Re \equiv 2vR/\nu$, where R is the length scale over which the velocity varies and ν is the kinematic viscosity coefficient. The geometry of the flow does not change—it is still the same fluid in a pipe—and the viscosity coefficient is still the same. In order to keep the flow dynamically similar when $R \longrightarrow 4R$, the Reynolds number must be kept constant, therefore it must be $v \longrightarrow v/4$.

6.2   Gases

Gases are composed of particles relatively far apart from each other, which can be assumed to interact with each other through instantaneous collisions instead of with complicated short range forces requiring detailed models as is the case with liquids and solids. On average, gas particles travel a distance λ (*mean free path*) between consecutive collisions. The particles may have an overall velocity $\vec{v}$ (*gas velocity*) onto which random velocities are superposed. If the gas is not extremely rarified, in a reference frame moving with the gas at velocity $\vec{v}$ the particles have a Maxwell–Boltzmann distribution characterized by the *gas temperature T*. On length scales much larger than the mean free path λ the gas can be described as a compressible fluid and we can employ as physical variables the mass density ρ, pressure P, temperature T, volume V, macroscopic velocity field $\vec{v}$, and the external force per unit volume $\vec{f}$ (for example, gravity). The fluid is usually viscous, in which case we have to introduce a simple model for internal dissipation. If instead one wants to study phenomena on length scales of the order of λ or smaller, the kinetic theory of gases [62, 9, 38, 10, 6] must be used.

The most important gas for the environmental scientist is of course air, which is a mixture of different gases, but other gases are also important, for example, greenhouse gases, especially CO_2, which participates in the carbon cycle, photosynthesis, and the greenhouse effect), ozone, which in the atmosphere shields living organisms from harmful ultraviolet rays and at ground level is a pollutant, methane produced by agricultural practices and the cattle industry, radioactive radon seeping from the ground, or various other pollutants emitted in gaseous form. Here we focus on the physical properties of gases—see Chapter 3 for applications to atmospheric physics.

1 **(A)** Determine the number of particles in $1.00\ \mathrm{m}^3$ of gas at temperature $10°C$ and pressure $P = 2$ atm.

Solution
The ideal gas law
$$PV = NkT$$

yields the number of particles

$$N = \frac{PV}{kT} = \frac{2 \cdot (1.01 \cdot 10^5\,\mathrm{Pa}) \cdot (1.00\,\mathrm{m}^3)}{(1.38 \cdot 10^{-23}\,\mathrm{J \cdot K^{-1}}) \cdot (283\,\mathrm{K})} = 5.17 \cdot 10^{25}.$$

2 **(C)** Define the ideal gas and write the equation of state that it obeys. Do real gases satisfy it? What corrections can be introduced to account for the behavior of a real gas?

Solution
The ideal gas is defined as one such that the volume of its particles is negligible in comparison to the volume occupied by the gas and the interatomic/intermolecular forces are weak. The molecules only interact through scattering.

The ideal gas equation of state is

$$PV = NkT, \tag{6.7}$$

where P, V, N, and T are, respectively, the pressure, volume, number of particles, and absolute temperature of the gas, and k is the Boltzmann constant. In a (V, P) plane and at constant temperature T, the ideal gas equation of state is represented by the hyperbola $P = \mathrm{const.}/V$ (ideal gas *isotherm*).

Real gases obey the ideal gas equation of state in conditions of relatively high temperature and low pressure and density. To take into

account the finite volume occupied by the gas particles and the interaction forces between particles, the Van der Waals equation of state
is often used,

$$\left(P + \frac{a}{V^2}\right) \cdot (V - b) = N\,k\,T, \tag{6.8}$$

where a and b are constants characteristic of the particular gas considered, with dimensions of an energy times a volume and of a volume,
respectively.

3 **(A)** The ideal gas law is

$$PV = NkT, \tag{6.9}$$

where N is the number of gas particles contained in the volume V, k
is the Boltzmann constant, and T is the absolute temperature of the
gas. Show that Eq. (6.9) can be rewritten as

$$P = n_d\,k\,T, \tag{6.10}$$

where n_d is the number density of gas particles, or as

$$PV = n\,R\,T \tag{6.11}$$

where n is the number of moles of gas in V and R is the universal
gas constant, or as

$$P = \rho\,\frac{k\,T}{m}, \tag{6.12}$$

where m is the mass of the gas particle and ρ is the mass density.

Solution
By dividing Eq. (6.9) by the volume V and introducing the number
density of particles $n_d \equiv N/V$, we immediately obtain Eq. (6.10).

By introducing the number of moles of gas in V, $n_{\text{mol}} = N/N_A$,
where $N_A = 6.022 \cdot 10^{23}\,\text{mol}^{-1}$ is Avogadro's number, we obtain

$$PV = \frac{N}{N_A}\,N_A kT = n\,RT,$$

where $R \equiv N_A k = 8.31\,\text{J} \cdot \text{mol}^{-1} \cdot \text{K}^{-1}$ is the *universal gas constant*.

Finally, by introducing the mass m of each gas particle and using
Eq. (6.10), we notice that the mass density of the gas is $\rho = m\,n_d$
and we obtain

$$P = mn_d\,\frac{kT}{m} = \rho\,\frac{kT}{m}.$$

4 (A) You see a lightning flash and hear the thunder 6.00 s later: how far away, approximately, did the lightning strike?

Solution
The distance to the point where the lightning struck is approximately

$$d = v_{\text{sound}} \, t \simeq (330 \, \text{m/s}) \cdot (6 \, \text{s}) = 1980 \, \text{m},$$

where $v_{\text{sound}} \simeq 330 \, \text{m/s}$ is the speed of sound at temperature of $0°C$ and pressure of 1 atm.

5 (A) Derive the law for adiabatic transformations of an ideal gas

$$T V^{\gamma-1} = \text{constant}, \tag{6.13}$$

where T is the temperature of the ideal gas, V is its volume, and $\gamma = c_P/c_V > 1$ is the ratio of specific heats at constant pressure and constant volume.

Solution
The equation of state of an ideal gas is $PV = n_d kT$, where n_d and k are the number density of gas particles and the Boltzmann constant, respectively. Adiabatic transformations obey Poisson's law

$$P V^{\gamma} = \text{const.},$$

from which we obtain

$$P V^{\gamma} = (PV) \, V^{\gamma-1} = n_d \, k \, T \, V^{\gamma-1} = \text{const.},$$

and therefore

$$T V^{\gamma-1} = \frac{\text{const.}}{n_d \, k} = C,$$

where C is a constant.

6 (B) Compute the expansivity, the isothermal compressibility, and the bulk modulus of an ideal gas.

Solution
The *expansivity* is the percent change in volume V as the temperature changes at constant pressure,

$$\beta \equiv \frac{1}{V} \left(\frac{\partial V}{\partial T} \right)_P.$$

The ideal gas equation of state

$$PV = nRT,$$

where n is the number of moles of the gas and R is the universal gas constant yields

$$\beta = \frac{P}{nRT} \left(\frac{\partial}{\partial T}\right)_P \left(\frac{nRT}{P}\right) = \frac{1}{T}.$$

The *isothermal compressibility* is the percent change in volume as the pressure changes at constant temperature, changed in sign,

$$\kappa \equiv -\frac{1}{V} \left(\frac{\partial V}{\partial P}\right)_T = \frac{-P}{nRT} \left(\frac{\partial}{\partial P}\right)_T \left(\frac{nRT}{P}\right) = \frac{1}{P}.$$

The *isothermal bulk modulus* is the inverse of the isothermal compressibility, $B \equiv \kappa^{-1} = P$.

7 **(A)** The decomposition of organic materials at the bottom of a lake 20m deep liberates gases in the form of bubbles that ascend to the surface of the lake. A bubble has radius $r_b = 2$ mm at the bottom of the lake, where the temperature is 4°C. What is the new radius of the bubble at the surface, where the temperature is 2°C? For simplicity, assume that the bubble ascends slowly enough that it always maintains thermal equilibrium with its surroundings.

Solution
By treating the gas in the bubble as an ideal gas and using the ideal gas equation of state, we obtain

$$P_b V_b = nRT_b$$

at the bottom and

$$P_s V_s = nRT_s$$

at the surface of the lake, where $V = 4\pi r^3/3$ is the bubble volume, n is the number of moles of gas in the bubble, and T is the Kelvin temperature. The pressure P_b at the bottom of the lake is the sum of the atmospheric pressure P_0 and of the hydrostatic pressure $\rho g h$, where ρ is the freshwater density, g is the acceleration of gravity, and h is the lake depth, or $P_b = P_0 + \rho g h$. Division of the equation of state at the surface and at the bottom yields

$$\left(\frac{r_s}{r_b}\right)^3 = \frac{T_s}{T_b} \left(1 + \frac{\rho g h}{P_0}\right).$$

The bubble radius at the surface of the lake is then

$$r_s = r_b \left[\frac{T_s}{T_b}\left(1 + \frac{\rho g h}{P_0}\right)\right]^{1/3} = (2\,\mathrm{mm})$$

$$\cdot \left\{ \left(\frac{275\,\text{K}}{277\,\text{K}} \right) \left[1 + \frac{\left(1.0 \cdot 10^3\,\text{kg/m}^3\right)\left(9.8\,\text{m/s}^2\right)\left(20\,\text{m}\right)}{1.013 \cdot 10^5\,\text{Pa}} \right] \right\}^{1/3}$$

$$= \quad 2.84\,\text{mm} \simeq 3\,\text{mm}.$$

8 **(B)** Compute the configuration work done when a volume V of an ideal gas is changed isothermally.

Solution
The ideal gas obeys the equation of state $PV = nRT$, where P, n, R, and T are the pressure, number of moles, universal gas constant, and Kelvin temperature, respectively. The configuration work done in an infinitesimal volume change dV is $dW = PdV$. Therefore, for a finite expansion from initial volume V_i to final volume V_f at constant temperature, it is

$$W = \int_{V_i}^{V_f} P(V)dV = nRT \int_{V_i}^{V_f} \frac{dV}{V} = nRT \ln\left(\frac{V_f}{V_i}\right).$$

If $V_f > V_i$ (expansion), then $W > 0$ and work is done *by* the gas.
If $V_f < V_i$ (compression), then $W < 0$ and work is done *on* the gas.

9 **(B)** Compute the specific (molar) work done when the specific (molar) volume v of a gas satisfying the Van der Waals equation of state

$$\left(P + \frac{a}{v^2}\right)(v - b) = RT$$

(where a and b are constants) is changed isothermally.

Solution
The configuration work done in an infinitesimal change dv of specific volume is $dw = Pdv$ and therefore, for a finite expansion from initial molar volume v_i to final molar volume v_f, it is $w = \int_{v_i}^{v_f} Pdv$. The pressure as a function of the other two thermodynamic variables v and T is obtained from the Van der Waals equation of state rewritten as

$$P(v, T) = \frac{RT}{v - b} - \frac{a}{v(v - b)} + \frac{ab}{v^2(v - b)}.$$

Since the temperature T is constant, the specific work done is

$$w = RT \int_{v_i}^{v_f} \frac{dv}{v - b} - a \int_{v_i}^{v_f} \frac{dv}{v(v - b)} + ab \int_{v_i}^{v_f} \frac{dv}{v^2(v - b)}.$$

By using the decompositions

$$\frac{1}{v\left(v-b\right)} = \frac{-1/b}{v} + \frac{1/b}{v-b} = \frac{1}{b}\left(\frac{1}{v-b} - \frac{1}{v}\right),$$

$$\frac{1}{v^2\left(v-b\right)} = \frac{-1/b - v/b^2}{v^2} + \frac{1/b^2}{v-b} = \frac{1}{b}\left[-\frac{1}{v^2} - \frac{1}{bv} + \frac{1}{b\left(v-b\right)}\right],$$

which are straightforward to find, we obtain

$$w = RT\int_{v_i}^{v_f}\frac{dv}{v-b} - \frac{a}{b}\left[\int_{v_i}^{v_f}\frac{dv}{v-b} - \int_{v_i}^{v_f}\frac{dv}{v}\right]$$

$$+a\left[\int_{v_i}^{v_f}dv\frac{-1}{v^2} - \frac{1}{b}\int_{v_i}^{v_f}\frac{dv}{v} + \frac{1}{b}\int_{v_i}^{v_f}\frac{dv}{v-b}\right]$$

$$= RT\ln\left(\frac{v_f-b}{v_i-b}\right) + \frac{a}{v_f} - \frac{a}{v_i}.$$

10 (A) Compute the root mean square speed of particles of mass m in a monoatomic gas at temperature T using the classical principle of equipartition of energy.

Solution
A monoatomic gas is composed of particles with only three degrees of freedom and no internal structure. The principle of equipartition of energy assigns the energy $kT/2$ to each degree of freedom, and the average kinetic energy of the gas particles is then

$$\frac{1}{2}m\bar{v^2} = \frac{3}{2}kT.$$

The root mean square molecular speed is then

$$v\text{rms} \equiv \left(\bar{v^2}\right)^{1/2} = \left(\frac{3kT}{m}\right)^{1/2}.$$

Chapter 7

EVAPOTRANSPIRATION, SOILS, AND HYDROLOGY

The success of any physical investigation depends on the judicious selection of what is observed as of primary importance, combined with a voluntary abstraction of the mind from those features which, however attractive they may appear, we are not sufficiently advanced in science to investigate with profit.
—James Clerk Maxwell

Even within the relatively limited range of temperatures experienced on planet Earth, phase transitions are commonly observed. For the environmental scientist it is phase transitions of water that are most important as evaporation and condensation determine the transfer of solar energy around the globe and precipitation.

Historically, hydrology developed as a branch of engineering devoted to the study of water management for human uses and only later was raised to the dignity of an earth science. The hydrologic cycle consists of perennial transfer of water: water is removed by evapotranspiration from the oceans (especially in tropical regions), land waters, soils, and the vegetation cover; it is transported as water vapor carrying large amounts of latent heat; and finally it returns to the land and oceans through precipitation. Water falling in liquid form on the land is removed as runoff into streams and rivers, or as groundwater or through evapotranspiration again. In this chapter we focus on soil physics of which water dynamics is an important part, and on groundwater hydrology.

7.1   Phase transitions, hygrometry, and evapotranspiration

Phase changes are very interesting in fundamental physics, cryogenics, thermodynamics, and statistical mechanics. Exercises on the thermodynamical aspects of phase transitions are presented in Chapter 5. From the environmental scientist's point of view, the phase transitions of water are the most important. *Evaporation* of liquid water is the part of the hydrologic cycle by which water is transferred into the atmosphere. *Transpiration* is the process of evaporation from the leaves of plants—the collective name *evapotranspiration* denotes both of these processes. During evaporation of a solution most but not all[1] chemicals and salts dissolved are left behind—evaporation is nature's powerful way of purifying water, making clean water available again. Most of the evaporation on the planet occurs from the oceans, mostly in tropical regions. An accurate knowledge of evapotranspiration is required, for example, to model runoff following rainfall, plan efficient irrigation of crops in agriculture (agriculture consumes much more water than industry and households), and to plan water reservoirs efficiently. Evaporation (sweating) is also important as a temperature-regulating factor in the human body.

In addition to evaporation, *sublimation* liberates water molecules in the atmosphere directly from the solid phase—ice or snow. The opposite phase transition from gas to solid is responsible for frost. *Condensation* is fundamental in meteorology because it is the process that allows precipitation to occur so that water is returned to the land and oceans. Dew is another manifestation of condensation.

The melting of ice and snow is important in the hydrologic cycle, and the freezing of water bodies and rivers is also significant.

1 **(C)** Discuss the anomalous behavior of the density of water during temperature changes and during the water/ice phase transition, and the consequences for marine and freshwater life.

Solution

Most substances decrease their density when the temperature rises but the density of water *increases* from 0°C to 4°C and is maximum at 4°C ($\rho = 1.000 \cdot 10^3 \, \text{kg/m}^3$). In addition, most substances increase their density during the transition from the liquid to the solid phase but water decreases its density during this phase transition. Ice is less dense than water (icebergs float on water). The anomalous behavior has important consequences for marine and freshwater life in cold

---

[1]DDT is a notable exception.

climates: water at 4°C tends to sink, and ice begins forming at the water surface. This fact allows life to continue in water underneath the ice when the air temperature drops below the freezing point. The layer of ice on top acts as an insulator for the water below and it may prevent freezing all the way to the bottom. If the density of water decreased monotonically with decreasing temperature, colder water would sink and freezing would begin from the bottom and continue to the top.

2 **(A)** Water seeping into cracks of rocks, roads, or walls of buildings causes substantial damage when it freezes and is a significant erosion agent in the mountains when the temperature oscillates around 0°C. Compute the pressure applied to the walls of a container filled with water when the latter turns into ice. The densities of water and ice are, respectively, $\rho_w = 1.00 \cdot 10^3 \text{ kg} \cdot \text{m}^{-3}$ and $\rho_{ice} = 9.17 \cdot 10^2 \text{ kg} \cdot \text{m}^{-3}$, and the bulk modulus of ice is $B = 1.13 \cdot 10^6 \text{ N} \cdot \text{m}^{-2}$. Why should a mountaineer be particularly careful about rockfall at sunrise and sunset?

Solution
The percent change in volume of ice under compression is

$$\frac{\delta V}{V_0} = \frac{1}{B}\frac{F}{A},$$

where F is the force on the unit normal area of the ice block. When a mass m of water turns into ice, the densities before and after the phase change satisfy the relation

$$m = \rho_w V_0 = \rho_{ice}\left(V_0 + \delta V\right).$$

Hence,

$$\frac{\rho_w}{\rho_{ice}} - 1 = \frac{\delta V}{V_0},$$

and the required pressure is

$$
\begin{aligned}
P &= \frac{F}{A} = B\frac{\delta V}{V_0} = B\left(\frac{\rho_w}{\rho_{ice}} - 1\right)\\
&= \left(1.13 \cdot 10^6 \text{ N} \cdot \text{m}^{-2}\right) \cdot \left(\frac{1.00 \cdot 10^3 \text{ kg} \cdot \text{m}^{-3}}{0.917 \cdot 10^3 \text{ kg} \cdot \text{m}^{-3}} - 1\right)\\
&= 1.0 \cdot 10^5 \text{ Pa}.
\end{aligned}
$$

A mountaineer should be especially alert for falling rocks at sunrise because the first heat from the rising Sun melts the ice that locks rocks in place, causing them to fall. At sunset water from melting snow will freeze and expand when the temperature drops, dislodging rocks that are in loose balance.

3 **(A)** The air temperature is 30°C and the hygrometer reads the relative humidity as 75%. What is the vapor pressure? The saturated vapor pressure at 30°C is 4.23 kPa.

Solution
The relative humidity is the ratio between the vapor pressure P_v and the saturated vapor pressure at that temperature P_{sat},

$$W \equiv \frac{P_v}{P_{sat}};$$

therefore,

$$P_v = W P_{sat} = 0.75 \cdot \left(4.23 \cdot 10^3 \, \text{Pa}\right) = 3.2 \, \text{kPa}.$$

4 **(B)** The Clausius–Clapeyron equation of thermodynamics applies to the liquid–vapor phase transition for water when the water vapor is saturated (i.e., in equilibrium with the liquid phase):

$$\frac{dP_{sat}}{dT} = \frac{L_v}{T\left(V_g - V_l\right)},$$

where P_{sat} is the saturated vapor pressure, L_v is the latent heat of vaporization, T is the Kelvin temperature, and V_g and V_l are the volumes of water in the gaseous and liquid phase, respectively. Derive a formula giving the saturated vapor pressure as a function of the temperature. Treat the vapor as an ideal gas, and comment on the physical significance of your result for atmospheric physics.

Solution
The volume change going from the liquid to the vapor phase is extremely large[2] and $V_g \gg V_l$: this justifies writing

$$\frac{dP_{sat}}{dT} \simeq \frac{L_v}{T V_g}.$$

By treating water vapor as an ideal gas and applying the ideal gas equation of state $P_{sat} V_g = nRT$, where n is the number of moles of steam, we obtain $V_g = nRT/P_{sat}$. Substituting this value in the

---

[2]During the phase transition 1 m³ of liquid water turns into 1600 m³ of steam.

previous equation yields

$$\frac{dP_{\text{sat}}}{dT} = \frac{L_v P_{\text{sat}}}{nRT^2},$$

or

$$\frac{1}{P_{\text{sat}}} \frac{dP_{\text{sat}}}{dT} = \frac{l_v}{RT^2},$$

where $l_v \equiv L_v/n$ is the molar latent heat of vaporization. In terms of differentials, we have

$$d\left(\ln P_{\text{sat}}\right) = -\frac{l_v}{R} d\left(\frac{1}{T}\right),$$

which integrates to

$$P_{\text{sat}}(T) = P_* \exp\left[\frac{1}{R}\int dT \frac{l_v(T)}{T^2}\right],$$

where P_* is an integration constant. If l_v can be considered approximately independent of the temperature (e.g., by considering only a restricted range of temperatures over which l_v does not change significantly), then

$$P_{\text{sat}}(T) = P_* \exp\left[\frac{-l_v}{R}\left(\frac{1}{T} - \frac{1}{T_*}\right)\right],$$

where T_* is another integration constant (see Fig. 7.1).

The saturated vapor pressure is strongly dependent on the temperature, and its slope

$$\frac{dP_{\text{sat}}}{dT} = \frac{L_v}{TV_g} = \frac{L_v P_{\text{sat}}}{nR} \frac{1}{T^2}$$

formally diverges as $T \to 0^+$. This means that the curve representing $P_{\text{sat}}(T)$ is very steep and P_{sat} is very sensitive to temperature changes. Then warm air can accommodate much more water vapor than colder air. This fact is important for meteorology because warm air in tropical regions can store large amounts of water vapor and the associated latent heat. The latter can be transported in the atmosphere, condense to form precipitation, or fuel hurricanes.

5 (B) More than two thirds of the evaporation from the surface of the Earth occurs in regions at latitudes comprised between 30°S and 30°N. What percentage of the surface of the Earth is covered by these

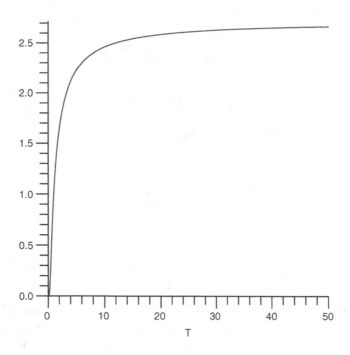

Figure 7.1.   Saturated vapor pressure versus temperature.

regions? Why there is so much evaporation in these regions?

Solutions
The region with latitude 30°S $\leq \lambda \leq$ 30°N corresponds to polar angles (colatitude θ)

$$\frac{\pi}{2} - \frac{\pi}{6} \leq \theta \leq \frac{\pi}{2} + \frac{\pi}{6},$$

or $\pi/3 \leq \theta \leq 2\pi/3$, and the area it covers is given by the surface integral

$$S_1 = R^2 \int_{\pi/3}^{2\pi/3} d\theta \int_0^{2\pi} d\varphi \sin\theta = 2\pi R^2 \left[-\cos\theta \right]_{\pi/3}^{2\pi/3} = 2\pi R^2;$$

this is half of the total area of the globe $4\pi R^2$. If heat is supplied at a constant rate, evaporation in air is described by Dalton's empirical law

$$m = c\,\frac{A\,(P_H - P_h)\,t}{P_0},$$

where m is the mass of the liquid that evaporates, A is its surface area, P_H and P_h are, respectively, the maximum vapor pressure at the temperature of the liquid and the pressure of the vapor already present in air, and P_0 is the atmospheric pressure, while t is time and c a coefficient. A large amount of water evaporates in the regions at latitudes $30°S \le \lambda \le 30°N$ because of the amount of solar energy available in these regions and because of the large surface area A that they cover. In addition, the latent heat of vaporization of water is comparatively lower in these regions because of the higher temperatures found there.

6 **(B)** It is estimated that 50% of the energy reaching the Earth from the Sun goes into evaporation of ocean water and evapotranspiration of land waters. Assume for simplicity that water is distributed uniformly around the surface of the planet, covering it completely. Calculate the thickness of the layer of water evaporated in one year from the surface of such fictitious water-covered planet.[3] The solar constant is $S = 1370\,\text{W/m}^2$, the average Earth radius is 6370 km, and the latent heat of vaporization of water is $L_v = 2.26 \cdot 10^6\,\text{J/kg}$.

Solution
The power received from the Sun per unit of normal area is the solar constant S. An element of the Earth surface dA with unit normal making an angle θ with the direction of propagation of the rays from the Sun (assumed to be all parallel to each other) presents a normal area $dA_\perp = dA\cos\theta = R^2 \sin\theta\,d\theta\,d\varphi$ to the rays, where R is the radius of the Earth and φ is an azimuthal angle. By integrating over the entire hemisphere illuminated by the Sun, we obtain the normal area presented by the Earth to the Sun:

$$A_\perp = \int_0^{\pi/2} d\theta \int_0^{2\pi} d\varphi\, R^2 \sin\theta\cos\theta = 2\pi R^2 \int_0^{\pi/2} d\theta\,\frac{\sin(2\theta)}{2}$$

$$= \pi R^2 \left[-\frac{\cos(2\theta)}{2}\right]_0^{\pi/2} = \pi R^2.$$

---

[3]Of this, 88% is water evaporated from the oceans while 12% is due to evapotranspiration from land waters.

The total power received by the planet from the Sun at any given time is

$$S\pi R^2 = \pi \left(1370\,\frac{\text{W}}{\text{m}^2}\right)\left(6.37\cdot 10^6\,\text{m}\right)^2 = 1.75\cdot 10^{17}\,\text{W}.$$

The total energy received in one year is

$$E = \left(1.75\cdot 10^{17}\,\text{W}\right)(365\,\text{days})\left(12\,\frac{\text{hours}}{\text{day}}\right)\left(3600\,\frac{\text{s}}{\text{hour}}\right)$$

$$= 2.76\cdot 10^{24}\,\text{J}.$$

Half of this energy, $Q = E/2 = 1.38\cdot 10^{24}\,\text{J}$, goes into evaporating water. The mass of water evaporated during one year is therefore $m = Q/L_v$, where L_v is the latent heat of evaporation of water,

$$m = \frac{Q}{l_v} = \frac{1.38\cdot 10^{24}\,\text{J}}{2.26\cdot 10^6\,\text{J/kg}} = 6.1\cdot 10^{17}\,\text{kg}.$$

This mass corresponds to a layer of thickness x around a fictitious planet entirely covered by water—then $m = \rho V = \rho\, 4\pi R^2 x$, where ρ is the density of water. This yields

$$x = \frac{m}{4\pi R^2\rho} = \frac{6.1\cdot 10^{17}\,\text{kg}}{4\pi\left(6.37\cdot 10^6\,\text{m}\right)^2\left(10^3\,\text{kg/m}^3\right)} = 1.2\,\text{m} \simeq 1\,\text{m}.$$

7 **(A)** Transpiration from an area covered with short vegetation produces the volume flux density (in $\text{m}^3/(\text{m}^2\cdot\text{s})$) of water vapor[4]

$$q'' = F(v)\left(P_{\text{sat}} - P_v\right), \tag{7.1}$$

where P_{sat} is the saturated vapor pressure at the temperature of the vegetated surface, P_v is the vapor pressure, and the coefficient $F(v)$ is a "conductance" that depends on the wind velocity[5] v.

a) Derive an electrical analogy for this transpiration law.

b) What is the flux density of latent heat associated with transpiration? Express it in terms of the relative humidity.

---

[4]Equation (7.1) is a form of Dalton's law of evaporation.
[5]Both P_v and v are measured at a conventional height, usually two meters above the ground with short vegetation.

Solution

a) In an electrical analogy, consider a resistor of resistance R subject to the potential difference $V_1 - V_2$ between its ends and traversed by an electrical current of intensity I according to Ohm's law

$$V_1 - V_2 = RI.$$

The given law for the flux of transpiration vapor has a similar form

$$P_{\text{sat}} - P_v = R_T I_T,$$

where the difference $P_{\text{sat}} - P_v$ between the saturated vapor pressure and the vapor pressure is the analogue of the potential difference $V_1 - V_2$, the quantity $I_T \equiv q''A$ (the flux density of vapor multiplied by the cross-sectional area A normal to the flow) is the analogue of the electric current intensity I, and $R_T \equiv 1/F$ is the "transpiration resistance," the analogue of the electrical resistance. The differential form of Ohm's law is $\vec{J} = -\sigma \vec{\nabla} V$, where $\vec{J}$ is the current density, σ is the electrical conductivity, and V is the potential. The finite difference form of this equation is $J = \sigma (V_1 - V_2)/l$, which is formally the same as the law given for transpiration.

b) Since the amount of heat required to evaporate a mass m of liquid is $Q = L_v m$, where L_v is the latent heat of vaporization, during transpiration latent heat is released from the stomata of the plants at the rate

$$\frac{dQ}{dt} = L_v \frac{dm}{dt}.$$

The heat flux density is the heat energy evaporated per unit of normal area and per unit time, while $m = \rho_v V_v$ is the mass of vapor released (ρ_v and V_v are the density and volume occupied by the vapor). Hence, the heat flux density is

$$q''_Q = L_v \rho_v q''_v = L_v \rho_v F(v) (P_{\text{sat}} - P_v).$$

By remembering that the relative humidity is $W = P_v/P_{\text{sat}}$, we obtain

$$q''_Q = L_v \rho_v F(v) P_{\text{sat}} (1 - W).$$

8 **(B)** A lake is formed in an Australian desert by a once-in-a-lifetime rainfall. Water evaporates from the surface of the lake of area A into a slab of air of height h with a vapor flux density q'' (in kg·m^{-2}·s^{-1}) given by Dalton's law[6]

$$q'' = c (P_{\text{sat}} - P_v),$$

---

[6] John Dalton discovered this empirical law in 1802.

where P_{sat} is the saturated vapor pressure at the temperature of the liquid surface, P_v is the vapor pressure in the air slab, and c is a constant coefficient.

a) By treating the vapor as an ideal gas and assuming that the temperatures of the air and of the liquid surface remain constant, derive an ordinary differential equation for the mass of liquid $m(t)$ left in the lake as a function of time t, by knowing that at the initial time $t = 0$ this mass of water is m_0.

b) Assume that the evaporated liquid is immediately diluted so that the vapor pressure in the air P_v remains constant, and solve the differential equation that you found for $m(t)$. What is the mass of water in the lake after a long time?

Solution
a) The flux density of vapor leaving the liquid surface is $-\frac{1}{A}\frac{dm}{dt}$, and the mass of vapor present in the air at time t is $m_v(t)$. This is given by the ideal gas equation of state $P_vV = nRT$, where the number of moles of the vapor is $n = m_v/M$, M is the molar mass of water, and T is the Kelvin temperature, hence $P_vAh = m_vRT/M$. The mass of water evaporated at time t is $m_v(t) = m_0 - m(t)$, hence

$$P_v = [m_0 - m(t)]\frac{RT}{MAh}.$$

Dalton's law can be written as

$$\frac{1}{A}\frac{d}{dt}[m_0 - m(t)] = c\left\{P_{sat} - [m_0 - m(t)]\frac{RT}{MAh}\right\}.$$

The saturated vapor pressure P_{sat} can be considered as constant because the temperature of the liquid surface is assumed to be constant, as well as the air temperature T, and we can write

$$\frac{dm}{dt} = -\frac{cRT}{Mh}\left[m(t) - \left(m_0 - \frac{P_{sat}MAh}{RT}\right)\right],$$

or

$$\frac{dm}{dt} = -\alpha\left[m(t) - \mu_0\right],$$

where

$$\alpha \equiv \frac{cRT}{Mh}, \qquad \mu_0 = m_0 - \frac{P_{sat}MAh}{RT}$$

are constants.

b) In order to solve this differential equation for $m(t)$, we divide both sides by $(m - \mu_0)$, obtaining

$$\frac{1}{m - \mu_0} \frac{dm}{dt} = \frac{d}{dt} \left(\ln \left| \frac{m - \mu_0}{\mu} \right| \right) = -\alpha,$$

where μ is an integration constant. This equation is immediately integrated, yielding

$$m(t) = \mu e^{-\alpha t} + \mu_0.$$

The initial condition $m(t = 0) = m_0$ determines the integration constant $\mu = m_0 - \mu_0$, hence

$$m(t) = \mu_0 + (m_0 - \mu_0) e^{-\alpha t}.$$

The late time state of the lake is described by the steady-state solution

$$m(t \to +\infty) \approx \mu_0 = m_0 - \frac{P_{sat} M A h}{RT}.$$

In this model, if $\mu_0 > 0$ there will be water left in the lake, while if $\mu_0 < 0$ the lake evaporates completely in a finite time. If μ_0 is exactly zero (an unlikely situation corresponding to a practically impossible fine-tuning), the lake will take a very long (formally infinite) time to dry up. If $\mu_0 > 0$, there will be water left in the lake, while if $\mu_0 < 0$ the lake will dry up completely in a finite time.

9 **(C)** Is melting the only cause for the disappearance of the snowpack in spring?

Solution
No: a large fraction of the snowpack disappears due to *sublimation*, the direct change from the solid phase (in this case ice—snow consists of ice crystals) to the gas phase (water vapor). At any temperature, molecules leave the surface of a solid and merge with air molecules. The opposite process (*deposition*) takes place when water molecules go directly from the vapor to the ice phase, forming frost.

10 **(C)** In polar regions water is locked in form of snow and ice. What causes the presence—albeit scarce—of water vapor in air in these regions given that precipitation is almost absent and, if there is any, it is in form of snow (ice crystals)?

Solution
The presence of water vapor in polar regions, which are cold deserts, is due to *sublimation*, the direct change of water from the solid to the vapor phase (cf. previous problem).

7.2   Soil physics

By drilling a borehole vertically into the ground one encounters, beginning from the surface, a region called *unsaturated soil*—sometimes called also *dry soil*, which is perhaps a misnomer because humidity is usually present. Below this there is a very thin transition layer called *capillary fringe* or *tension-saturated zone* and, further below, there is the *saturated soil*, which constitutes the subject of groundwater hydrology (or geohydrology).

The unsaturated soil is practically irrelevant for groundwater hydrology because water residing in it cannot be effectively pumped out. Water in unsaturated soils is called *soil moisture* or *vadose water*—the simultaneous presence of water, oxygen, microorganisms, and organic materials make unsaturated soils interesting from the point of view of biochemistry and biology and, of course, agriculture. Plant roots also penetrate unsaturated soil.

1 **(B)** Consider a soil composed of three phases: solid particles (with total mass m_s and volume V_s), liquid water (with total mass m_l and volume V_l), and gaseous air (with negligible mass and volume V_g). The total mass of the soil is $m = m_s + m_l$ and its total (or bulk) volume is $V_t = V_s + V_l + V_g$, while the soil *porosity* is defined as

$$\epsilon \equiv \frac{V_l + V_g}{V_t}$$

and the *void ratio* is

$$e \equiv \frac{V_l + V_g}{V_s}.$$

Express the porosity as a function of the solid particles density $\rho_s \equiv m_s/V_s$ and the (dry) bulk density[7] $\rho_b \equiv m_s/V_t$. Derive the relation $\epsilon(e)$ between porosity and void ratio and its inverse $e(\epsilon)$.

Solution

By using the fact that the total volume of the soil is $V_t = V_s + V_l + V_g$, the porosity can be written as

$$\epsilon \equiv \frac{V_l + V_g}{V_t} = \frac{V_t - V_s}{V_t} = 1 - \frac{V_s}{V_t}.$$

The definitions of bulk and particle densities $\rho_b \equiv m_s/V_t$ and $\rho_s \equiv m_s/V_s$ yield $V_t = m_s/\rho_b$ and $V_s = m_s/\rho_s$, which, upon substitution

---

[7]This is determined by weighing a sample of soil dried in an oven.

into the previous equation, yield

$$\epsilon = 1 - \frac{V_s}{V_t} = 1 - \frac{m_s/\rho_s}{m_s/\rho_b}$$

and finally

$$\epsilon = 1 - \frac{\rho_b}{\rho_s},$$

which is the desired expression. Similarly we can express the void ratio as

$$e = \frac{V_l + V_g}{V_s} = \frac{V_t - V_s}{V_s} = \frac{V_t}{V_s} - 1 = \frac{m_s/\rho_b}{m_s/\rho_s} - 1 = \frac{\rho_s}{\rho_b} - 1.$$

By using the fact that $\rho_b/\rho_s = 1 - \epsilon$, we obtain

$$e(\epsilon) = \frac{1}{1 - \epsilon} - 1 = \frac{\epsilon}{1 - \epsilon}.$$

This relation can be easily inverted to obtain

$$\epsilon(e) = \frac{e}{1 + e}.$$

2 **(B)** The *volumetric water content* of a soil is defined as $\theta \equiv V_l/V_t$, where V_l is the volume of liquid water in it and V_t is the total (bulk) volume given by $V_t = V_l + V_s + V_g$. Here V_s and V_g are the volumes occupied by solid particles and gaseous air in the soil, respectively. The *gravimetric water content* of the soil is defined as $\theta_m \equiv m_w/m_s$, where m_w is the mass of liquid water in the soil and m_s is the mass of the solid particles. Prove that the range of values of θ is $0 \le \theta \le \epsilon$, where ϵ is the soil porosity. Prove the relation between θ and θ_m

$$\theta = \theta_m \frac{\rho_b}{\rho_w},$$

where $\rho_b \equiv m_s/V_t$ is the (dry) bulk density and ρ_w is the density of water.

Another measure of the water content used in swelling soils is the *liquid ratio* defined by $\theta_r \equiv V_l/V_s$. Prove that

$$\theta_r = \theta_m \frac{\rho_s}{\rho_w},$$

where $\rho_s = m_s/V_s$ is the density of solid particles, and prove that $\theta_r = \theta(1 + e)$, where e is the void ratio.

Solution

Since $0 \leq V_l \leq V_t - V_s$ and $\theta \equiv V_l/V_t$ it is also

$$0 \leq \frac{V_l}{V_t} \leq \frac{V_t - V_s}{V_t} = 1 - \frac{V_s}{V_t} = \frac{V_l + V_g}{V_t} \equiv \epsilon \leq 1,$$

or $0 \leq \theta \leq \epsilon$.

The relation between θ and θ_m is obtained by using

$$\theta \equiv \frac{V_l}{V_t} = \frac{m_w/\rho_w}{m_s/\rho_b} = \frac{m_w}{m_s}\frac{\rho_b}{\rho_w} = \theta_m \frac{\rho_b}{\rho_w}.$$

The liquid ratio is

$$\theta_r \equiv \frac{V_l}{V_s} = \frac{m_w/\rho_w}{m_s/\rho_s} = \frac{m_w}{m_s}\frac{\rho_s}{\rho_w} \equiv \theta_m \frac{\rho_s}{\rho_w}.$$

The relation between θ_r and θ is obtained by considering that

$$\theta_r = \theta_m \frac{\rho_s}{\rho_w} = \left(\theta \frac{\rho_w}{\rho_b}\right)\left(\frac{\rho_s}{\rho_w}\right) = \theta \frac{\rho_s}{\rho_b}$$

by using the fact that $\theta = \theta_m \, \rho_b/\rho_w$. Now use the relations

$$\epsilon = 1 - \frac{\rho_b}{\rho_s} = \frac{e}{1+e}$$

to obtain $\rho_b/\rho_s = (1+e)^{-1}$ and

$$\theta_r = \theta \frac{\rho_s}{\rho_b} = \theta\,(1+e).$$

3 **(B)** A possible way of measuring the water content of a soil is to study the attenuation of γ-rays propagating through it, which is due to both solid particles in the soil (with density $\rho_b = m_s/V_t$, where m_s is the mass of solid particles and V_t is the total volume) and water contained the soil pores (with mass m_w). The attenuation of γ-rays is exponential—if N_m denotes the count rate for γ-rays transmitted from a given source to a detector in moist soil (see Fig. 7.2), the following law is obeyed:

$$N_m(x) = N_0 \exp\left[-\left(\mu_s \rho_b + \mu_w \theta\right) x\right],$$

where x is the distance traveled in the soil, N_0 is the count rate for γ-rays propagating in free air, θ is the water content in volume of the soil ($\theta \equiv V_l/V_t$, where V_l is the volume occupied by the liquid water in the soil), while μ_s and μ_w are attenuation coefficients for

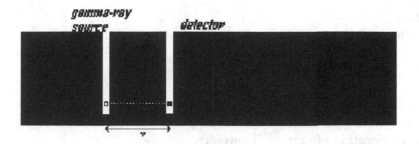

Figure 7.2. Attenuation of γ-rays in a soil.

solid particles and water, respectively. To measure the water content of a soil one drills two vertical boreholes of equal depth and places a source of γ-rays in one hole and a γ-ray detector, at the same depth, in the second hole. The count rate N_m in the soil is then measured. This measurement is repeated on a sample of soil taken to the lab and dried up, determining the count rate N_d. Derive a formula that allows one to determine the water content θ of the moist soil in terms of the known attenuation coefficient μ_w and of the measured count rates N_m and N_d.

Solution
For the moist soil we have

$$N_m(x) = N_0 \exp\left[-\left(\mu_s\rho_b + \mu_w\theta\right)x\right],$$

while for the dry soil

$$N_d(x) = N_0 \exp\left(-\mu_s\rho_b\, x\right);$$

by dividing term to term we obtain

$$\frac{N_d}{N_m} = \frac{N_0 \exp\left(-\mu_s\rho_b x\right)}{N_0 \exp\left[-\left(\mu_s\rho_b + \mu_w\theta\right)x\right]} = e^{\mu_w\theta x},$$

and therefore,

$$\theta = \frac{1}{\mu_w\, x} \ln\left(\frac{N_d}{N_m}\right).$$

4 **(A)** The suction head in a gravelly sandy soil is $\psi = 1.0$ cm, while in a silty soil it is $\psi = 1.0$ m. Estimate the corresponding values of the effective radius of pores, i.e., the radius of a fictitious vertical tube in the soil carrying water by capillarity. The surface tension of water at 20°C is $\gamma = 7.3 \cdot 10^{-3}$ N/m.

Solution
Water in a capillary tube of radius r rises to the height

$$h = \frac{2\gamma \cos \theta}{\rho g r},$$

where θ is the contact angle, which can be taken equal to zero for most substances in contact with water, ρ is the water density, and g is the acceleration of gravity. Therefore,

$$r \simeq \frac{2\gamma}{\rho g h}$$

$$\simeq \frac{2\left(7.3 \cdot 10^{-3}\,\mathrm{N/m}\right)}{\left(1.00 \cdot 10^{3}\,\mathrm{kg} \cdot \mathrm{m}^{-3}\right)\left(9.8\,\mathrm{m} \cdot \mathrm{s}^{-2}\right)\left(1.0 \cdot 10^{-2}\,\mathrm{m}\right)}$$

$$\simeq 1.5 \cdot 10^{-4}\,\mathrm{m} = 150\,\mu\mathrm{m} \quad \text{for gravelly sandy soil,}$$

$$\simeq \frac{2\left(7.3 \cdot 10^{-3}\,\mathrm{N/m}\right)}{\left(1.00 \cdot 10^{3}\,\mathrm{kg} \cdot \mathrm{m}^{-3}\right)\left(9.8\,\mathrm{m} \cdot \mathrm{s}^{-2}\right)\left(1.0\,\mathrm{m}\right)}$$

$$\simeq 1.5 \cdot 10^{-6}\,\mathrm{m} = 1.5\,\mu\mathrm{m} \quad \text{for silty soil.}$$

5 **(B)** Particle size analysis in the classification of soils with finer particles employs the rate of settling for sedimentation in water, while sieves are employed for coarser particles. According to Stokes' law (which can be applied to particles with radius $< 40\mu m$ [18]) the friction force encountered by a sphere of radius r moving with speed v in a fluid with dynamic viscosity coefficient η is $F_v = 6\pi\eta\,rv$. Particles of different sizes will settle at different speeds and the determination of the amount of soil settled after a given time will provide the abundance of particles of a given size in the soil sample.

Assuming that the soil particles are spherical, find the terminal velocity of the particles of a clayey soil with radius $r = 1\,\mu$m and density $\rho_s = 2.65 \cdot 10^{3}\,\mathrm{kg/m}^3$. Do finer or coarser particles settle first? Estimate the time taken for a soil sample mostly made of particles of the same size to settle in a 30.0cm deep tank. The density and dynamic viscosity coefficient of water at 10°C are $\rho_w = 1.00 \cdot 10^{3}\,\mathrm{kg/m}^3$ and $\eta = 1.30 \cdot 10^{-3}\,\mathrm{Pa} \cdot \mathrm{s}$.

Solution
During the sedimentation process a soil particle is subject to three vertical forces: its own weight $mg = \frac{4\pi}{3}r^{3}\rho_s g$ pointing downward,

the buoyant force $F_b = \frac{4\pi}{3} r^3 \rho_w g$ pointing upward, and the viscous drag $F_v = 6\pi\eta r v$ also directed upward. When the particle reaches its terminal velocity these forces balance, giving zero net force. In this regime the particle experiences zero acceleration and

$$F_b + F_v - mg = 0$$

or

$$\frac{4\pi}{3} r^3 \rho_w g - \frac{4\pi}{3} r^3 \rho_s g + 6\pi\eta r v = 0.$$

This equation yields

$$v = \frac{2}{9} \frac{r^2 g}{\eta} (\rho_s - \rho_w).$$

The deposition rate is proportional to the square of the particle size and therefore finer particles settle much more slowly than larger particles. Numerically,

$$v = \frac{2}{9} \frac{(1.00 \cdot 10^{-6}\,\text{m})^2 (9.81\,\text{m/s}^2)}{1.30 \cdot 10^{-3}\,\text{Pa} \cdot \text{s}} \left(2.65 \cdot 10^3\, \frac{\text{kg}}{\text{m}^3} - 1.00 \cdot 10^3\, \frac{\text{kg}}{\text{m}^3} \right)$$

$$= 2.77 \cdot 10^{-6}\, \frac{\text{m}}{\text{s}}.$$

A homogeneous soil sample will settle approximately in the time

$$t \approx \frac{l}{v} = \frac{0.30\,\text{m}}{2.77 \cdot 10^{-6}\,\text{m/s}} = 1.08 \cdot 10^5\,\text{s} \simeq 30\,\text{hours}.$$

6 **(A)** The Darcy law of groundwater hydrology in saturated soils can be applied with some caution also to unsaturated soils.[8] Consider an unsaturated soil in which the flow of water only occurs in the vertical direction. The hydraulic potential ϕ can be written as the sum of the *suction head* $\psi = P/(\rho g)$ and of the depth z, $\phi = \psi + z$. Here P, ρ, and g are the water pressure and density and the acceleration of gravity, respectively. The z-axis is vertical and points upward, and the origin $z = 0$ is at a conventional reference level, usually sea level. Derive a condition on the vertical gradient of the suction head ψ characterizing

---

[8]Essentially the hydraulic conductivity K becomes a function of the water content that is changing with the flow in unsaturated soils.

a) a regime in which the transfer of soil moisture is predominantly due to evaporation,

b) a regime in which the transfer of soil moisture is predominantly due to infiltration into the soil following rainfall or irrigation.

Solution
Darcy's law of groundwater hydrology states that the specific discharge vector $\vec{q}$ (volume of water flowing through the unit of normal area per unit time) is proportional to the gradient of the hydraulic potential ϕ,

$$\vec{q} = -K \vec{\nabla} \phi,$$

where $K(z)$ is the hydraulic conductivity (called *capillary conductivity* in unsaturated soils). Since the flow is purely vertical $\vec{q} = (0, 0, q_z)$ or $q_z = |\vec{q}|$ and

$$q_z = -K \frac{d\phi}{dz},$$

where $\phi = \phi(z)$ and $\psi = \psi(z)$ only. By introducing the suction head given by $\phi = \psi + z$, we have

$$q_z = -K \left(\frac{d\psi}{dz} + 1 \right).$$

During evaporation the soil moisture moves upward in the soil in the positive z direction and $q_z > 0$. Hence,

$$\frac{d\psi}{dz} < -1$$

is the desired condition on the suction head ψ. During infiltration following rainfall or irrigation the flow is downward, $q_z < 0$, which yields the opposite condition

$$\frac{d\psi}{dz} > -1.$$

7 **(B)** Microorganisms in the top region of a homogeneous and isotropic soil produce CO_2 at the rate $\dot{\chi}$ ($kg \cdot m^{-3} \cdot s^{-1}$) per unit volume and per unit time. This gas diffuses upward along the positive z-axis through the soil until it reaches a steady-state equilibrium at which its concentration at the surface $z = 0$ is $C(t, 0) = C_0$. At depth $z = -d$ there is a layer of soil impassable to the gas. Solve the one-dimensional diffusion equation for the concentration $C(t, z)$ of the gas.

Solution

The one-dimensional diffusion equation for CO_2 is

$$\frac{\partial C}{\partial t} = D \frac{\partial^2 C}{\partial z^2} + \dot{\chi}.$$

In steady state $\partial C / \partial t = 0$, and the boundary-value problem to solve is

$$\frac{d^2 C}{dz^2} + \frac{\dot{\chi}}{D} = 0,$$

$$C(0) = C_0,$$

$$-\frac{dC}{dz}\bigg|_{z=-d} = 0,$$

where $\dot{\chi}$ and D are constants with respect to both t and z and the second boundary condition at $z = -d$ expresses the fact that the flux of gas vanishes there. The general solution of the ordinary differential equation is obtained by two elementary integrations,

$$\frac{dC}{dz} = -\frac{\dot{\chi}}{D} z + \alpha,$$

and

$$C(z) = -\frac{\dot{\chi}}{2D} z^2 + \alpha z + \beta,$$

where α and β are integration constants to be determined by imposing the boundary conditions. The condition $C(0) = C_0$ implies that $\beta = C_0$, while the second boundary condition at $z = -d$ yields $\alpha = -\dot{\chi} d/D$. The concentration of the diffusing gas is therefore

$$C(t, z) = C_0 - \frac{\dot{\chi}}{2D} z (z + 2d).$$

8 **(B)** Consider an unconfined aquifer and assume that all the water vapor leaving the water table evaporates out of the soil with flux density q_v (in $m^3 \cdot m^{-2} \cdot s^{-1}$), and that this process determines a groundwater flow that takes place only in the vertical direction. Find an expression for the depth of the water table as a function of q_v, the hydraulic conductivity K, and the suction head ψ. Assume that $\psi = \psi(z)$ only and that the groundwater has nonzero salinity. What happens to the dissolved salts during evaporation from the soil?

Solution

Darcy's law yields the specific discharge in the z- direction

$$q_z = -K \frac{d\phi}{dz},$$

where $\phi = \psi + z$ is the hydraulic potential, ψ is the suction head, and z is the level of the water table. Therefore,

$$q_z = -K \left(\frac{d\psi}{dz} + 1 \right).$$

If the vertical flow of groundwater is entirely due to evaporation from the surface of the soil, then $q_z = q_v$ and

$$-\frac{q_v}{K} = \frac{d\psi}{dz} + 1.$$

By integrating between the surface of the soil $z = 0$ and the water table at level z, we obtain

$$z + \int_0^z \frac{d\psi}{ds} \, ds = -\frac{q_v}{K} z$$

and

$$z + \psi(z) - \psi(0) = -\frac{q_v}{K} z,$$

or

$$z = K \frac{\psi(0) - \psi(z)}{K + q_v}.$$

Since $\psi(z)$ will be larger than $\psi(0)$, the level of the water table will be $z < 0$, as it should be if the origin of the z-axis is at the surface.

If the groundwater has nonzero salinity, during evaporation from the soil the dissolved salts are left behind and deposited at the surface where they concentrate. This gives rise to the salinity problem frequently encountered with irrigation in agriculture.

9 **(B)** Infiltration in a homogeneous soil following rainfall or irrigation can be described by the changing volumetric water content of the soil $\theta(t, z)$ as a function of depth z. A possible model for the infiltration process is the one-dimensional diffusion problem

$$\frac{\partial \theta}{\partial t} = D \frac{\partial^2 \theta}{\partial z^2} \qquad (7.2)$$

with the boundary conditions

$$\theta(t, 0) = \theta_0 \qquad (7.3)$$

at the surface $z = 0$,

$$\lim_{z \to +\infty} \theta\,(t, z) = \bar{\theta} \qquad (7.4)$$

at infinite depth $z \to +\infty$ (modeling the soil as a semi-infinite slab), and with the initial condition

$$\theta\,(0, z) = \bar{\theta} \qquad (z > 0), \qquad (7.5)$$

with $\bar{\theta} < \theta_0$ corresponding to sudden wetting at $t = 0$. Look for a solution for $z \in [\,0, +\infty\,)$ in the form of a function of the new variable

$$\bar{z} \equiv \frac{z}{\sqrt{t}}$$

introduced by Boltzmann himself [5].

Solution
By using the new variable, we have

$$\frac{\partial \bar{z}}{\partial t} = -\frac{z}{2t\sqrt{t}},$$

$$\frac{\partial}{\partial t} = \left(\frac{\partial}{\partial \bar{z}}\right)\left(\frac{\partial \bar{z}}{\partial t}\right) = -\frac{z}{2t\sqrt{t}}\frac{\partial}{\partial \bar{z}},$$

$$\frac{\partial}{\partial z} = \left(\frac{\partial}{\partial \bar{z}}\right)\left(\frac{\partial \bar{z}}{\partial z}\right) = \frac{1}{\sqrt{t}}\frac{\partial}{\partial \bar{z}},$$

and

$$\frac{\partial^2}{\partial z^2} = \frac{1}{t}\frac{\partial^2}{\partial \bar{z}^2}.$$

In terms of the new variable $\bar{z}$ the diffusion equation becomes

$$\frac{d^2\theta}{d\bar{z}^2} + \frac{\bar{z}}{2D}\frac{d\theta}{d\bar{z}} = 0,$$

where the total derivative replaces the partial derivatives because θ depends only on $\bar{z}$. The original partial differential equation is reduced to an ordinary differential equation. By setting $u \equiv d\theta/d\bar{z}$ this ODE reduces to

$$\frac{du}{d\bar{z}} = -\frac{\bar{z}\,u}{2D},$$

which is immediately integrated, yielding

$$u = \frac{d\theta}{dz} = C_1\,e^{-\frac{\bar{z}^2}{4D}},$$

where C_1 is an arbitrary integration constant. Then

$$\theta\left(\bar{z}\right) = C_1 \int_0^{\bar{z}} d\zeta\, e^{-\frac{\zeta^2}{4D}} + C_2 = C_1\sqrt{4D} \int_0^{\frac{\bar{z}}{\sqrt{4D}}} d\xi\, e^{-\xi^2} + C_2$$

$$= C_3\, \mathrm{erf}\left(\frac{\bar{z}}{\sqrt{4D}}\right) + C_2,$$

where $\xi \equiv \bar{z}/\sqrt{4D}$ and the definition of the error function

$$\mathrm{erf}\left(s\right) \equiv \frac{2}{\sqrt{\pi}} \int_0^s dx\, e^{-x^2}$$

has been used. Therefore,

$$\theta\left(t, z\right) = C_3\, \mathrm{erf}\left(\frac{z}{2\sqrt{Dt}}\right) + C_2$$

is the desired solution of the diffusion equation (Fig. 7.3).

Upon use of the property $\mathrm{erf}(0) = 0$, the boundary condition $\theta\left(t, 0\right) = \theta_0$ fixes the value of the integration constant $C_2 = \theta_0$. The second boundary condition $\lim_{z \to +\infty} \theta\left(t, z\right) = \bar{\theta}$ yields, using the other property $\lim_{s \to +\infty} \mathrm{erf}(s) = 1$, the condition $C_3 + C_2 = \bar{\theta}$. Therefore, the solution of the problem is

$$\theta\left(t, z\right) = \left(\bar{\theta} - \theta_0\right)\mathrm{erf}\left(\frac{z}{2\sqrt{Dt}}\right) + \theta_0.$$

The property of the error function $\mathrm{erf}(+\infty) = 1$ guarantees that

$$\lim_{t \to 0^+} \theta\left(t, z\right) = \bar{\theta}$$

and therefore that also the initial condition is satisfied.

The physical interpretation of the solution found is as follows. If we model the soil as a semi-infinite slab, the sudden wetting at the surface $z = 0$ and the subsequent diffusion process increase the volumetric water content θ of the soil. The latter decays exponentially fast with the depth z to its average constant value before rainfall or irrigation. However, if we wait sufficiently long (formally, as $t \to +\infty$), the water content θ at a fixed depth z will reach the value θ_0 found at the surface because

$$\lim_{t \to +\infty} \mathrm{erf}\left(\frac{z}{2\sqrt{Dt}}\right) = \mathrm{erf}(0) = 0$$

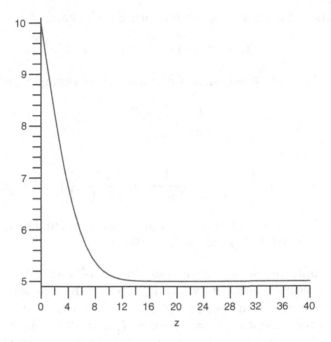

$Figure\ 7.3.$ The solution of the infiltration problem (7.2)–(7.5).

and $\theta\,(t \to +\infty, z) = \theta_0$.

10 **(B)** Consider water-carrying pipes buried at a depth z. The solution of the one-dimensional heat equation with the periodic boundary condition

$$T(t,0) = \bar{T} + T_0 \cos(\omega t)$$

modeling seasonal changes (with $\bar{T}$ the year average at large depths) is

$$T(t,z) = \bar{T} + T_0\, e^{-Az} \cos(kz - \omega t).$$

How deep should one bury the pipes in order to avoid freezing of the water in them? In the geographical location under consideration, the daily thermal excursion is $20.0°C$, the average ground temperature at large depths is constant and equal to $3.50°C$, and the penetration depth is $A^{-1} = 1\,\mathrm{m}$.

Solution

One wants to bury the pipes at a depth z such that $T(t, z) > 0°C$, or

$$T_0\,e^{-Az}\cos(kz - \omega t) > -\bar{T};$$

since $\cos(kz - \omega t) \geq -1$, this is guaranteed by taking z such that

$$T_0\,e^{-Az} \leq \bar{T},$$

yielding

$$z \geq -\frac{1}{A}\ln\left(\frac{\bar{T}}{T_0}\right) = -\frac{1}{1\,m^{-1}}\ln\left(\frac{3.50°C}{10.0°C}\right) = 1.05\,m,$$

where we used the fact that the temperature excursion is $T_{max} - T_{min} = 2T_0 = 20.0°C$, yielding $T_0 = 10.0°C$.

11 **(B)** Calculate the steady-state heat flux density in a homogeneous soil with thermal conductivity $k = 1.1\,W{\cdot}m^{-1}{\cdot}K^{-1}$ by knowing that the surface temperature is $T_s = 32°C$ and the temperature measured at the reference depth $z_{ref} = 65$ cm is $T_{ref} = 19°C$. Assuming that initially all this heat goes into evaporation of water and that vapor can escape freely from the top portion of the soil, how much water (in $kg{\cdot}m^{-2}\cdot s^{-1}$) is evaporated in one hour from the portion of the soil between the surface $z = 0$ and the depth z_{ref}? The latent heat of evaporation of water is $L_v = 2.26 \cdot 10^6$ J/kg.

Solution

The heat flux density due to conduction satisfies Fourier's law

$$\vec{q}'' = -k\vec{\nabla}T.$$

For a homogeneous soil and assuming that $\vec{q}''$ is constant at equilibrium, the equation becomes

$$q'' = -k\frac{dT}{dz},$$

where the z-axis is vertical and pointing downward into the soil. This equation is integrated, obtaining

$$q''z = -k\,[T(z) - T_0],$$

where T_0 is an integration constant determined by imposing latthe boundary condition $T(z = 0) = T_s$, which yields $T_0 = T_s$. At the depth z_{ref}, we have $q'' z_{\text{ref}} = -k\left(T_{\text{ref}} - T_s\right)$ and the required heat flux density is therefore

$$q'' = \frac{-k\left(T_{\text{ref}} - T_s\right)}{z_{\text{ref}}} = \frac{-\left(1.1\,\text{W}\cdot\text{m}^{-1}\cdot\text{K}^{-1}\right)(19 - 32)\,\text{K}}{0.65\,\text{m}} = 22\,\frac{\text{W}}{\text{m}^2}.$$

If the heat dQ goes into evaporating water, the mass of water leaving the soil in the time dt is given by $dQ = L_v\,dm$, where L_v is the latent heat of evaporation of water. For an area A of soil

$$\frac{1}{A}\frac{dQ}{dt} = q'' = \frac{L_v}{A}\frac{dm}{dt}.$$

The rate of evaporation from the top part of the soil with $0 \le z \le z_{\text{ref}}$ is

$$\frac{1}{A}\frac{dm}{dt} = \frac{q''}{L_v} = \left(22\,\frac{\text{W}}{\text{m}^2}\right)\left(2.26\cdot 10^6\,\frac{\text{J}}{\text{kg}}\right)^{-1} = 9.7\cdot 10^{-6}\,\frac{\text{kg}}{\text{m}^2\cdot\text{s}}.$$

12 (A, B) A flux density q_r'' (in $\text{m}^3/\left(\text{m}^2\cdot\text{s}\right)$) of rain at temperature T_r falls on a snowpack at subzero temperature T_s. What happens to the rainwater and to the snowpack? Derive an expression for the flux density q_Q'' of heat supplied to the snowpack as a function of q_r'' and the temperature difference $T_r - T_s$. Assume that all the rain penetrates the snowpack without running off. What is the heat flux supplied to a snowpack by 3.8 cm of rain at 3.5°C falling on it during one day of heavy rain? The specific heat of ice at this temperature is $0.500\,\text{Kcal}\cdot\text{kg}^{-1}\cdot\text{K}^{-1}$, and the latent heat of fusion is $L_f = 3.34\cdot 10^5\,\text{J/kg}$.

Solution
Rainwater cools to zero degrees Celsius and then freezes. The heat lost by rainwater warms up the snowpack: this is the sum of the heat lost by the water during its cooling from T_r to zero degrees Celsius, plus the latent heat released when rainwater freezes.

The heat lost by a mass m of rainwater during cooling by an infinitesimal amount dT is $dQ_1 = c\,m\,dT$. Since the range of temperatures involved is rather small, we can assume that the specific heat does not change over this small range and we integrate dQ_1 between T_r and °C, obtaining the heat $Q_1 = c\,m\,(0°\text{C} - T_r)$. The mass of rainwater is $m = \rho V$, where ρ is the water density and V its volume, and it is supplied at the rate $dm/dt = \rho\,dV/dt$ over an area A of the snowpack. The mass flux density is

$$q_m'' = \frac{\rho}{A}\frac{dV}{dt} = \rho\,q_r''.$$

The flux density of heat supplied to the snowpack due to the cooling of rainwater is

$$q_1'' = \frac{1}{A}\left|\frac{dQ_1}{dt}\right| = \frac{c}{A}\frac{dm}{dt}\left(T_r - 0\,°C\right) = c\,\rho\,q_r''\left(T_r - 0\,°C\right).$$

The latent heat released when the mass m of rainwater freezes is $Q_2 = L_f\,m$, where L_f is the latent heat of fusion of water and its flow rate is

$$q_2'' = \frac{1}{A}\frac{dQ_2}{dt} = L_f\,\rho\,q_r''.$$

The total heat flux supplied to the snowpack is therefore

$$q_Q'' = q_1'' + q_2'' = \rho\left[c\left(T_r - 0\,°C\right) + L_f\right]q_r''.$$

With the given data, this amounts to

$$q_Q'' = \left(1.00\cdot 10^3\,\frac{\text{kg}}{\text{m}^3}\right)\left[\left(0.500\,\frac{\text{kcal}}{\text{kg}\cdot\text{K}}\cdot 4187\,\frac{\text{J}}{\text{kcal}}\right)(3.5\,\text{K})\right.$$
$$\left. + 3.34\cdot 10^5\,\frac{\text{J}}{\text{kg}}\right]\cdot\left(\frac{0.038\,\text{m}}{24\cdot 3600\,\text{s}}\right) = 150\,\frac{\text{W}}{\text{m}^2}.$$

7.3   Groundwater hydrology

Groundwater hydrology studies the flow of underground water and the dynamics of aquifers. Perhaps the word "flow" is exaggerated because although there is a net flow of groundwater, which is extremely slow in comparison with the faster flow in a stream or river, the process is more similar to a slow diffusion of water finding its way through the system of pores between solid particles in a random walk than to the organized (or even disorganized and turbulent) flow in a river. Concepts from hydrostatics such as hydrostatic pressure are fundamental to understand the various potentials used in groundwater hydrology and soil physics and use is made of hydrodynamical quantities and equations as well.

1 **(A)** An unconfined aquifer in a gravelly soil has porosity $\epsilon = 0.25$ and effective porosity $\epsilon_{eff} = 0.22$. A borehole isolates a cylindrical column of saturated soil with radius $r = 15$ cm and 30m high. What mass of water is contained in this column? What mass of water could in practice be pumped from such a column in the aquifer?

Solution
The volume of the cylindrical column is $V_c = \pi r^2 h$, where h is the height of the column. Since the porosity is defined as the ratio between the volume of the pores filled with water in a saturated soil

and the total volume of the material, the volume of water contained in the saturated column is equal to the total volume of pores

$$V_{\text{pores}} = \epsilon V_c = \epsilon \pi r^2 h = 0.25\pi \, (0.15\,\text{m})^2 \, (30\,\text{m}) = 0.53\,\text{m}^3.$$

The mass of water contained in the column is $m_w = \rho_w V_{\text{pores}}$, where ρ_w is the density of freshwater (assume zero salinity). Numerically,

$$m_w = \rho_w V_{\text{pores}} = \left(1.00 \cdot 10^3 \, \frac{\text{kg}}{\text{m}^3}\right) (0.53\,\text{m}^3) = 5.3 \cdot 10^2 \, \text{kg}.$$

The volume of water that can practically be extracted by such a column of water is smaller because $\epsilon_{\text{eff}} < \epsilon$, due to dead-end pores and the fact that a film of groundwater adsorbs on the solid particles in the soil. This volume is given by

$$V_{\text{eff}} = \epsilon_{\text{eff}} V_c = 0.22\pi \, (0.15\,\text{m})^2 \, (30\,\text{m}) = 0.47\,\text{m}^3.$$

The maximum mass of water that can be pumped from the column is

$$m_{\text{eff}} = \rho_w V_{\text{eff}} = \left(1.00 \cdot 10^3 \, \frac{\text{kg}}{\text{m}^3}\right) (0.47\,\text{m}^3) = 4.7 \cdot 10^2 \, \text{kg}.$$

2 **(A, B)** A soil is composed of gravel and sand and has hydraulic conductivity $K = 400\,\text{m/day}$. Two wells are drilled 300m apart along a line parallel to the groundwater flow. The hydraulic potential (or *groundwater head*) is $\phi_1 = 26$m in the well upstream and $\phi_2 = 22$m in the well downstream. Estimate the magnitude of the specific discharge vector (volume of groundwater passing through the unit of normal area per unit time) and the flow rate through a cross section of the aquifer 20m^2 wide. Is the kinetic energy of water going to be important in the study of groundwater flow?

Solution
The specific discharge vector is given by Darcy's law

$$\vec{q}'' = -K\,\vec{\nabla}\phi.$$

Since the line joining the two wells is parallel to the average direction of $\vec{q}''$ and the two wells are relatively close, the gradient $\vec{\nabla}\phi$ in Darcy's law can be approximated as

$$\left|\vec{q}''\right| \approx K\left|\frac{\Delta\phi}{\Delta x}\right| = K\left|\frac{\phi_1 - \phi_2}{\Delta x}\right| = \left(400\,\frac{\text{m}}{\text{day}}\right)\left(\frac{26\,\text{m} - 22\text{m}}{300\,\text{m}}\right)$$

$$= 5.3\,\frac{\text{m}}{\text{day}} = 6.2 \cdot 10^{-5}\,\frac{\text{m}}{\text{s}} :$$

this number shows that groundwater flow is very slow in terms of everyday velocities. Therefore, the kinetic energy of groundwater is usually orders of magnitude smaller than the other forms of energy and work involved in groundwater flow, and it is usually neglected in the energy balance.

The flow rate through a cross section of the aquifer of area A is

$$\left| \vec{q''} \right| A = \left(6.2 \cdot 10^{-5} \, \frac{m^3}{m^2 \cdot s} \right) (20 \, m^2) = 1.2 \cdot 10^{-3} \, \frac{m^3}{s}.$$

3 (A) Consider the continuity equation for steady flow of incompressible water and derive an average velocity for groundwater flow from Darcy's law—this is called *Darcy velocity* v_D. Argue that the Darcy velocity actually underestimates the real velocity of the flow in the soil pores. Find a relation among the true velocity of groundwater, the Darcy velocity, and the porosity of the soil.

Hint: Consider the definition of porosity.

Solution
For the steady flow of an incompressible fluid, the continuity equation assumes the simple form

$$\frac{V}{t} = Av = \text{constant},$$

where V is the volume of water flowing through the normal area A during the time t and v is the fluid velocity. Darcy's law gives the specific discharge vector (volume of fluid passing through the unit of normal area per unit time) $\vec{q''} = -K \vec{\nabla}\phi$ in terms of the hydraulic conductivity K and of the hydraulic potential ϕ. The specific discharge coincides with the flow rate V/t. Hence the Darcy velocity is

$$v_D = \frac{V}{At} = \left| \vec{q''} \right| = K \left| \vec{\nabla}\phi \right|.$$

In fact, only a part A_p of the cross section A of an aquifer is occupied by pores through which groundwater can flow—the rest is occupied by solid particles. By considering an imaginary tube in the aquifer parallel to the flow and with cross-sectional area A and length l, we find that the actual average velocity of the flow v is given by

$$\frac{V}{t} = A v_D = A_p v,$$

which yields

$$v = \frac{A}{A_p} v_D = \frac{Al}{A_p l} v_D = \frac{V_t}{V_{\text{pores}}} v_D.$$

The porosity ϵ of the soil is defined as the ratio of the volume V_{pores} occupied by the pores and the bulk volume V_t of the soil. Hence,

$$v = \frac{v_D}{\epsilon}.$$

Because $\epsilon < 1$, it is $v > v_D$ and the Darcy velocity can seriously underestimate the average groundwater flow velocity. The effect of the solid particles in the soil is to reduce the cross-sectional area and therefore increase the speed of the flow, as described by the continuity equation. This is analogous to squeezing a garden hose to increase the speed of the water coming out of it.

4 **(B)** In the absence of major perturbing factors (torrential rainfalls, droughts, etc.) groundwater flow reaches an equilibrium in a steady-state regime. Prove that in this regime and in the absence of wells and sources, the hydraulic potential ϕ in a homogeneous aquifer satisfies the Laplace equation $\nabla^2 \phi = 0$. Assume that groundwater has the same temperature and zero salinity everywhere in the aquifer.

Solution
Conservation of the groundwater mass is expressed by the continuity equation

$$\frac{\partial \rho}{\partial t} + \vec{\nabla} \cdot (\rho \vec{v}) = 0,$$

where ρ and $\vec{v}$ are the density and velocity field of the water. In steady state $\partial \rho / \partial t = 0$. Furthermore, the normal velocity v is given by the Darcy velocity divided by the soil porosity ϵ, $\vec{v} = \vec{q''}/\epsilon$, where $\vec{q''}$ is the specific discharge vector given by Darcy's law $\vec{q''} = -K\,\vec{\nabla}\phi$, and K is the hydraulic conductivity. Therefore,

$$\vec{\nabla} \cdot \left(\rho\, \frac{\vec{q''}}{\epsilon} \right) = -\vec{\nabla} \cdot \left(\rho\, \frac{K}{\epsilon}\, \vec{\nabla}\phi \right) = 0.$$

The aquifer is homogeneous, hence $\partial K / \partial x^i = 0$ and $\partial \epsilon / \partial x^i = 0$. Since all the points of the aquifer are at the same temperature and there is no salinity, the density of groundwater is also constant, or $\partial \rho / \partial x^i = 0$. Then we can write

$$\frac{\rho K}{\epsilon}\, \vec{\nabla} \cdot \vec{\nabla}\phi = 0.$$

Finally, using the vector identity $\vec{\nabla} \cdot \vec{\nabla} = \nabla^2$, we obtain the Laplace equation for the hydraulic head

$$\nabla^2 \phi = 0.$$

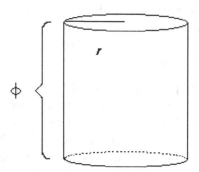

Figure 7.4. Groundwater through a cylinder of radius r and height $\phi(r)$.

5 **(A)** The level of the water table decreases near a pumped well in
an unconfined aquifer forming the *cone of depression*. Assume that
the aquifer is homogeneous and in steady-state equilibrium, that the
original water table (i.e., without the well) was horizontal, and that
the flow to the well is approximately horizontal[9] and radial in cylindrical coordinates centered on the well. Compute the profile $\phi(r)$ of
the hydraulic head as a function of the horizontal distance r from the
axis of the well, knowing the rate Q $(\mathrm{m^3/s})$ at which water is pumped
from the well. Assume that the well reaches the bottom of the aquifer.

Solution
According to Darcy's law the specific discharge vector (volume of
water flowing through the unit of normal area per unit time) is
$\vec{q}'' = -K\,\vec{\nabla}\phi$, where ϕ is the hydraulic potential (hydraulic head)
and K is the hydraulic conductivity. The water arriving from the
distance r to the well and pumped out of the well passes through a
cylinder of radius r and height ϕ (see Fig. 7.4) coaxial with the well,
because the well reaches the bottom of the aquifer. The lateral area

---

[9]This is the *Dupuit approximation.*

of this cylinder is $2\pi r \phi(r)$, and the flow rate through it is

$$Q = 2\pi r \phi(r) \left| \vec{q''} \right| = 2\pi r \phi(r) K \frac{d\phi}{dr};$$

hence, we have the ordinary differential equation

$$\phi \frac{d\phi}{dr} = \frac{Q}{2\pi K r}.$$

Integration yields

$$\int_{\phi_1}^{\phi} \phi' d\phi' = \frac{Q}{2\pi K} \int_{r_1}^{r} \frac{dr}{r},$$

where Q is constant because steady state is assumed and $\partial K / \partial r = 0$ because the aquifer is homogeneous. Here the integration constant $r_1 > r$ has the meaning of an arbitrary reference value at which the hydraulic head $\phi_1 \equiv \phi(r_1)$ will be measured with an observation well. Hence,

$$\frac{\phi^2}{2} - \frac{\phi_1^2}{2} = \frac{Q}{2\pi K} \ln\left(\frac{r}{r_1}\right)$$

and

$$\phi(r) = \left[\phi_1^2 - \frac{Q}{\pi K} \ln\left(\frac{r_1}{r}\right)\right]^{1/2}$$

is the profile of the depression cone. This expression makes sense for $r > r_0$, where r_0 is the lower limit on r at which the argument of the square root in the expression of $\phi(r)$ is nonnegative, i.e.,

$$r_0 = r_1 \exp\left(-\frac{\pi K}{Q} \phi_1^2\right).$$

In practice the expression for $\phi(r)$ is valid almost all the way to the well if the argument of the exponential is in absolute value much larger than unity, or $\phi_1 \gg \sqrt{\frac{Q}{\pi K}}$. This means that

$$\left(\frac{\phi}{\phi_1}\right)^2 - 1 = \frac{Q}{\pi K \phi_1^2} \ln\left(\frac{r}{r_1}\right) \ll 1,$$

or that the slope of the cone of depression

$$\frac{d\phi}{dr} = \frac{Q}{2\pi K \phi r}$$

is small. In other words, the well does not significantly disturb the level of the water table.

6 **(B)** In steady state and outside sources and wells, the hydraulic potential obeys the Laplace equation $\nabla^2 \phi = 0$. Show that the approximate solution for ϕ found in the previous exercise satisfies the Laplace equation approximately but not exactly, due to the approximations made. Argue that the solution for multiple wells, all satisfying the assumptions of the previous exercise, is the sum of the solutions for each individual well were all the others absent.

Solution
The approximate solution found

$$\phi(r) = \left[\phi_1^2 + \frac{Q}{\pi K} \ln \left(\frac{r}{r_1} \right) \right]^{1/2}$$

depends only on the cylindrical radius. Therefore, its Laplacian in cylindrical coordinates reduces to

$$\nabla^2 \phi = \frac{1}{r} \frac{d}{dr} \left(r \frac{d\phi}{dr} \right).$$

We find

$$\frac{d\phi}{dr} = \frac{Q}{2\pi K} \frac{1}{r\phi},$$

and

$$\frac{1}{r} \frac{d}{dr} \left(r \frac{d\phi}{dr} \right) = \frac{1}{r} \frac{d}{dr} \left(\frac{Q}{2\pi K} \frac{1}{\phi} \right) = \frac{Q}{2\pi K} \frac{1}{r} \left(-\frac{1}{\phi^2} \frac{d\phi}{dr} \right)$$

$$= -\frac{Q}{2\pi K} \frac{1}{r\phi^2} \frac{Q}{2\pi K} \frac{1}{r\phi} = -\left(\frac{Q}{2\pi Kr} \right)^2 \frac{1}{\phi^3} \approx 0$$

because the approximation used to find the solution $\phi(r)$ in the previous exercise is that the slope of the depression cone is small, i.e.,

$$\left| \frac{d\phi}{dr} \right| = \left| \frac{Q}{2\pi Kr\phi} \right| \ll 1.$$

Therefore, the Laplace equation is satisfied approximately but not exactly.

If there are n wells, each satisfying the small drawdown assumption, let $\phi^{(i)}$ be the solution corresponding to each well without all the others. Because the Laplace equation is linear, a superposition principle holds and the approximate solution in the presence of n wells is simply the sum of the individual approximate solutions $\phi^{(i)}$,

$$\phi = \sum_{i=1}^{n} \phi^{(i)}.$$

7 **(B)** a) Derive an electrical analogy between Darcy's law in a homogeneous isotropic aquifer and Ohm's law.

b) Consider an anisotropic soil composed of n horizontal layers, each of which is homogeneous and isotropic and has hydraulic conductivity K_i, placed on top of each other. Assume that groundwater flows only in the horizontal (x) direction, and prove that the effective hydraulic conductivity of this layered soil is the sum of the hydraulic conductivities of each layer,

$$K = \sum_{i=1}^{n} K_i.$$

Solution
The differential form of Ohm's law is $\vec{J} = \sigma \vec{E}$, where $\vec{J}$ is the current density vector (charge flowing per unit of normal area and per unit time), σ is the electrical conductivity of the ohmic material, and $\vec{E} = -\vec{\nabla}V$ is the electric field, while V is the electric potential. Darcy's law states that the specific discharge vector $\vec{q''}$ (volume of water flowing per unit of normal area and per unit time) is $\vec{q''} = -K\vec{\nabla}\phi$, where K is the hydraulic conductivity and ϕ is the hydraulic potential. The analogy between Darcy's law and Ohm's law in the form $\vec{J} = -\sigma \vec{\nabla}V$ is evident. The following quantities are analogous:

- electric charge Q and volume of water V_w
- electric current $I \equiv dQ/dt$ and flow rate dV_w/dt
- current density $\vec{J}$ and specific discharge $\vec{q''}$
- electric potential V and hydraulic potential ϕ
- electric conductivity σ and hydraulic conductivity K
- electric resistivity $\rho = \sigma^{-1}$ and "hydraulic resistivity" $\rho_h = K^{-1}$
- wire cross-sectional area A and aquifer cross-sectional area A
- length of wire l and length of aquifer l
- electric resistance R and "hydraulic resistance" $R_h = \frac{l}{KA}$
- Ohm's law (differential form) $\vec{J} = -\sigma \vec{\nabla}V$ and Darcy's law $\vec{q''} = -K\vec{\nabla}\phi$
- Ohm's law $V_1 - V_2 = RI$ and $\phi_1 - \phi_2 = K\, dV_w/dt$.

The finite form of Ohm's law applied to a resistor of resistance R traversed by a current of intensity I and subject to the potential difference $V_1 - V_2$ between its ends is $V_1 - V_2 = RI$. The finite form of Darcy's law is obtained as follows. Consider a parallelepiped parallel to the groundwater flow, with cross-sectional area A and length l. Since the soil is homogeneous, K is constant and

$$\frac{d\phi}{dx} = \frac{\Delta\phi}{\Delta x} = \frac{\phi_2 - \phi_1}{x_2 - x_1} = \frac{\phi_2 - \phi_1}{l};$$

then

$$q'' A = K A \frac{\phi_2 - \phi_1}{l}$$

or

$$\frac{dV_w}{dt} = \frac{KA}{l} (\phi_1 - \phi_2).$$

We can write this relation as

$$\phi_1 - \phi_2 = \frac{l}{KA} \frac{dV_w}{dt} \equiv R_h \frac{dV_w}{dt},$$

where $R_h \equiv l/(KA)$. The electrical analogy is not perfect because groundwater hydrology considers the flux density of the *volume* of water $\vec{q''}$ instead of the flux density of the *mass* of water as would be more intuitive.

b) Consider an aquifer composed of n horizontal, homogeneous and isotropic layers on top of each other, each characterized by the hydraulic conductivity K_i $(i = 1, ..., n)$. The horizontal flow of water across a vertical cross section of the aquifer is

$$\vec{q''} = \sum_{i=1}^{n} \vec{q''}_i,$$

where $\vec{q''}_i$ is the specific discharge across the ith layer and the total flow rate is the sum of the flow rates $dV_w^{(i)}/dt$ across each layer:

$$\frac{dV_w}{dt} = \sum_{i=1}^{n} \frac{dV_w^{(i)}}{dt} = \sum_{i=1}^{n} (\phi_1 - \phi_2) \frac{K_i A_i}{l} = (\phi_1 - \phi_2) \frac{1}{l} \sum_{i=1}^{n} K_i A_i,$$

where the difference $(\phi_1 - \phi_2)$ is the same for each horizontal layer because the flow is horizontal and the hydraulic potential is independent of the vertical coordinate: $\phi = \phi(x)$ only. The last equation can be rewritten as

$$\phi_1 - \phi_2 = R_h \frac{dV_w}{dt},$$

where $R_h = l/\sum_{i=1}^{n} K_i A_i$. If we further assume that all the layers have the same cross-sectional area $A_i = A$, then

$$R_h = \frac{l}{A} \sum_{i=1}^{n} K_i = \sum_{i=1}^{n} R_i.$$

The "hydraulic resistances" R_i then add up as electrical resistances in a parallel connection.

8 **(B)** A circular pond of radius r_p is surrounded by higher soil, which is homogeneous and isotropic. The hydraulic head ϕ is higher in the soil wall than in the pond (where it takes the value $\phi(r_p) \equiv \phi_p$). Assume that groundwater only flows horizontally (*Dupuit approximation*) and radially to the pond; hence, $\phi = \phi(r)$, with flow rate Q (in m^3/day). Assuming steady state, compute the flow rate Q in terms of the hydraulic heads $\phi(r)$ at r and ϕ_p, and in terms of the hydraulic conductivity K.

Solution
Consider a vertical cylinder of soil coaxial with the pond and with radius r. Groundwater flows across the side of this cylinder moving radially to the pond. The flow rate is $Q = 2\pi r \phi(r) q''$, where $\vec{q}'' = -K \vec{\nabla} \phi$ is the specific discharge vector, which points radially. Hence,

$$Q = 2\pi r \phi(r) K \frac{d\phi}{dr}$$

at any radius $r \geq r_p$. By noting that

$$\phi \frac{d\phi}{dr} = \frac{d}{dr}\left(\frac{\phi^2}{2}\right),$$

we can write

$$\frac{1}{2}\frac{d}{dr}\left(\phi^2\right) = \frac{Q}{2\pi K r}.$$

Integration with respect to r between the radius of the pond r_p and r yields

$$\int_{\phi_p^2}^{\phi^2} d\left(\phi^2\right) = \frac{Q}{\pi K} \int_{r_p}^{r} \frac{dr'}{r'},$$

or

$$\phi^2(r) - \phi_p^2 = \frac{Q}{\pi K} \ln\left(\frac{r}{r_p}\right),$$

from which the desired expression of Q follows:

$$Q = \pi K \, \frac{\phi^2(r) - \phi_p^2}{\ln{(r/r_p)}}.$$

Chapter 8

POLLUTION

Prediction is very difficult, especially about the future.
—Attributed to Mark Twain

While natural forms of pollution (e.g., volcanic eruptions and natural forest fires) exist, the human impact on the environment is very visible around us. Pollution resulting from human activities is one of the most urgent problems of the modern world. Power plants pollute while there is increasing demand for energy, manufacturing industries pollute while world economies push for larger production rates, agriculture pollutes and consumes large amounts of water while the demand for food and higher crop yields to feed an ever-increasing world population rises, households pollute and require more energy every year to sustain the high standard of living expected in developed countries, vehicles, airplanes, and ships pollute while the number of vehicles increases to satisfy the demands of an increasing population in developing countries, and the extraction and exploitation of oil, gas, and mineral resources contribute a great deal to pollution. The emission of CO_2 and other greenhouse gases threatens global warming while the ozone layer is being destroyed by manmade pollutants. Power plants and the industry also generate thermal pollution, while attention is also being drawn to radioactive, electromagnetic, acoustic, light, indoor, and visual pollution. In this chapter we focus on the way pollutants are transported in the environment, in particular in air and in water.

The best practice would be to limit, if not eliminate, pollution. However, complete elimination of pollution is a utopic ideal, and sometimes limiting the emission of a certain chemical increases the emission of an-

other one, or the consumption of energy. It is a fact that there will always
be manmade pollution and we have to deal with it. One way is to dilute
pollution. For example, organic waste is eliminated naturally (mainly
by aerobic organisms) if it is sufficiently diluted—however, this does not
apply, e.g., to heavy metals. Another way to deal with toxic waste is to
store it away in "safe" places so that it does not enter the environment—
this is done with radioactive waste from nuclear reactors. When these
approaches fail, the pollutant enters the environment and the last re-
source is environmental clean-up. This is usually a long and expensive
process, and often impossible. Examples are attempts to clean up oil
spills or the remediation of a contaminated aquifer. There are many
general and specialized references for this chapter—[69, 29, 63, 4, 39, 58]
are recommended.

8.1   Transport equations

Common ideas and methods underlie different models of the transport of pollutants in different media in the environment. The spreading
of a pollutant in still air or water is described by diffusion processes.
Diffusion is usually very slow and most of the transport is carried out
by convective motion when the fluid flows. However, in the presence of
turbulence, random motion in the turbulent fluid can be described again
by diffusion-type processes, provided that the parameters are changed
with respect to proper diffusion. The spreading of pollutants in a flowing medium is described by the Navier–Stokes equations of fluid dynamics and by the theory of turbulent diffusion. Special solutions of the
advection-diffusion equation called *Gaussian plume models* are particularly important due to their wide range of applications in atmospheric
pollution and hydrology. In general, one can consider the transport of
mass (of a fluid or of a pollutant), the transport of energy (e.g., heat
conduction already considered in Chapter 5, or the transport of kinetic
energy in a moving fluid, or of potential energy in the form of latent
heat stored in water vapor in the atmosphere), and the transport of
momentum due to viscosity or turbulent eddies. The relevant transport
and conservation equations are, of course, applied to many other areas
of physics.

1 **(C)** Discuss the basic ideas of transport theory: flux density, flux,
electrical analogies, and conservation equations.

Solution
Transport theory studies the transport of a certain quantity Q, e.g.,

mass, volume of liquid, energy, momentum, or angular momentum, etc. The *flux density* is the amount of the quantity Q that is transported per unit time across the unit of area normal to the direction of motion. Flux densities are usually denoted with the symbol $\vec{q}''$ (a vector pointing in the direction of motion) and have the dimensions $[q''] = [Q] / [\text{m}^2 \cdot \text{s}]$.

The amount of quantity Q transported per unit time across a certain surface S is the *flux* of Q across that surface, usually denoted with the symbol q:

$$q = \int\!\!\int_S \vec{q}'' \cdot d\vec{S}.$$

If $\vec{q}''$ is homogeneous across a surface S of area A, then we have simply $\vec{q}'' = \vec{q}A$. It is a good thing to remember this relation as it gives the dimensions of q: $[q] = [q''] [\text{m}^2]$.

In most applications of transport theory a linear approximation is used that assumes the flux density to be proportional to the gradient of a quantity that is the cause of the transport. In heat conduction this approximation takes the form of Fourier's law $\vec{q}'' = -k\vec{\nabla}T$, in diffusion theory we find Fick's law $\vec{q}'' = -D\vec{\nabla}C$, in groundwater hydrology Darcy's law $\vec{q}'' = -k\vec{\nabla}\phi$, etc. This linear approximation resembles the differential form of Ohm's law $\vec{J} = \sigma \vec{E} = -\sigma\vec{\nabla}V$ between the current density $\vec{J}$ (flux density of electric charge), the conductivity σ, and the electric field $\vec{E} = -\vec{\nabla}V$, which is minus the gradient of the electrostatic potential V. This is the basis of many electrical analogies encountered in the theory of heat conduction, diffusion, turbulent diffusion, fluid mechanics, and geohydrology. Therefore, it is not surprising to also see analogues of the finite form of Ohm's law

$$I = \frac{1}{R}\Delta V$$

between electric current I (flux of electric charge across the cross section of a wire), the electrical resistance R, and the potential difference ΔV across a resistor. It is common to find the words *conductance* to refer to the analogue of R^{-1} and *conductivity* for the analogue of σ. For example, one speaks of thermal or hydraulic conductivity, or of the conductance of a soil with respect to the evaporation of water.

A conservation equation for the quantity Q transported with velocity $\vec{v}$ expresses the fact that, in the absence of sources or sinks of Q, the amount of Q that enters a volume V in the unit time is equal to the amount of Q that exits V per unit time. In practice, the differential form of the conservation equation is more useful than its finite (or integral) form and is written as

$$\frac{\partial \rho_Q}{\partial t} + \vec{\nabla} \cdot (\rho_Q \vec{v}) = 0,$$

where ρ_q is the volume density of the quantity Q.

2 **(C)** There is an analogy between heat conduction and diffusion: Fourier's law for the heat flux $\vec{q''} = -k \vec{\nabla} T$ resembles Fick's law for the flux of a diffusing substance, $\vec{F} = -D \vec{\nabla} C$; and the heat equation for the temperature T,

$$\frac{\partial T}{\partial t} = a \nabla^2 T, \qquad (8.1)$$

is formally the same as the diffusion equation for the concentration C of a pollutant,

$$\frac{\partial C}{\partial t} = D \nabla^2 T. \qquad (8.2)$$

What is the physical origin of this formal analogy?

Solution
The formal analogy is due to the fact that both diffusion and heat conduction are due to the same kind of microscopic phenomena: random motions and collisions of particles. Molecules and atoms vibrate, move, and scatter randomly in a medium, while electrons in a conductor move and undergo scatterings against other electrons and ions of the metal. It is not surprising that the macroscopic descriptions of these phenomena are also similar.

3 **(A)** Derive the dimensions

- of the Fourier coefficient a by using the heat equation,
- of the diffusion coefficient D by using the diffusion equation,
- of the thermal conductivity k by using Fourier's law,
- of the diffusion coefficient D by using Fick's law.

Solution
The heat equation

$$\frac{\partial T}{\partial t} = a \nabla^2 T$$

and the diffusion equation

$$\frac{\partial C}{\partial t} = D \nabla^2 C$$

give $[a] = [D] = [\mathrm{m}^2 \cdot \mathrm{s}^{-1}]$.

Fourier's law $\vec{q}'' = -k\vec{\nabla}T$ and Fick's law $\vec{F}_C = -D\vec{\nabla}C$, plus the expression $\vec{F} = C\vec{v}$, give

$$[k] = [\mathrm{W} \cdot \mathrm{m}^{-1} \cdot \mathrm{K}^{-1}]$$

and

$$[D] = [\mathrm{m}^2 \cdot \mathrm{s}^{-1}].$$

4 **(C)** Explain why the diffusion coefficient D is smaller for heavier molecules.

Solution

From the microscopic point of view, molecular diffusion is due to the random walk of molecules through the medium. Molecules cannot cover macroscopic distances quickly because of the continuous scattering with other molecules. A more massive, larger molecule has smaller root mean square speed

$$v_{\mathrm{rms}} = \left(\frac{3kT}{m}\right)^{1/2}$$

and a larger cross section. Therefore, its diffusion will be slower than the diffusion of a lighter molecule. Macroscopically, this fact is reflected in smaller values of the diffusion coefficient D.

5 **(B)** The continuity equation for the concentration C (mass density) of a pollutant is

$$\frac{\partial C}{\partial t} + \vec{\nabla} \cdot \vec{F} = 0$$

away from sources or sinks of the pollutant, where the flux density $\vec{F}$ of the pollutant is given by Fick's law

$$\vec{F} = -D\vec{\nabla}C$$

and D is the diffusion coefficient.

a) Derive the continuity equation and comment on its physical content. What can you say about the number of particles in a volume of space V?

b) How is the continuity equation modified if diffusion takes place in a medium (water or air) that is flowing with velocity $\vec{v}$?

Solution

a) Consider a fixed volume of space V enclosed by the surface S (see Fig. 5.3). The mass of the pollutant contained in V is

$$\int\int\int_V dV\, C\,(t, \vec{x})\,,$$

and its rate of variation is

$$\frac{d}{dt}\left(\int\int\int_V dV\, C\right) = \int\int\int_V dV\, \frac{\partial C}{\partial t}.$$

The flux leaving the volume V through the surface S in the unit of time is

$$\int\int_S \vec{F}\cdot\vec{n}\,dS,$$

where $\vec{n}$ is the outward unit normal to S. In the absence of sources or sinks of the pollutant, the mass disappearing from V in the unit time is equal to the mass leaving V in the same time and flowing through S, i.e.,

$$-\frac{d}{dt}\left(\int\int\int_V dV\, C\right) = \int\int_S \vec{F}\cdot\vec{n}\,dS;$$

by applying Gauss' law, we obtain

$$\int\int\int_V dV\, \frac{\partial C}{\partial t} + \int\int\int_V dV\, \vec{\nabla}\cdot\vec{F} = 0.$$

Since the integration volume V is arbitrary, it must be

$$\frac{\partial C}{\partial t} + \vec{\nabla}\cdot\vec{F} = 0.$$

The physical content of the continuity equation is simply mass conservation for the pollutant.

The concentration can be written as $C = m\, n_d$, where m is the mass of the polluting particles and $n_d = C/m$ is the number density of particles. The total number of particles in a volume V is

$$N = \int\int\int_V d^3\vec{x}\, n_d = \frac{1}{m}\int\int\int_V d^3\vec{x}\, C,$$

and its time derivative is

$$\frac{dN}{dt} = \frac{1}{m} \int \int \int_V d^3\vec{x}\, \frac{\partial C}{\partial t} = -\frac{1}{m} \int \int \int_V d^3\vec{x}\, \vec{\nabla}\cdot\vec{F}$$

$$= -\frac{1}{m} \int \int_S \vec{F}\cdot\vec{n}\, dS.$$

If there is no flux of particles across the boundary S of the volume V and there are no chemical reactions, then the number of particles in V is constant, $dN/dt = 0$.

b) In a medium moving with velocity $\vec{v}$, the flux density becomes

$$\vec{F} = -D\vec{\nabla}C + C\vec{v},$$

where the second term in the right-hand side is usually much larger than the first one. The continuity equation becomes

$$\frac{\partial C}{\partial t} + C\vec{\nabla}\cdot\vec{v} + \vec{v}\cdot\vec{\nabla}C = \vec{\nabla}\cdot\left(D\vec{\nabla}C\right)$$

and, if the fluid motion is solenoidal and the diffusion coefficient D does not depend on the position, then

$$\frac{\partial C}{\partial t} + \vec{v}\cdot\vec{\nabla}C = D\nabla^2 C;$$

this is called the *advection-diffusion equation*.

6 **(B)** A pollutant diffuses in the x direction in a homogeneous medium at rest. Prove that *in stationary regime* the flux density of the pollutant does not depend on the coordinate x.

Solution
Let $C(t,x)$ be the concentration (mass density) of the pollutant. C and the flux $\vec{F}$ of the pollutant obey the continuity equation

$$\frac{\partial C}{\partial t} + \vec{\nabla}\cdot\vec{F} = 0. \tag{8.3}$$

In stationary regime $\partial C/\partial t = 0$ and, since the flux is unidirectional, $\vec{F} = (F, 0, 0)$, it follows from Eq. (8.3) that

$$\vec{\nabla}\cdot\vec{F} = \frac{\partial F}{\partial x} = 0.$$

7 **(B)** Show that the change of the time coordinate $t \longrightarrow \tau \equiv Dt$ changes the diffusion equation

$$\frac{\partial C}{\partial t} = D \, \nabla^2 C \qquad\qquad (8.4)$$

for the concentration $C(t, x)$ of a pollutant in a fluid at rest into

$$\frac{\partial C}{\partial \tau} = \nabla^2 C.$$

Does the new variable τ have the dimensions of a time?

Solution
We have

$$\frac{\partial}{\partial t} = \frac{\partial}{\partial \tau}\frac{d\tau}{dt} = D\frac{\partial}{\partial \tau},$$

and the diffusion equation (8.4) is changed into

$$\frac{\partial C}{\partial \tau} = \nabla^2 C.$$

The dimensions of the variable τ are

$$[\tau] = [D][T] = \left[\mathrm{m}^2 \cdot \mathrm{s}^{-1} \cdot \mathrm{s}\right] = \left[\mathrm{m}^2\right],$$

so τ has the dimensions of a length squared, not of a time.

8 **(B)** Verify that

$$C(t, x) = \frac{C_0}{2\sqrt{\pi D t}} \exp\left(-\frac{x^2}{4Dt}\right)$$

is the solution of the one-dimensional diffusion equation

$$\frac{\partial C}{\partial t} = D\frac{\partial^2 C}{\partial x^2} \qquad\qquad (8.5)$$

with the boundary condition

$$C(t, x) \to 0 \qquad \text{as } |x| \to +\infty$$

and with the initial condition

$$C(0, x) = C_0 \, \delta(x)$$

representing a single pointlike puff or spill of pollutant in still air or water. Consider the solution at an arbitrarily small time $t > 0$

and an arbitrarily large value of $|x|$ and discuss the implications for causality. Discuss the graphical interpretation of this solution in the limit $t \to 0^+$. Various modifications of this solution in the case of flowing media are the basis of Gaussian plume models of pollution.

Solution
We have

$$\frac{\partial C}{\partial t} = \frac{C_0}{4\sqrt{\pi D}\, t^{3/2}} \, e^{-\frac{x^2}{4Dt}} \left(\frac{x^2}{2Dt} - 1 \right),$$

$$\frac{\partial C}{\partial x} = -\frac{C_0\, x}{4\sqrt{\pi D^3}\, t^3} \, e^{-\frac{x^2}{4Dt}},$$

and

$$\frac{\partial^2 C}{\partial x^2} = \frac{C_0}{4\sqrt{\pi D^3}\, t^3} \, e^{-\frac{x^2}{4Dt}} \left(\frac{x^2}{2Dt} - 1 \right).$$

Hence, $D \partial^2 C / \partial x^2 = \partial C / \partial t$ and the diffusion equation is satisfied.

The boundary condition is easily verified, since $\lim_{x \to \pm\infty} e^{-\alpha^2 x^2} = 0$. To verify the initial condition, introduce the quantity $\alpha \equiv (2\sqrt{Dt})^{-1}$; then $\alpha \to +\infty$ as $t \to 0^+$ and, using the representation of the Dirac delta [17]

$$\delta(x) = \frac{1}{\sqrt{\pi}} \lim_{\alpha \to +\infty} \alpha\, e^{-\alpha^2 x^2},$$

we verify that

$$\lim_{t \to 0^+} C(t, x) = \lim_{\alpha \to +\infty} \frac{C_0}{\sqrt{\pi}} \, \alpha\, e^{-\alpha^2 x^2} = C_0\, \delta(x).$$

The initial condition corresponds to the instantaneous release of a certain amount of pollutant at $x = 0$ at the initial time $t = 0$. At any later time t and position x, with t arbitrarily small and x arbitrarily large, the mass density $C(t, x)$ of the pollutant is strictly positive: this fact implies that Eq. (8.5) describes *instantaneous* diffusion.[1] The initial rate of change of C at $x = 0$ (the only point where $C \neq 0$ at $t = 0$) is

$$\left. \frac{\partial C}{\partial t} \right|_{x=0} \to \infty \quad \text{as} \quad t \to 0^+.$$

This feature of the diffusion equation does not constitute a problem for most practical applications because the relativistic corrections

---

[1] This phenomenon is analogous to the heat paradox.

that would be needed to take into account the finite velocity of propagation are usually completely negligible—Eq. (8.5) is an excellent approximation for everyday situations.

The limit $t \to 0^+$ has a graphical interpretation: as $t \to 0^+$, the Gaussian (8.5) is more and more peaked on $x = 0$; the area between the x-axis and the graph of $C(t, x)$, given by the integral

$$A = \int_{-\infty}^{+\infty} dx\, C(t, x) = \frac{C_0}{\sqrt{\pi Dt}} \int_0^{+\infty} dx\, e^{-x^2/4Dt} = 2C_0 \int_0^{+\infty} d\xi\, e^{-\xi^2},$$

where $\xi \equiv x/(2\sqrt{Dt})$, does not depend on time. At $t = 0$ we have

$$A = C_0 \int_{-\infty}^{+\infty} dx\, \delta(x) = C_0.$$

The total mass of pollutant present over all space does not depend on time—it is initially concentrated at $x = 0$ and later is spread over the entire x-axis. This is consistent with the continuity equation

$$\frac{\partial C}{\partial t} + \vec{\nabla} \cdot \vec{F} = 0$$

(where $\vec{F} = -D\,\vec{\nabla} C$ is the flux of the pollutant) expressing the conservation of the pollutant's mass.

8.2   Water pollution

Water pollution is one of the major problems of environmental science. Pollutants are released in various ways into the environment and contaminate groundwater, streams, rivers, lakes, and oceans. One primary concern is the disappearance of safe drinking water for human use, followed by the destruction or alteration of ecosystems and plant and animal life. Water pollution comes from industry, households, transportation, and agriculture (which in addition to using pesticides is responsible for increasing nitrogen levels due to the heavy use of fertilizers, which cause eutrophication). Thermal pollution may also be of concern near power plants.

In the absence of advection, the transport of pollutants in water is described by molecular diffusion or by similar processes—when net flows are present in the water, convection and advection dominate the transport. If the flow is turbulent, which is usually the case in streams and rivers, the advection-diffusion equation with eddy coefficients replacing the molecular diffusion coefficients is a very useful tool to model the transport of pollutants.

1 **(A)** You take sugar in your morning tea or coffee. Why do you stir it? To answer, compute the average time it takes a $C_6H_{12}O_6$ (glucose) molecule to diffuse 1 cm in water. The diffusion coefficient is $D = 6.7 \cdot 10^{-10}\,\mathrm{m^2 \cdot s^{-1}}$.

Solution
The root mean square distance traveled by glucose molecules is given by $\sigma = \sqrt{2Dt}$, and the average time required to travel a distance σ is

$$t = \frac{\sigma^2}{2D} = \frac{\left(1.0 \cdot 10^{-2}\,\mathrm{m}\right)^2}{2 \cdot (6.7 \cdot 10^{-10}\,\mathrm{m^2 \cdot s^{-1}})} = 7.5 \cdot 10^4\,\mathrm{s} = 21\,\text{hours}.$$

Obviously no one wants to wait so long to drink a coffee, and stirring causes turbulent mixing. In practice, spontaneous convective motions accelerate the process even without stirring.

2 **(B)** The volume of water in a lake is constant and is given by the balance of the flow rates (expressed in $\mathrm{m^3/s}$)

$$\mathcal{F}_R + \mathcal{F}_P + \mathcal{F}_E + \mathcal{F}_O = 0, \tag{8.6}$$

where $\mathcal{F}_R, \mathcal{F}_P, \mathcal{F}_E$, and $\mathcal{F}_O$ are the constant flow rates due to runoff into the lake ($\mathcal{F}_R > 0$), precipitation ($\mathcal{F}_P > 0$), evaporation ($\mathcal{F}_E < 0$), and outflow from the lake ($\mathcal{F}_O < 0$), respectively. Eutrophication in the lake is limited by the inflow of nitrogen, the nutrient least available to algae, and it is crucial to know its concentration $C(t)$ (in $\mathrm{kg/m^3}$) in the lake as a function of time. Derive an ordinary differential equation for the nitrogen concentration $C(t)$ by knowing the concentration C_R in the water entering the lake. Find the solution of this ODE by knowing that at the time $t = 0$ the measured concentration has the value C_*.

Hint: Consider the mass of nitrogen $m(t)$ in the lake at the instant t and compute its rate of change dm/dt, neglecting the contributions to $C(t)$ from precipitation and evaporation.

Solution
The mass of nitrogen flowing into the lake per unit time is $C_R\mathcal{F}_R$, the mass leaving the lake is $C\mathcal{F}_O$ (the concentration of nitrogen in the outflow from the lake is the same as the concentration in the lake), while precipitation and evaporation do not contribute to the balance of nitrogen mass. The rate of change of the nitrogen mass in the lake is

$$\frac{dm}{dt} = C_R\mathcal{F}_R + C\mathcal{F}_O$$

and $m = CV$, where V is the volume of water in the lake, which does not vary in time. Therefore,

$$\frac{dC}{dt} - \frac{\mathcal{F}_O}{V} C = \frac{C_R \mathcal{F}_R}{V},$$

which is the desired ODE. Its general solution is the sum of the general solution of the associated homogeneous equation and of a particular solution of the inhomogeneous equation. These are easily determined to be $\alpha e^{\mathcal{F}_O t/V}$ (where α is an arbitrary integration constant) and $-C_R \mathcal{F}_R / \mathcal{F}_O$. The general solution of the inhomogeneous equation is

$$C(t) = \alpha \exp\left(-\frac{|\mathcal{F}_O| t}{V}\right) + \frac{C_R \mathcal{F}_R}{|\mathcal{F}_O|}.$$

The integration constant α is determined by imposing the initial condition $C(0) = C_*$, which yields $\alpha = C_* - C_R \mathcal{F}_R / |\mathcal{F}_O|$. The solution of the problem is therefore the nitrogen concentration

$$C(t) = C_1 e^{-t/\tau} + C_2,$$

where

$$\tau = \frac{V}{|\mathcal{F}_O|},$$

$$C_1 = C_* - \frac{C_R}{\mathcal{F}_R} |\mathcal{F}_O|,$$

$$C_2 = \frac{C_R \mathcal{F}_R}{|\mathcal{F}_O|}.$$

The late time state of the lake is one with nitrogen concentration equal to the asymptotic value C_2. The timescale for changes in C is τ. In order to limit the eutrophication of the lake, one must keep the asymptotic concentration of the nitrogen—the nutrient for the algae—below a certain threshold C_{max}, or

$$C_2 = \frac{C_R \mathcal{F}_R}{|\mathcal{F}_O|} < C_{max}.$$

If we use the water balance equation for the lake (8.6), this inequality becomes

$$C_R < C_{max} \left[1 - \frac{|\mathcal{F}_E| - \mathcal{F}_P}{\mathcal{F}_R}\right].$$

3 **(B)** Phosphate (a major agent of eutrophication) enters a lake at a constant rate a (expressed in kg/s) through a stream that collects waste from a factory producing detergents. The outflow of water from the lake occurs through a stream that carries a constant flow $\mathcal{F}_{\text{out}}$ (expressed in m^3/s). Derive an ordinary differential equation describing the concentration C (in kg/m^3) of phosphate in the lake and solve it. What is the late time state of the lake? Is it stable? Neglect evaporation, precipitation, and groundwater flow into and out of the lake, and assume that rapid mixing occurs.

Solution

The rate at which the phosphate accumulates in the lake is simply the input rate a minus the output rate. Because rapid mixing is assumed, the concentration of the pollutant in the water leaving the lake is the same as in the lake and the outflow of the pollutant (in kg/s) is $C\mathcal{F}_{\text{out}}$. Therefore, the rate of change of the phosphate concentration in the lake is given by

$$\frac{d(CV)}{dt} = a - C\mathcal{F}_{\text{out}},$$

where V is the volume of the lake, which is assumed to be constant. We write

$$\frac{dC}{dt} = a - \frac{C}{\tau},$$

where $\tau \equiv V/\mathcal{F}_{\text{out}}$ has the dimensions of a time. This linear, first-order, inhomogeneous ODE can be solved using the method of variation of parameters. The complementary equation $dC/dt = -C/\tau$ has general solution $\alpha\,e^{-t/\tau}$, where α is an integration constant. Therefore, we look for a solution of the inhomogeneous equation in the form

$$C(t) = u(t)\,e^{-t/\tau}.$$

By substituting into the nonhomogeneous equation, we obtain

$$\frac{du}{dt} = a\,e^{t/\tau},$$

which is immediately integrated, obtaining

$$u(t) = a\tau\,e^{t/\tau} + \beta,$$

with β an integration constant, and

$$C(t) = a\tau + \beta e^{-t/\tau}.$$

The constant β is determined by imposing an initial condition $C(0) = C_0$ at the initial time $t = 0$, which yields

$$C(t) = a\tau + (C_0 - a\tau)\,e^{-t/\tau}.$$

The late time state of the lake (as $t \to +\infty$) is the steady state $C\,(t \to +\infty) \approx a\tau$. This is an exact solution corresponding to equilibrium $(dC/dt = 0)$, which is also obtained by inspection of the original ODE for $C(t)$. In order to study its stability we write $C(t)$ as the equilibrium state $a\tau$ plus a perturbation $\delta C(t)$,

$$C(t) = a\tau + \delta C(t),$$

and assume that at the initial time $t = 0$ the perturbation has magnitude ϵ, which can be of either sign. By inserting this expression into the ODE for C, we obtain the evolution equation satisfied by the perturbation δC

$$\frac{d\,(\delta C)}{dt} = -\frac{\delta C}{\tau},$$

which is immediately integrated to yield

$$\delta C(t) = \epsilon\,e^{-t/\tau}.$$

Any perturbation decays exponentially fast irrespective of its sign or of its initial amplitude.[2] The late time state of the lake is asymptotically stable.

4 **(B)** A pollutant is released in a long pond of still water. One side of the pond (at $x = 0$) faces a rock wall completely impassable to the pollutant, which does not stick to it. At the opposite side $(x = L)$ the pollutant is so diluted that its concentration there is measured to be zero. At an initial time $t = 0$ a measurement has given a non-vanishing distribution $C_0(x)$ for the pollutant. Find the solution of the diffusion equation at a later time $t > 0$.
Hint: Use separation of variables.

Solution
The diffusion equation

$$\frac{\partial C}{\partial t} = D\,\nabla^2 C$$

---

[2]Note that we did not assume that the perturbation is small: the ODE for the perturbation is exact. This is possible because the equation for $C(t)$ is linear.

must be solved for $0 \leq x \leq L$ and $t \geq 0$ with the boundary conditions

$$-D \frac{\partial C}{\partial x}(t, 0) = 0 \tag{8.7}$$

(no flux of the pollutant at $x = 0$) and

$$C(L) = 0, \tag{8.8}$$

and with the initial condition $C(0, x) = C_0(x)$. Separation of variables

$$C(t, x) = T(t) X(x) \tag{8.9}$$

yields

$$X \frac{dT}{dt} = DT \frac{d^2 X}{dx^2} \tag{8.10}$$

and division by $C = TX$ gives

$$\frac{1}{T} \frac{dT}{dt} = \frac{D}{X} \frac{d^2 X}{dx^2}. \tag{8.11}$$

The last equation can only be satisfied if the left-hand side and the right-hand side, which depend on different variables, are constant:

$$\frac{1}{T} \frac{dT}{dt} = -\lambda^2, \tag{8.12}$$

and

$$\frac{D}{X} \frac{d^2 X}{dx^2} = -\lambda^2, \tag{8.13}$$

where $-\lambda^2$ is a separation constant, which is chosen to be negative on physical grounds (see the discussion ahead). The equations

$$\frac{dT}{dt} + \lambda^2 T(t) = 0,$$

$$\frac{d^2 X}{dx^2} + \frac{\lambda^2}{D} X(x) = 0,$$

have the solutions

$$T(t) = e^{-\lambda^2 t},$$

$$X(x) = A \cos\left(\frac{\lambda}{\sqrt{D}} x\right) + B \sin\left(\frac{\lambda}{\sqrt{D}} x\right),$$

where A and B are integration constants determined by the boundary and initial conditions. The boundary condition (8.7) yields $B = 0$, while Eq. (8.8) leads to $X(L) = \cos\left(\lambda L/\sqrt{D}\right) = 0$ and

$$\lambda_n = \frac{\pi\sqrt{D}}{2L}\,(2n+1) \qquad\qquad (n = 0, 1, 2, 3, \cdots).$$

Only discrete values of the separation constant are allowed. The general solution is a superposition of elementary solutions,

$$C(t, x) = \sum_{n=0}^{+\infty} A_n\,\cos\left[(2n+1)\,\frac{\pi x}{2L}\right]\exp\left[-(2n+1)^2\,\frac{\pi^2 D}{4L^2}\,t\right].$$

The initial concentration profile $C_0(x)$ is also expanded as a Fourier series

$$C_0(x) = \sum_{n=0}^{+\infty} C_n\,\cos\left[(2n+1)\,\frac{\pi x}{2L}\right]$$

with Fourier coefficients

$$C_n = \frac{2}{L}\int_0^L dx\,C_0(x)\,\cos\left[(2n+1)\,\frac{\pi x}{2L}\right].$$

By imposing the initial condition $C(0, x) = C_0(x)$, we find $A_n = C_n$ for $n = 0, 1, 2, 3, \cdots$ and

$$C(t, x) = \sum_{n=0}^{+\infty} C_n\,\cos\left[(2n+1)\,\frac{\pi x}{2L}\right]\exp\left[-(2n+1)^2\,\frac{\pi^2 D}{4L^2}\,t\right].$$

$$(8.14)$$

We now discuss the sign of the separation constant. If $\lambda = 0$, we obtain the trivial solution

$$T(t) = \text{const.}, \qquad X(x) = \alpha x + \beta,$$

and the boundary conditions (8.7) and (8.8) imply $\alpha = \beta = 0$, which does not satisfy the initial condition $C(0, x) = C_0(x) \neq 0$.

If the separation constant is positive $(+\lambda^2$ instead of $-\lambda^2)$, we have the solutions

$$T(t) = T_0\,e^{\lambda^2 t}, \qquad X(x) = \alpha\,e^{\lambda x/\sqrt{D}} + \beta\,e^{-\lambda x/\sqrt{D}};$$

the exponential increase with time of the concentration at a fixed point is clearly unphysical. From the mathematical point of view, the boundary conditions (8.7) and (8.8) give $\alpha = \beta = 0$, leaving as the only possible solution the trivial one, which is incompatible with the initial condition (8.9).

5 **(B)** Solve the previous problem with the initial condition given by the linear concentration

$$C_0(x) = \left(3 \, \frac{\text{kg}}{\text{m}}\right) \cos\left(\frac{\pi x}{2L}\right).$$

Solution
The only nonvanishing Fourier coefficient appearing in the series (8.14) is

$$C_0 = 3 \, \frac{\text{kg}}{\text{m}},$$

corresponding to $n = 0$. Therefore, the solution is

$$C(t, x) = \left(3 \, \frac{\text{kg}}{\text{m}}\right) \exp\left[-\left(\frac{\pi^2 D}{4L^2} t\right)\right] \cos\left(\frac{\pi x}{2L}\right).$$

This solution represents a concentration that oscillates in x with a long wavelength $\lambda = 4L$ (only a quarter of the wavelength fits in the length of the pond) and decreasing exponentially fast in time.

6 **(B)** Compute the total derivative with respect to time of the concentration of a pollutant $C = C(t, \vec{x})$ in a river flowing with velocity field $\vec{v}$.

Solution
The total derivative of the concentration $C(t, \vec{x})$ is

$$\frac{dC}{dt} = \frac{\partial C}{\partial t} + \sum_{i=1}^{3} \frac{\partial C}{\partial x^i} \frac{dx^i}{dt} = \frac{\partial C}{\partial t} + \sum_{i=1}^{3} \frac{\partial C}{\partial x^i} v^i, \qquad (8.15)$$

or, in compact form,

$$\frac{dC}{dt} = \frac{\partial C}{\partial t} + \vec{v} \cdot \vec{\nabla} C. \qquad (8.16)$$

This expression of the total derivative explicitly takes into account the fact that, in addition to the explicit dependence of C on time, the particles of the pollutant moving with the fluid have time-dependent coordinates $\vec{x}(t)$ as well. The concentration at a fixed point of space changes because of an explicit dependence on t (described by $\partial C/\partial t$) and because the flow with velocity $\vec{v}$ spreads the pollutant, as described by the term $\vec{v} \cdot \vec{\nabla} C$.

7 (B) The spreading of pollutants in groundwater in a homogeneous but anisotropic aquifer is described by the advection-diffusion equation

$$\frac{\partial C}{\partial t} = K_x \frac{\partial^2 C}{\partial x^2} + K_y \frac{\partial^2 C}{\partial y^2} + K_z \frac{\partial^2 C}{\partial z^2} - \vec{u} \cdot \vec{\nabla} C,$$

where C is the concentration of the pollutant and $\vec{u}$ is the velocity field of the groundwater flow. The hydraulic conductivity is represented by the diagonal matrix

$$(K_{ij}) = \begin{pmatrix} K_x & 0 & 0 \\ 0 & K_y & 0 \\ 0 & 0 & K_z \end{pmatrix}.$$

Find a coordinate transformation that reduces the advection-diffusion equation to the corresponding one in a fictitious isotropic aquifer.

Solution
Redefine the spatial coordinates $\{x^i\}$ according to

$$x \longrightarrow \bar{x} \equiv \frac{x}{\sqrt{K_x}}, \qquad y \longrightarrow \bar{y} \equiv \frac{y}{\sqrt{K_y}}, \qquad z \longrightarrow \bar{z} \equiv \frac{z}{\sqrt{K_z}}.$$

Then

$$\frac{\partial}{\partial x^i} = \sum_{j=1}^{3} \left(\frac{\partial}{\partial \bar{x}^j} \right) \frac{\partial \bar{x}^j}{\partial x^i} = \frac{1}{\sqrt{K_i}} \frac{\partial}{\partial \bar{x}^i},$$

and by applying this operator twice,

$$\frac{\partial^2}{\partial (x^i)^2} = \frac{1}{K_i} \frac{\partial^2}{\partial (\bar{x}^i)^2}, \tag{8.17}$$

$$\sum_{i=1}^{3} K_i \frac{\partial^2}{\partial (x^i)^2} = \sum_{i=1}^{3} K_i \frac{1}{K_i} \frac{\partial^2}{\partial (\bar{x}^i)^2} = \bar{\nabla}^2, \tag{8.18}$$

where $\bar{\nabla}^2$ is the Laplacian operator in the new coordinates $\{\bar{x}^i\}$.

The components of the velocity vector field $\vec{u}$ transform according to the law

$$\bar{u}^i = \sum_{j=1}^{3} \frac{\partial \bar{x}^i}{\partial x^j} u^j = \sum_{j=1}^{3} \frac{1}{\sqrt{K_i}} \delta_{ij} u^j = \frac{u^i}{\sqrt{K_i}},$$

and the operator $\vec{u} \cdot \vec{\nabla}$ becomes, in the new coordinates,

$$\vec{u} \cdot \vec{\nabla} = \sum_{i=1}^{3} u_i \frac{\partial}{\partial x^i} = \sum_{i=1}^{3} \sqrt{K_i}\, \bar{u}_i \frac{1}{\sqrt{K_i}} \frac{\partial}{\partial \bar{x}^i} = \vec{\bar{u}} \cdot \vec{\bar{\nabla}},$$

where $\vec{\bar{\nabla}}$ is the gradient with respect to $\{\bar{x}^i\}$. The advection-diffusion equation is rewritten in the new coordinates as

$$\frac{\partial C}{\partial t} = \bar{\nabla}^2 C - \vec{\bar{u}} \cdot \vec{\bar{\nabla}} C.$$

The hydraulic conductivity in the fictitious aquifer corresponding to the new unphysical coordinates $\{\bar{x}^i\}$ coincides with the identity tensor represented by the identity matrix $\mathrm{diag}(1,1,1)$. Note that the new coordinates do not have the dimensions of a length but instead have dimension $[\bar{x}^i] = [T^{1/2}]$.

8 **(A)** Consider a factory spilling a mass $m = 1.0 \cdot 10^3$ kg of a certain pollutant into a turbulent river. What is the distribution of the average concentration (in kg/m) of the pollutant after six hours? How fast is the peak of this average concentration diluted as time goes by? Assume that the spill is instantaneous, the turbulent diffusion coefficient is $1.0 \cdot 10^{-3}\,\mathrm{m^2/s}$, and the average velocity of the water in the river is $\bar{v} = 5.0\,\mathrm{m/s}$.

Solution
The average concentration $\bar{C}$ of the pollutant satisfies the one-dimensional dispersion equation

$$\frac{\partial \bar{C}}{\partial t} + \bar{v} \frac{\partial \bar{C}}{\partial x} = K \frac{\partial^2 \bar{C}}{\partial x^2},$$

where x is a coordinate along the river with $x = 0$ at the polluting source and K is the turbulent diffusion (or dispersion) coefficient. The one-time, instantaneous spill is modeled by a delta-like source at $t = 0$. The solution of the dispersion equation is

$$\bar{C}(t, x) = \frac{m}{\sqrt{2\pi}\,\sigma} \exp\left[-\frac{(x - \bar{v}t)^2}{2\sigma^2} \right],$$

where the dispersion is $\sigma = \sqrt{2Kt}$. Hence,

$$\bar{C}(t, x) = \frac{m}{\sqrt{4\pi K t}} \exp\left[-\frac{(x - \bar{v}t)^2}{4Kt} \right].$$

Six hours after the spill, the average concentration is

$$\bar{C}(x) = \frac{(1.0 \cdot 10^3 \,\text{kg})}{\sqrt{4\pi \,(1.0 \cdot 10^{-3}\,\text{m}^2/\text{s}) \cdot 6 \cdot (3600\,\text{s})}}$$

$$\exp\left[-\frac{[x - (5.0\,\text{m/s}) \cdot 6 \cdot (3600\,\text{s})]^2}{4 \cdot (1.0 \cdot 10^{-3}\,\text{m}^2/\text{s}) \cdot 6 \cdot (3600\,\text{s})}\right]$$

$$= \left(61\,\frac{\text{kg}}{\text{m}}\right)\exp\left\{-1.35 \cdot 10^8 \left[\left(\frac{x}{110\,\text{km}}\right) - 1\right]^2\right\}.$$

The dilution away from the front $x = \bar{v}t$ is exponential and the peak of the (linear) average concentration is at $x = \bar{v}t$ and has value

$$\bar{C}_0 = \left(150\,\frac{\text{kg}}{\text{m}}\right)\left(\frac{t}{1\,\text{hour}}\right)^{-1/2}.$$

9 **(A)** A mass $m_1 = 1.00 \cdot 10^5$ kg of water at 65.0°C used for cooling a power plant is discharged into a nearby pond containing $3.00 \cdot 10^4$ m^3 of water at 15.50°C (thermal pollution). What is the temperature change of the pond after thermal equilibrium is established?

Solution
The pond contains the mass of water

$$m_2 = \rho V = \left(1.00 \cdot 10^3\,\frac{\text{kg}}{\text{m}^3}\right) \cdot (3.00 \cdot 10^4\,\text{m}^3) = 3.00 \cdot 10^7\,\text{kg},$$

where ρ is the density of freshwater. The temperature rise of the lake is given by the heat balance

$$c\,m_1\,(T - T_1) + c\,m_2\,(T - T_2) = 0,$$

where c is the specific heat of water and T is the final temperature after thermal equilibrium between the two masses of water is reached. Simple algebra yields

$$T = \frac{m_1 T_1 + m_2 T_2}{m_1 + m_2}$$

$$= \frac{(1.00 \cdot 10^5\,\text{kg}) \cdot (65.0°\text{C}) + (3.00 \cdot 10^7\,\text{kg}) \cdot (15.5°\text{C})}{(1.00 \cdot 10^5\,\text{kg} + 3.00 \cdot 10^7\,\text{kg})}$$

$$= 15.7°\text{C}.$$

The temperature change of the pond is 0.2°C.

8.3   Air pollution

Atmospheric pollution is another major problem of environmental science. Apart from global problems such as the emission of CO_2 and other greenhouse gases in the atmosphere since the advent of the Industrial Revolution, or the ozone hole problem, there are more local problems. Examples are the well-known London smog from coal burning and the photochemical Los Angeles smog. These exemplify pollution problems caused by pollutants of different chemical nature. Pollutants are released by industry, power generation plants, transportation, and domestic activities. Here we propose a few exercises on zero-dimensional models and Gaussian plume models based on the advection-diffusion equation.

1 **(B)** A crude but simple zero-dimensional model for air pollution in a city is the following. Imagine the city enclosed in a parallelepiped of sides L_1 and L_2 and height h equal to the mixing length in the atmosphere. A pollutant is emitted at a constant rate P (in kg/s) and its concentration $C(t)$ (in kg/m^3) is uniform and instantly distributed over the city. A horizontal wind with uniform and constant velocity v enters the side of the box of length L_1 perpendicular to it, and moves along the side of length L_2 (one can always orient the parallelepiped to achieve this geometrical configuration) and exits from the opposite side, thus diluting the pollutant.

a) Derive an ordinary differential equation for the concentration $C(t)$ of the pollutant as a function of time and solve it by knowing that the initial concentration at time $t = 0$ is C_0. What happens to your solution if there is no wind?

b) Assume now that the emission rate P is not constant in time but varies periodically as $P(t) = P_0 + P_1 \cos(\omega t)$, where $P_{0,1}$ are constants (for example, the pollutant could be carbon monoxide released by automobile traffic, which peaks at rush hour). Find the corresponding solution $C(t)$.

Solution

a) Conservation of the pollutant mass CV is expressed by the balance equation for the rate of change of the concentration C

$$\frac{d(CV)}{dt} = P - \mathcal{M},$$

where $V = L_1 L_2 h$ is the volume of the imaginary parallelepiped enclosing the city and $\mathcal{M}$ is the outflow of air from it (in kg/s) due to the wind. The mass of the pollutant leaving the box in the time dt is $C\,dV_{\text{out}}$, where the volume of air leaving the box during dt is $dV_{\text{out}} = L_1 h\,dx = L_1 h v\,dt$. Therefore, $\mathcal{M} = C\,dV_{\text{out}}/dt = L_1 h v C$ and mass conservation of the pollutant is described by

$$\frac{dC}{dt} + \frac{v}{L_2} C = \frac{P}{L_1 L_2 h}.$$

The general solution of this equation is the sum of the general solution of the associated homogeneous equation and of a particular solution of the inhomogeneous one. Those are easily found to be

$$C(t) = \alpha \exp\left(-\frac{t}{\tau}\right) + \frac{P}{L_1 h v},$$

where $\tau = L_2/v$ is the time scale for dilution and α is an integration constant determined by the initial condition. By imposing $C(0) = C_0$, we obtain $\alpha = C_0 - P/(L_1 v h)$, and the solution is

$$C(t) = \left(C_0 - \frac{P}{L_1 h v}\right) e^{-t/\tau} + \frac{P}{L_1 h v}.$$

The asymptotic state as $t \to +\infty$ is the steady-state concentration $C = P/(L_1 h v)$. This goes with the inverse of the transverse size of the box L_1, the inverse of the atmospheric mixing length h, and the inverse of the wind speed v. The dilution timescale τ also goes with the inverse of v and is of course proportional to L_2, the distance that the wind has to travel to cross the entire city. In this model if there is no wind $(v = 0)$, there is no dilution of the pollutant that is constantly produced, $\mathcal{M} = 0$, and the ODE is modified to

$$\frac{dC}{dt} = \frac{P}{L_1 L_2 h},$$

which has the linear solution

$$C(t) = \frac{P}{L_1 L_2 h} t + C_0$$

satisfying the initial condition $C(0) = C_0$. The pollutant keeps accumulating, and its concentration diverges linearly in time. This solution can also be obtained as the limiting case of the general one as $v \to 0$ by Taylor-expanding in powers of v:

$$\exp\left(-\frac{t}{\tau}\right) = \exp\left(-\frac{v}{L_2} t\right) = 1 - \frac{v}{L_2} t + \dots$$

and

$$C(t) = \left(C_0 - \frac{P}{L_1 h v}\right)\left(1 - \frac{v}{L_2}t + \dots\right) + \frac{P}{L_1 h v}$$

$$= C_0 + \frac{P}{L_1 L_2 h}t + \dots .$$

b) If the production rate P of the pollutant changes sinusoidally with time as $P(t) = P_0 + P_1 \cos(\omega t)$, the conservation equation for the mass $m(t)$ of the pollutant becomes

$$\frac{dC}{dt} + \frac{v}{L_2}C = \frac{P_0}{L_1 L_2 h} + \frac{P_1}{L_1 L_2 h}\cos(\omega t).$$

The general solution of the associated homogeneous equation is the exponential $\alpha e^{-t/\tau}$. We look for a particular solution of the inhomogeneous equation in the form

$$C_1(t) = \beta\cos(\omega t) + \gamma\sin(\omega t) + \delta,$$

with β, γ, and δ constants. Substitution into the ODE yields

$$\left(-\omega\beta + \frac{v}{L_2}\gamma\right)\sin(\omega t) + \left(\omega\gamma + \frac{v}{L_2}\beta\right)\cos(\omega t) + \frac{v}{L_2}\delta$$
$$= \frac{P_1}{L_1 L_2 h}\cos(\omega t) + \frac{P_0}{L_1 L_2 h}.$$

Then it must be

$$-\omega\beta + \frac{v}{L_2}\gamma = 0,$$

$$\omega\gamma + \frac{v}{L_2}\beta = \frac{P_1}{L_1 L_2 h},$$

$$v\delta = \frac{P_0}{L_1 h}.$$

This linear system has the solution

$$\beta = \frac{P_1 v}{L_1 h\left(\omega^2 L_2^2 + v^2\right)},$$

$$\gamma = \frac{\omega L_2 P_1}{L_1 h\left(\omega^2 L_2^2 + v^2\right)},$$

$$\delta = \frac{P_0}{L_1 h v}.$$

Hence, the general solution of the ODE is

$$C(t) \;=\; \alpha\,e^{-t/\tau} + \frac{P_1}{L_1 h\left(\omega^2 L_2^2 + v^2\right)}\left[v\cos\left(\omega t\right) + \omega L_2 \sin\left(\omega t\right)\right]$$

$$+\,\frac{P_0}{L_1 h v}.$$

By imposing the initial condition $C(0) = C_0$, we determine the integration constant

$$\alpha = C_0 - \frac{P_1 v}{L_1 h\left(\omega^2 L_2^2 + v^2\right)} - \frac{P_0}{L_1 h v}.$$

The final solution is

$$C(t) \;=\; C_0\,e^{-t/\tau} + \frac{P_1 v}{L_1 h\left(\omega^2 L_2^2 + v^2\right)}\left[\cos\left(\omega t\right) - e^{-t/\tau}\right]$$

$$+\frac{\omega L_2 P_1}{L_1 h\left(\omega^2 L_2^2 + v^2\right)}\sin\left(\omega t\right) + \frac{P_0}{L_1 v h}\left(1 - e^{-t/\tau}\right).$$

The late-time solution is the oscillating concentration

$$C\left(t \to +\infty\right) \approx \frac{P_1}{L_1 h\left(\omega^2 L_2^2 + v^2\right)}\left[v\cos\left(\omega t\right) + \omega L_2, \sin\left(\omega t\right)\right] + \frac{P_0}{L_1 h v}.$$

2 **(B)** Consider a homogeneous fluid at rest and suppose that at time $t = 0$ an amount C_0 of pollutant is released instantaneously in the $x = 0$ plane. The pollutant diffuses through the fluid in both the positive and negative x directions. What is the root mean square distance traveled by pollutant particles at a time $t > 0$?

Solution
The solution of the one-dimensional diffusion equation with the appropriate initial condition $C(x,0) = C_0\,\delta(x)$ is the Gaussian function

$$C(t, x) = \frac{C_0}{2\sqrt{\pi D t}}\,e^{-\frac{x^2}{4Dt}},$$

where $C(t, x)$ is the concentration of the pollutant and D is the diffusion coefficient. The root mean square distance traveled by polluting particles at time t is the variance of the Gaussian $\sigma = \sqrt{2Dt}$.

3 **(A)** Fifty grams of poisonous gas are instantaneously released when a container is accidentally dropped on the floor of a laboratory. What

is the gas concentration 10m from the point of the accident, after one hour? The air is still and the diffusion coefficient is $D = 9.0 \cdot 10^{-5} \mathrm{m^2 \cdot s^{-1}}$. Can diffusion be the main agent in spreading the gas?

Hint: Treat the emission as pointlike.

Solution

The solution of the diffusion equation for the concentration $C(t, \vec{x})$ in the case of a point source in three dimensions is

$$C(t, \vec{x}) = \frac{Q}{\left(\sqrt{2\pi}\,\sigma\right)^3} \, e^{-\frac{r^2}{2\sigma^2}}, \qquad \sigma = \sqrt{2D\,t}.$$

We have $\sigma = \sqrt{2 \cdot (9.0 \cdot 10^{-5} \mathrm{m^2 \cdot s^{-1}}) \cdot (3600\,\mathrm{s})} = 0.80\mathrm{m}$ and

$$C(1\,\mathrm{h}, 10\,\mathrm{m}) = \frac{(5 \cdot 10^{-2}\,\mathrm{kg})}{\left(\sqrt{2\pi}\,0.80\,\mathrm{m}\right)^3} \exp\left[-\left(\frac{10\,\mathrm{m}}{\sqrt{2} \cdot (0.80\,\mathrm{m})}\right)^2\right]$$

$$= 7.3 \cdot 10^{-37} \, \frac{\mathrm{kg}}{\mathrm{m^3}}.$$

This extremely small number clearly shows that diffusion is not an efficient process in transporting pollutants over macroscopic scales in short times. The main agents spreading the poisonous gas are convective motions and advection.

4 (B) Verify that

$$C(t, \vec{x}) = \frac{Q}{\left(\sqrt{2\pi}\,\sigma\right)^3} \, e^{-\frac{r^2}{2\sigma^2}},$$

where $\sigma = \sqrt{2D\,t}$, solves the diffusion equation

$$\frac{\partial C}{\partial t} = D\,\nabla^2 C.$$

It represents the solution for a instantaneous point source in three dimensions (a single puff in still air).

Solution

Explicit calculation of the derivatives yields

$$\frac{\partial C}{\partial t} = \frac{Q}{\sigma\left(\sqrt{2\pi}\,\sigma\right)^3} \, e^{-\frac{r^2}{2\sigma^2}} \left(-3 + \frac{r^2}{\sigma^2}\right) \frac{\partial \sigma}{\partial t}$$

$$= \frac{Q}{\sigma\left(\sqrt{2\pi}\,\sigma\right)^3} \, e^{-\frac{r^2}{2\sigma^2}} \left(-3 + \frac{r^2}{\sigma^2}\right) \sqrt{\frac{D}{2t}} = \frac{C}{\sigma^2}\left(-3 + \frac{r^2}{\sigma^2}\right) D,$$

and

$$\vec{\nabla}C = \frac{-Q}{\sigma^2 \left(\sqrt{2\pi}\,\sigma\right)^3}\, e^{-\frac{r^2}{2\sigma^2}}\, \vec{x} = -\frac{C}{\sigma^2}\, \vec{x}.$$

The Laplacian of C is easily computed by using the identity

$$\nabla^2 C = \vec{\nabla}\cdot\vec{\nabla}C = \frac{-1}{\sigma^2}\left(\vec{\nabla}C\cdot\vec{x} + C\vec{\nabla}\cdot\vec{x}\right) = \frac{1}{\sigma^2}\left(-\frac{C}{\sigma^2}\vec{x}\cdot\vec{x} + 3C\right)$$

$$= \frac{C}{\sigma^2}\left(\frac{r^2}{\sigma^2} - 3\right).$$

Hence, we have

$$\frac{\partial C}{\partial t} = D\,\frac{C}{\sigma^2}\left(\frac{r^2}{\sigma^2} - 3\right) = D\,\nabla^2 C.$$

5 **(C)** Describe the principles of smoke precipitators and electrostatic air cleaning.

Solution
Smoke precipitators are based on electrostatics. Air containing particulate matter pollutants from industrial processes passes through a positively charged grid at a potential of the order of 30 kV that gives the particles a positive charge. They are then attracted by a second grid kept at a negative potential with respect to the first one and are collected on it, thus effectively removing them from the air flow.

Air cleaners for household use are based on the same principle to remove particles of smoke, dust, pollen, and other polluting or irritating materials.

6 **(A)** Beginning at time $t = 0$, a pointlike source of CO_2 emits continuously at the rate of $1\,\text{kg}\cdot\text{s}^{-1}$ in uniform wind with velocity $5.0\,\text{m}\cdot\text{s}^{-1}$.

a) In a coordinate system with the source at the origin and the x-axis in the direction of the wind, what is the concentration C_A of CO_2 (in kg/m^3) at the point $A = (x, y, z) = (1\,\text{km}, 0, 0)$ on the x-axis?

b) Consider a point $B = (x, y, z) = (1\,\text{km}, y, 0)$ with $|y| \ll 1$ km: for what values of y has the concentration in B gone down by two e-folds, i.e., its value is $e^{-2}C_A$?
In both cases, compute the concentration a long time after the source began emitting and take $D = 1.64 \cdot 10^{-5}\text{m}^2\cdot\text{s}^{-1}$ as the value of the diffusion coefficient.

Solution

a) The solution of the advection-diffusion equation

$$\frac{\partial C}{\partial t} + \vec{v} \cdot \vec{\nabla} C = D \nabla^2 C \qquad (8.19)$$

for a continuously emitting pointlike source, with the geometry of the problem and in the approximation $t \to +\infty$ (a long time after the source begins to emit), is (e.g., Ref. [4])

$$C(\vec{x}) = \frac{q}{4\pi Dr} e^{-\frac{v(r-x)}{2D}}, \qquad (8.20)$$

where $r = \sqrt{x^2 + y^2 + z^2}$ and $q = 1\,\mathrm{kg/s}$. At point A it is $r = x$, $e^{-v(r-x)} = 1$ and

$$C_A = \frac{q}{4\pi Dr} = \frac{\left(1\,\mathrm{kg \cdot s^{-1}}\right)}{4\pi \cdot (1.64 \cdot 10^{-5}\,\mathrm{m^2 \cdot s^{-1}}) \cdot (1.0 \cdot 10^3\,\mathrm{m})} = 4.85 \frac{\mathrm{kg}}{\mathrm{m^3}}.$$

b) At point B (with y coordinate to be determined), we have, in the approximation $|y_B| \ll x_A$,

$$\frac{q}{4\pi Dr} e^{-\frac{v(r-x)}{2D}} = e^{-2} C_A,$$

which yields

$$\frac{v(r-x)}{2D} = 2,$$

or $\sqrt{x^2 + y^2} = x + 4D/v$. Hence,

$$y = \pm \left[\left(x + \frac{4D}{v} \right)^2 - x^2 \right]^{1/2} = \pm \frac{4D}{v} \sqrt{1 + \frac{vx}{2D}}$$

$$= \pm \frac{4 \cdot (1.64 \cdot 10^{-5}\,\mathrm{m^2 \cdot s^{-1}})}{(5.0\,\mathrm{m \cdot s^{-1}})} \sqrt{1 + \frac{(5.0\,\mathrm{m \cdot s^{-1}}) \cdot (1.0 \cdot 10^3\,\mathrm{m})}{2 \cdot (1.64 \cdot 10^{-5}\,\mathrm{m^2 \cdot s^{-1}})}}$$

$$= \pm 0.16\,\mathrm{m}.$$

Appendix A
Physical constants

Fundamental constants

gravitational constant $G = 6.673 \cdot 10^{-11} \, \text{N} \cdot \text{m}^2 \cdot \text{kg}^{-2}$

speed of light in vacuum $c = 2.998 \cdot 10^8 \, \text{m/s}$

molar gas constant $R = 8.315 \, \text{J} \cdot \text{mol}^{-1} \cdot \text{K}^{-1}$

Avogadro's number $N_A = 6.022 \cdot 10^{23} \, \text{mol}^{-1}$

molar volume[1] $V_m = 22.414 \cdot 10^{-3} \, \text{m}^3 \cdot \text{mol}^{-1}$

universal gas constant $R = 8.314 \, \text{J} \cdot \text{K}^{-1} \cdot \text{mol}^{-1}$

Boltzmann constant $k = R/N_A = 1.381 \cdot 10^{-23} \, \text{J/K}$

Stefan–Boltzmann constant $\sigma = 5.671 \cdot 10^{-8} \, \text{W} \cdot \text{m}^{-2} \cdot \text{K}^{-4}$

permittivity of vacuum $\epsilon_0 = 8.854 \cdot 10^{-12} \, \text{F/m}$

permeability of vacuum $\mu_0 = 4\pi \cdot 10^{-7} \, \text{H/m}$

Microscopic physics

electron charge $e = 1.602 \cdot 10^{-19} \, \text{C}$

electron mass $m_e = 9.109 \cdot 10^{-31} \, \text{kg} = 511.0 \, \text{keV}$

proton mass $m_p = 1.673 \cdot 10^{-27} \, \text{kg} = 938.3 \, \text{MeV}$

atomic mass unit $1 \, \text{a.m.u.} = 1.661 \cdot 10^{-27} \, \text{kg} = 931.5 \, \text{MeV}$

Planck constant $h = 6.626 \cdot 10^{-34} \, \text{J} \cdot \text{s}$

reduced Planck constant $\hbar = \frac{h}{2\pi} = 1.0546 \cdot 10^{-34} \, \text{J} \cdot \text{s}$

fine structure constant $\alpha = \mu_0 c e^2/(2h) = 7.297 \cdot 10^{-3}$

Rydberg constant $\mathcal{R} = 1.0974 \cdot 10^7 \, \text{m}^{-1}$

Compton wavelength of the electron $\lambda_c = 2.426 \cdot 10^{-12} \, \text{m}$

Wien's displacement law constant $b = \lambda_{max} T = 2.8978 \cdot 10^{-3} \, \text{m} \cdot \text{K}$

Astronomical constants

mass of the Earth $M_E = 5.978 \cdot 10^{24} \, \text{kg}$

mass of the Sun $M_\odot = 1.989 \cdot 10^{30} \, \text{kg}$

---

[1]Ideal gas at $T = 273.15$ K and 1 atm.

mass of the Moon $M_M = 7.35 \cdot 10^{22}\,\text{kg}$
average radius of the Earth $R_E = 6.370 \cdot 10^6\,\text{m}$
equatorial radius of the Earth $R = 6.378 \cdot 10^6\,\text{m}$
radius of the Sun $R_\odot = 6.96 \cdot 10^8\,\text{m}$
radius of the Moon $R_M = 1.74 \cdot 10^6\,\text{m}$
average Sun–Earth distance   1 A.U.$= 1.496 \cdot 10^{11}\,\text{m}$
solar constant $S = 1.370 \cdot 10^3\,\text{W/m}^2$
average acceleration of gravity on the Earth's surface $g = 9.81\,\text{m/s}^2$
(bolometric) luminosity of the Sun $L_\odot = 3.826 \cdot 10^{26}\,\text{W}$

Air

density of air at 10°C $\rho = 1.247\,\text{kg/m}^3$
density of air at 20°C $\rho = 1.205\,\text{kg/m}^3$
specific heat of dry air at constant pressure $c_p = 1004\,\text{J} \cdot \text{K}^{-1} \cdot \text{kg}^{-1}$
Fourier coefficient $a = 2.25 \cdot 10^{-5}\,\text{m}^2/\text{s}$
thermal conductivity at 27°C $k = 0.026\,\text{W} \cdot \text{m}^{-1} \cdot (°\text{C})^{-1}$
speed of sound at 1 atm and 0°C $v_s = 330\,\text{m/s}$

Water

density of freshwater at 20°C $\rho = 0.998 \cdot 10^3\,\text{kg/m}^3$
density of sea water at 20°C $\rho = 1.03 \cdot 10^3\,\text{kg/m}^3$
latent heat of fusion $L_f = 3.34 \cdot 10^5\,\text{J/kg} = 79.7\,\text{cal/g}$
latent heat of vaporization $L_v = 2.26 \cdot 10^6\,\text{J/kg} = 539\,\text{cal/g}$
coefficient of thermal expansion at 20°C $2.10 \cdot 10^{-4}\,(°\text{C})^{-1}$
specific heat at 25°C $4187\,\text{J} \cdot \text{kg}^{-1} \cdot (°\text{C})^{-1} = 1.00\,\text{cal/g}$
surface tension at 20°C $\gamma = 7.28 \cdot 10^{-2}\,\text{N/m}$
dynamic viscosity coefficient of freshwater at 20°C $\eta = 1.00 \cdot 10^{-3}\,\text{Pa} \cdot \text{s}$

Conversion factors

radian $1\,\text{rad} = 57.3\,°$
Armstrong $1\,\overset{\circ}{\text{A}} = 10^{-10}\,\text{m}$
fermi $1\,\text{fm} = 10^{-15}\,\text{m}$
foot $1\,\text{ft} = 0.3048\,\text{m}$
arcminute $1' = 2.909 \cdot 10^{-4}\,\text{rad}$
arcsecond $1" = 4.848 \cdot 10^{-6}\,\text{rad}$
electronvolt $1\,\text{eV} = 1.602 \cdot 10^{-19}\,\text{J}$
kilowatthour $1\,\text{kWh} = 3.6 \cdot 10^6\,\text{J}$
kilocalorie $1\,\text{Kcal} = 4187\,\text{J}$
calorie $1\,\text{cal} = 10^{-3}\,\text{Kcal} = 4.187\,\text{J}$
British thermal unit $1\,\text{BTU} = 1.055 \cdot 10^3\,\text{J}$
horsepower $1\,\text{HP} = 746\,\text{W}$
atmosphere $1\,\text{atm} = 1.01325 \cdot 10^5\,\text{Pa}$
torr (millimeter of mercury) $1\,\text{torr} = 133.3224\,\text{Pa}$
foot of water[2] $1\,\text{f.w.} = 2.989 \cdot 10^3\,\text{Pa}$

---

[2] At 4°C.

gallon (UK)   1 gl $= 4.546 \cdot 10^{-3}\,\mathrm{m}^3$
gallon (US)   1 gl $= 3.785 \cdot 10^{-3}\,\mathrm{m}^3$
gauss   1 gauss $= 10^{-4}$ T

Appendix B
Mathematical identities

Trigonometric identities

$$\tan\theta \equiv \frac{\sin\theta}{\cos\theta}$$

$$\cot\theta \equiv (\tan\theta)^{-1}$$

$$\sec\theta \equiv (\cos\theta)^{-1}$$

$$\csc\theta \equiv (\sin\theta)^{-1}$$

$$\sin(\alpha \pm \beta) = \sin\alpha\cos\beta \pm \sin\beta\cos\alpha$$

$$\cos(\alpha \pm \beta) = \cos\alpha\cos\beta \mp \sin\alpha\sin\beta$$

$$\tan(\alpha \pm \beta) = \frac{\tan\alpha \pm \tan\beta}{1 \mp \tan\alpha\tan\beta}$$

$$\cos\alpha\cos\beta = \frac{1}{2}\left[\cos(\alpha+\beta) + \cos(\alpha-\beta)\right]$$

$$\sin\alpha\sin\beta = \frac{1}{2}\left[\cos(\alpha-\beta) - \cos(\alpha+\beta)\right]$$

$$\sin\alpha\cos\beta = \frac{1}{2}\left[\sin(\alpha+\beta) + \sin(\alpha-\beta)\right]$$

$$\sin^2\theta + \cos^2\theta = 1$$

$$\sin(2\theta) = 2\sin\theta\cos\theta$$

$$\cos(2\theta) = \cos^2\theta - \sin^2\theta$$

$$\tan(2\theta) = \frac{2\tan\theta}{1 - \tan^2\theta}$$

$$\cos^2\theta = \frac{1 + \cos(2\theta)}{2}$$

$$\sin^2\theta = \frac{1 - \cos(2\theta)}{2}$$

Hyperbolic identities

$$\sinh x \;\equiv\; \frac{e^x - e^{-x}}{2}$$

$$\cosh x \;\equiv\; \frac{e^x + e^{-x}}{2}$$

$$\tanh x \;\equiv\; \frac{\sinh x}{\cosh x}$$

$$\coth x \;\equiv\; (\tanh x)^{-1}$$

$$\sinh(\alpha \pm \beta) = \sinh\alpha\cosh\beta \pm \sinh\beta\cosh\alpha$$

$$\cosh(\alpha \pm \beta) = \cosh\alpha\cosh\beta \pm \sinh\alpha\sinh\beta$$

$$\tanh(\alpha \pm \beta) = \frac{\tanh\alpha \pm \tanh\beta}{1 \pm \tanh\alpha\tanh\beta}$$

$$\cosh^2 x - \sinh^2 x = 1$$

$$\sinh(2x) = 2\sinh x\cosh x$$

$$\cosh(2x) = \cosh^2 x + \sinh^2 x$$

$$\tanh(2x) = \frac{2\tanh x}{1 + \tanh^2 x}$$

$$\cosh^2 x = \frac{1 + \cosh(2x)}{2}$$

$$\sinh^2 x = \frac{\cosh(2x) - 1}{2}$$

Error function

The *error function* is defined as

$$\mathrm{erf}(x) \equiv \frac{2}{\sqrt{\pi}} \int_0^x ds\, e^{-s^2}$$

and satisfies the properties

$$\mathrm{erf}(0) = 0, \quad \mathrm{erf}(+\infty) = 1, \quad \mathrm{erf}(-x) = -\mathrm{erf}(x).$$

The *complementary error function* is defined as

$$\mathrm{erfc}(x) \equiv 1 - \mathrm{erf}(x).$$

Appendix C
Differential operators in various coordinate systems

Gradient

The gradient of a function $f(\vec{x})$ in Cartesian coordinates (x, y, z) is

$$\vec{\nabla} f = \left(\frac{\partial f}{\partial x}, \frac{\partial f}{\partial y}, \frac{\partial f}{\partial z} \right).$$

The expression of the gradient of f in cylindrical coordinates (r, φ, z) is

$$\vec{\nabla} f = \frac{\partial f}{\partial r}\, \vec{e_r} + \frac{1}{r} \frac{\partial f}{\partial \varphi}\, \vec{e_\varphi} + \frac{\partial f}{\partial z}\, \vec{e_z}.$$

The gradient of f in spherical polar coordinates (r, θ, φ) is

$$\vec{\nabla} f = \frac{\partial f}{\partial r}\, \vec{e_r} + \frac{1}{r} \frac{\partial f}{\partial \theta}\, \vec{e_\theta} + \frac{1}{r \sin \theta} \frac{\partial f}{\partial \varphi}\, \vec{e_\varphi}.$$

Divergence

The divergence of a vector field $\vec{a}(\vec{x}) = (a_x, a_y, a_z)$ in Cartesian coordinates is

$$\vec{\nabla} \cdot \vec{a} = \frac{\partial a_x}{\partial x} + \frac{\partial a_y}{\partial y} + \frac{\partial a_z}{\partial z}.$$

The expression of the divergence of $\vec{a}$ in cylindrical coordinates is

$$\vec{\nabla} \cdot \vec{a} = \frac{1}{r} \frac{\partial}{\partial r}\, (r a_r) + \frac{1}{r} \frac{\partial a_\theta}{\partial \theta} + \frac{\partial a_z}{\partial z}.$$

The divergence of $\vec{a}$ in spherical polar coordinates is

$$\vec{\nabla} \cdot \vec{a} = \frac{1}{r^2} \frac{\partial}{\partial r}\, (r^2 a_r) + \frac{1}{r \sin \theta} \frac{\partial}{\partial \theta}\, (a_\theta \sin \theta) + \frac{1}{r \sin \theta} \frac{\partial a_\varphi}{\partial \varphi}.$$

Laplacian

The Laplacian (or Laplace operator) applied on a function $f(\vec{x})$ in Cartesian coordinates is

$$\nabla^2 f = \frac{\partial^2 f}{\partial x^2} + \frac{\partial^2 f}{\partial y^2} + \frac{\partial^2 f}{\partial z^2}.$$

The expression of the Laplacian of f in cylindrical coordinates is

$$\nabla^2 f = \frac{1}{r}\frac{\partial}{\partial r}\left(r\frac{\partial f}{\partial r}\right) + \frac{1}{r^2}\frac{\partial^2 f}{\partial \varphi^2} + \frac{\partial^2 f}{\partial z^2}.$$

The Laplacian of f in spherical polar coordinates is

$$\nabla^2 f = \frac{1}{r^2}\frac{\partial}{\partial r}\left(r^2\frac{\partial f}{\partial r}\right) + \frac{1}{r^2 \sin\theta}\frac{\partial}{\partial\theta}\left(\sin\theta\frac{\partial f}{\partial\theta}\right) + \frac{1}{r^2 \sin^2\theta}\frac{\partial^2 f}{\partial\varphi^2}.$$

d'Alembert's operator

The d'Alembert operator (Dalambertian) applied on a function $f(t,\vec{x})$ is

$$\Box f = \nabla^2 f - \frac{1}{v^2}\frac{\partial^2 f}{\partial t^2}.$$

Differential operator identities

$$\vec{\nabla}(fg) = f\vec{\nabla}g + g\vec{\nabla}f$$

$$\vec{\nabla}\cdot(f\vec{a}) = \left(\vec{\nabla}f\right)\cdot\vec{a} + f\vec{\nabla}\cdot\vec{a}$$

$$\vec{\nabla}\times(f\vec{a}) = \left(\vec{\nabla}f\right)\times\vec{a} + f\vec{\nabla}\times\vec{a}$$

$$\vec{\nabla}\cdot\vec{\nabla}f = \nabla^2 f$$

$$\vec{\nabla}\times\left(\vec{\nabla}f\right) = 0$$

$$\vec{\nabla}\cdot\left(\vec{\nabla}\times\vec{a}\right) = 0$$

$$\vec{\nabla}\times\left(\vec{\nabla}\times\vec{a}\right) = \vec{\nabla}\left(\vec{\nabla}\cdot\vec{a}\right) - \vec{\nabla}^2\vec{a} \quad \text{(Cartesian coordinates only)}$$

$$\vec{\nabla}\left(\vec{a}\cdot\vec{b}\right) = \vec{a}\times\left(\vec{\nabla}\times\vec{b}\right) + \left(\vec{a}\cdot\vec{\nabla}\right)\vec{b} + \vec{b}\times\left(\vec{\nabla}\times\vec{a}\right) + \left(\vec{b}\cdot\vec{\nabla}\right)\vec{a}$$

$$\vec{\nabla}\cdot\left(\vec{a}\times\vec{b}\right) = \vec{b}\cdot\left(\vec{\nabla}\times\vec{a}\right) - \vec{a}\cdot\left(\vec{\nabla}\times\vec{b}\right)$$

$$\vec{\nabla}\times\left(\vec{a}\times\vec{b}\right) = \vec{a}\left(\vec{\nabla}\cdot\vec{b}\right) - \vec{b}\left(\vec{\nabla}\cdot\vec{a}\right) + \left(\vec{b}\cdot\vec{\nabla}\right)\vec{a} - \left(\vec{a}\cdot\vec{\nabla}\right)\vec{b}$$

References

[1] Batchelor, G.K. (1967), *Introduction to Fluid Mechanics.* Cambridge: Cambridge University Press.

[2] Bear, J. (1979), *Hydraulics of Groundwater.* New York: McGraw-Hill.

[3] Bear, J. (1988), *Dynamics of Fluids in Porous Materials.* New York: Dover.

[4] Boeker, E. and van Grondelle, R. (1999), *Environmental Physics.* Chicester: J. Wiley & Sons.

[5] Boltzmann, L. (1894), *Ann. Phys. (Leipzig)* **53**, 959.

[6] Boltzmann, L. (1995), *Lectures on Gas Theory.* New York: Dover.

[7] Campbell, I.M. (1977), *Energy and the Atmosphere—A Physical-Chemical Approach.* London: J. Wiley & Sons.

[8] Campbell, G.S. and Norman, J.M. (1998), *An Introduction to Environmental Biophysics*, 2nd edition. New York: Springer-Verlag.

[9] Carter, A.H. (2001), *Classical and Statistical Thermodynamics.* Upper Saddle River, NJ: Prentice Hall.

[10] Collie, C.H. (1982), *Kinetic Theory and Entropy.* London: Longman.

[11] Csanady, G.T. (1973), *Turbulent Diffusion in the Environment.* Dordrecht: Reidel.

[12] Curzon, F. and Ahlborn, B. (1975), *Am. J. Phys.* **43**, 22.

[13] Defant, A. (1961), *Physical Oceanography.* Oxford: Pergamon Press.

[14] Dingman, S.L. (1994), *Physical Hydrology.* Englewood Cliffs, NJ: Prentice Hall.

[15] Fermi, E. (1956), *Thermodynamics.* New York: Dover.

[16] Fowler, J.M. (1984), *Energy and the Environment.* New York: McGraw-Hill.

[17] Gel'fand, I.M. and Shilov, G.E. (1964), *Generalized Functions*. New York: Academic Press.

[18] Gibbs, R.J., Matthews, M.D. and Link, D.A. (1971), *J. Sed. Petrol.* **41**, 7.

[19] Gill A.E. (1982), *Atmosphere-Ocean Dynamics*. New York: Academic Press.

[20] Goldstein, H. (1972), *Classical Mechanics*, 2nd edition. Reading, MA: Addison-Wesley.

[21] Gradshteyn, I.S. and Ryzhik, I.M. (1980), *Tables of Integrals, Series and Products*. New York: Academic Press.

[22] Griffiths, D.J. (1999), *Introduction to Electrodynamics*, 3rd edition. Upper Saddle River, NJ: Prentice Hall.

[23] Griffiths, D.J. (2005), *Introduction to Quantum Mechanics*. Upper Saddle River, NJ: Prentice Hall.

[24] Guyot, G. (1998), *Physics of the Environment and Climate*. Chicester: J. Wiley & Sons/Praxis Publishing.

[25] Hanski, L.A. (1991), *Bio. J. Linn. Soc.* **42**, 17.

[26] Harr, M.E. (1991), *Groundwater and Seepage*. New York: Dover.

[27] Harte, J. (1988), *Consider a Spherical Cow: A Course in Environmental Problem Solving*. Sausalito, CA: University Science Books.

[28] Harte, J. (2001), *Consider a Cylindrical Cow: More Adventures in Environmental Problem Solving*. Sausalito, CA: University Science Books.

[29] Heinsohn, R.J. and Kabel, R.L. (1999), *Sources and Control of Air Pollution*. Englewood Cliffs, NJ: Prentice Hall.

[30] Henry, J.G. and Heinke, G.W. (1989), *Environmental Science and Engineering*. Englewood Cliffs, NJ: Prentice Hall.

[31] Holton, J.R. (1992), *An Introduction to Dynamic Meteorology*. London: Academic Press.

[32] Hille, E. (1969), *Lectures on Ordinary Differential Equations*. Reading, MA: Addison-Wesley.

[33] Hillel, D. (1980), *Fundamentals of Soil Physics*. New York: Academic Press.

[34] Houghton, J.T. (1986), *The Physics of Atmospheres*. Cambridge: Cambridge University Press.

[35] Howes, R. and Fainberg, A. (eds.) (1991), *The Energy Sourcebook*. New York: American Institute of Physics.

[36] Huntley, H.E. (1967), *Dimensional Analysis*. New York: Dover.

[37] Jackson, J.D. (1975), *Classical Electrodynamics*, 2nd edition. New York: J. Wiley & Sons.

[38] Kauzmann, W. (1966), *Kinetic Theory of Gases*. New York: W.A. Benjamin.

[39] LaGrega, M.D., Buckingham, P.L. and Evans, J.C. (1994), *Hazardous Waste Management*. New York: McGraw-Hill.

[40] Lamb, H. (1932), *Hydrodynamics*. Cambridge: Cambridge University Press.

[41] Landau, L.D. and Lifshitz. E.M. (1965), *Quantum Mechanics (Non-Relativistic Theory)*. Reading, MA: Addison-Wesley.

[42] LeBlond, P.H. and Mysak, L.A. (1978), *Waves in the Ocean*. Amsterdam: Elsevier Science.

[43] Levins, R. (1969), *Bull. Entomol. Soc. Am.* **15**, 237.

[44] Main, I.G. (1993), *Vibrations and Waves in Physics*, 3rd edition. Cambridge: Cambridge University Press.

[45] Mason, N.J. and Hughes, P. (2001), *Introduction to Environmental Physics: Planet Earth, Life, and Climates*. New York: Taylor & Francis.

[46] Masters, G.M. (1974), *Introduction to Environmental Science and Technology*. New York: J. Wiley & Sons.

[47] McFarland, E.L., Hunt, J.L. and Campbell, J.L. (1994), *Energy, Physics and the Environment*. Winnipeg: Wuerz Publishing.

[48] McLellan, H.J. (1965), *Elements of Physical Oceanography*. Oxford: Pergamon Press.

[49] Messiah, A. (1961), *Quantum Mechanics*. Amsterdam: North-Holland.

[50] Meyer, R.E. (1971), *Introduction to Mathematical Fluid Dynamics*. New York: Wiley-Interscience.

[51] Moiseiwitsch, B.L. (1966), *Variational Principles*. London: Interscience.

[52] Monteith, J.L. and Unsworth, M.H. (1990), *Principles of Environmental Physics*. London: Butterworth-Heinemann.

[53] Muir, J. (1911), *My First Summer in the Sierra*. Boston: Houghton and Mifflin.

[54] Pain, H.J. (2005), *The Physics of Vibrations and Waves*, 6th edition. Chichester: J. Wiley & Sons.

[55] Pickard, G.L. and Emery, W.J. (1982), *Descriptive Physical Oceanography*, 4th edition. Oxford: Pergamon Press.

[56] Pond, S. and Pickard, G.L. (1993), *Introductory Dynamical Oceanography*. Oxford: Pergamon Press.

[57] Prandtl, L. (1952), *The Essentials of Fluid Dynamics*. London: Blackie.

[58] Princeton University Water Resources Program (1994), *Groundwater Contamination from Hazardous Wastes*. Englewood Cliffs, NJ: Prentice Hall.

[59] Reitz, J.R. and Milford, F.J. (1960), *Foundations of Electromagnetic Theory.* Reading, MA: Addison-Wesley.

[60] Ridley, B.K. (1979), *The Physical Environment.* Chicester: Ellis Horwood Publishers.

[61] Schiff, L.I. (1968), *Quantum Mechanics*, 3rd edition. New York: McGraw-Hill.

[62] Sears, F.W. and Salinger, G.L. (1975), *Thermodynamics, Kinetic Theory, and Statistical Thermodynamics.* Reading, MA: Addison-Wesley.

[63] Seinfeld, J.H. (1986), *Atmospheric Chemistry and Physics of Air Pollution.* New York: J. Wiley & Sons.

[64] Smith, C. (2001), *The Physical Environment.* New York: Taylor and Francis.

[65] Spain, B. (2003), *Tensor Calculus: A Concise Course.* New York: Dover.

[66] Stommel, H. (1987), *A View of the Sea.* Princeton: Princeton University Press.

[67] Stull, R.B. (1995), *Meteorology Today for Scientists and Engineers.* Minneapolis/St. Paul, MN: West Publishing Company.

[68] Svanberg, S. (1992), *Atomic and Molecular Spectroscopy, Basic Aspects and Practical Applications*, 2nd edition. New York: Springer-Verlag.

[69] Tchobanoglous, G. and Schroeder, E.D. (1985), *Water Quality.* Reading, MA: Addison-Wesley.

[70] Tennekes, H. and Lumley, J.L. (1972), *A First Course in Turbulence.* Cambridge, MA: MIT Press.

[71] Trewartha, G.T. and Horn, L.H. (1980), *An Introduction to Climate.* New York: McGraw-Hill.

[72] Wainwright, J. and Mulligan, M. (2004), *Environmental Modelling: Finding Simplicity in Complexity.* Chicester: J. Wiley & Sons.

[73] Wald, R.M. (1984), *General Relativity.* Chicago: Chicago University Press.

[74] Zemansky, M.W. (1968), *Heat and Thermodynamics*, 5th edition. New York: McGraw-Hill.

Index